vironmental and Climate Change in South and Southeast Asia

Climate and Culture

Series Editors

Carmen Meinert
Claus Leggewie

VOLUME 2

The titles published in this series are listed at *brill.com/clac*

Environmental and Climate Change in South and Southeast Asia

How are Local Cultures Coping?

Edited by

Barbara Schuler

BRILL

LEIDEN | BOSTON

Cover illustration: The bodhi tree in front of the famous Phra Pathom Chedi in Nakhon Pathom. Photo by Martin Seeger.

KULTURWISSENSCHAFTLICHES INSTITUT ESSEN
FORSCHUNGSKOLLEG DER UNIVERSITÄTSALLIANZ METROPOLE RUHR

Publication of this book was conceived and organised at the *Kulturwissenschaftliches Institut* Essen (KWI), Germany (Institute for Advanced Study in the Humanities).

Library of Congress Cataloging-in-Publication Data

Environmental and climate change in South and Southeast Asia : how are local cultures coping? / edited by Barbara Schuler.
pages cm. -- (Climate and culture ; volume 2)
Includes bibliographical references and index.
ISBN 978-90-04-24588-4 (hardback : alk. paper) -- ISBN 978-90-04-27322-1 (e-book) 1. Human beings--Effect of environment on--South Asia. 2. Human beings--Effect of environment on--Southeast Asia. 3. Climatic changes--Social aspects--South Asia. 4. Climatic changes--Social aspects--Southeast Asia. 5. Environmental degradation--Social aspects--South Asia. 6. Environmental degradation--Social aspects--Southeast Asia. 7. Human ecology--South Asia. 8. Human ecology--Southeast Asia. I. Schuler, Barbara, editor of compilation.

GF71.E68 2014
304.2'50954--dc23

This publication has been typeset in the multilingual 'Brill' typeface. With over 5,100 characters covering Latin, IPA, Greek, and Cyrillic, this typeface is especially suitable for use in the humanities. For more information, please see brill.com/brill-typeface.

ISSN 2213-0519
ISBN 978-90-04-24588-4 (hardback)
ISBN 978-90-04-27322-1 (e-book)

This book is printed on acid-free paper.

Contents

Foreword

Man influences the environment and climate and the consequences are now felt around the globe. National or regional efforts to restrict or at least contain the damage can only be insufficient: in principle environmental and climate protection needs a global concept.

Paradoxically, the way we *perceive* environmental and climate change and handle damage is closely linked to local or regional patterns of perception. It is these particularistic perceptions that often lead to different, in many instances opposing reactions to preventive and curative environmental and climate protection measures.

This local view is grounded not only in different ways of socio-economic development in different regions of the world, but also in differences in *cultural patterns*. Think for example, of the strongly varying notions of the actual problems at hand, or of different policy styles and politico-social environments. Also, the disturbance of the environment and climate causes relatively rapid social changes, in which the interpretation of symbols for the relationship between man and nature plays an important part. The history of climate and culture, patterns of perception of environmental and climate change and an informed assessment of the future direction of environmental and climate policy in various parts of the world have to be taken into account in order to get to grips with the problem.

From a variety of angles, such as the history of ideas, historiography, the study of civilisation, and the political sciences, the monographs and edited volumes in *Climate and Culture* will all deal with the following questions:

- How do local and regional cultures perceive changes in the environment and climate in past and present?
- How did and do they adjust to them?
- How do their various representatives and spokesmen introduce their respective views to the global debate and into emerging international negotiating systems?

The following titles will be published in the series:

VOLUME 1: *Nature, Environment and Culture in East Asia: The Challenge of Climate Change*, edited by Carmen Meinert, 2013.

VOLUME 2: *Environmental and Climate Change in South and Southeast Asia: How Are Local Cultures Coping?*, edited by Barbara Schuler, 2014.

VOLUME 3: *Cultural Dimensions of Climate Change and the Environment in North America*, edited by Bernd Sommer, 2014/2015 (forthcoming).

Carmen Meinert
Claus Leggewie

Preface

The present volume explores issues of relevance to the current climate change discourse over a range of cultural, socio-political, local, historical and geographical contexts in South and Southeast Asia. Though the title of the book specifies two broad regions, the volume concentrates mainly on India, Thailand and Indonesia. This selection is based on the fact that the three countries are at present (and will likely remain so well into the future) major contributors to global warming. The contributors to this volume, coming from various scholarly disciplines – including Indian, Thai and Southeast Asian studies, archaeology, anthropology, geography, political sciences, and media studies – approach these issues from the angle of their respective areas of research. The volume as a whole argues for taking bidirectionality as a key concept when approaching an issue of such critical importance for humans and nature as climate change: the premise, that is, that not only the climate shapes cultures but that, equally, cultures shape the climate.

This publication is a further volume in the BRILL book series *Climate and Culture*. It had its beginnings in an international symposium at the Institute for Advanced Study in the Humanities (KWI), Essen, in June 2011 organised by the editor of this volume upon an invitation by Claus Leggewie and Carmen Meinert. The contributions in it (both selected essays that go back to this occasion and supplementary ones) represent a great diversity of research areas and methodologies. They focus on three broad domains: the religious; the social and political; and the scientific and technological. The volume itself is divided into four parts: Part 1 looks at "Ideas", Part 2 is dedicated to an exploration of the "Past", Part 3 deals with the "Present", and the essays in Part 4 examine the "Prospects".

The volume would have not been possible without the KWI and the generous support of BRILL Publishers. I would like to acknowledge, too, my appreciation to Iain Sinclair, who did a final proofreading of the texts, and the anonymous peer reviewers as well, for their helpful suggestions. Last but not least, I am immensely grateful to Nele Fabian for her painstaking editorial assistance.

Barbara Schuler
Hamburg
Autumn 2013

Abbreviations

3R	reuse, reduce, recycle
AASA	Association of Academies of Sciences in Asia
ABC	atmospheric brown cloud
ABE	Advisory Body on Energy (India)
AEC	ASEAN Economic Community
AKB	Advokasi Kelembagaan Bencana (Indonesia)
AOSIS	Alliance of Small Island States
APSC	ASEAN Political Security Community
AR4	Fourth Assessment Report (of the IPCC)
ASEAN	Association of Southeast Asian Nations
ASSC	ASEAN Socio-Cultural Community
BASIC	Brazil, South Africa, India, and China
BAU	business-as-usual
BC	black carbon
BCE	before the common era
BE	Buddhist era
BJP	Bharatiya Janata Party
BNPB	Badan Nasional Penanggulangan Bencana (Indonesia)
BRICS	Brazil, Russia, India, China, South Africa
BRT	Bus Rapid Transport
CAIT	Climate Analysis Indicator Tool
CASE	Commission for Additional Sources of Energy (India)
CBDRM	Penanggulangan Bencana Berbasis Komunitas (Indonesia)
CCAFS	Climate Change Agriculture and Food Security
CCS	carbon capture and storage
CDASC	Child Disaster Awareness for Schools and Communities (Indonesia)
CE	common era
CNG	compressed natural gas
COP	Conference of Parties
CPO	crude palm oil
CSE	Centre for Science and Environment (Delhi)
CSO	Civil Society Organisation
DDA	Delhi Development Authority
DIY	Daerah Istimewa Yogyakarta (Indonesia)
DNES	Department of Non-conventional Energy Sources (India)
DST	Department of Science and Technology (India)

ENSO	El Niño/Southern Oscillation
EPCA	Environment Pollution (Prevention and Control) Authority
FBR	Betawi Brotherhood Forum
FICCI	Federation of Indian Chambers of Commerce and Industry
FPC	Fuel Policy Committee (India)
GDP	gross domestic product
GEF	Global Environment Facility
GHG	greenhouse gas
GIGA	German Institute of Global and Area Studies
GLOFs	glacial lake outburst floods
GNKL	Gerakan Nasional Kehutanan dan Lingkungan (Indonesia)
HGU	Hak Guna Usaha (Indonesia)
HKH	Hindu Kush-Karakoram-Himalaya
HT	Hizbut Tahrir
HTI	Hizbut Tahrir Indonesia
ICAS	Islamic College for Advanced Studies (Jakarta)
IFC	International Finance Corporation
IHDP	International Human Dimensions Programme on Global Environmental Change
IMF	International Monetary Fund
IPCC	Intergovernmental Panel on Climate Change
IRRI	International Rice Research Institute (Philippines)
ISC	Islamic Studies Center (Indonesia)
LLH	Lembaga Lingkungan Hidup (Indonesia)
LPBI NU	Lembaga Penanganan Bencana dan Perubahan Iklim Nahdlatul Ulama (Indonesia)
LSPLH	Lembaga Studi dan Pemberdayaan Lingkungan (Indonesia)
M7YAP	Muslim Seven Year Action Plan
MDMC	Muhammadiyah Disaster Management Center (Indonesia)
MLH	Majelis Lingkungan Hidup (Indonesia)
MNES	Ministry of Non-conventional Energy Sources (India)
MUI	Majelis Ulama Indonesia (Indonesia)
MW	megawatt
NCR	National Capital Region (India)
NCT	National Capital Territory (India)
NDTV	New Delhi Television
NGO	non-governmental organisation
n.p.	no place
NU	Nahdlatul Ulama (Indonesia)

OECD	Organization for Economic Co-operation and Development
PBNU	Pengurus Besar Nahdlatul Ulama (Indonesia)
PIL	public interest litigation
PKB	Partai Kebangkitan Bangsa (Indonesia)
PKS	Partai Keadilan Sejahtera (Indonesia)
PM	Prime Minister
PNAS	Proceedings of the National Academy of Sciences of the United States of America
PPP	purchasing power parity
PUC	Pollution Under Control
PV	photovoltaic
PVMTI	Photovoltaic Market Transformation Initiative (Kenya, Morocco, India)
R&D	research and development
REP	Religious Environmentalist Paradigm
Ri-MAP	Risk of Mortality/Morbidity due to Air Pollution
RMI	Rabithah Ma'ahid Islamiyah (Indonesia)
SDC	Sustainable Development Commission
SEZ	Special Economic Zone
SHS	solar home systems
Skt.	Sanskrit
SPM	suspended particulate matter
SPP	Solar Photovoltaic Programme (India)
TAR	Third Assessment Report (of the IPCC)
TERI	The Energy and Resources Institute (India)
TOI	*Times of India*
TSP	total suspended particulate
UN	United Nations
UNCED	UN Conference on Environment and Development
UNDP	United Nations Development Program
UNFCCC	United Nations Framework Convention on Climate Change
UN-REDD	Reducing Emissions from Deforestation and Forest Degradation (United Nations)
WBGU	German Advisory Council on Global Change
WCED	World Commission on Environment and Development
WGEP	Working Group on Energy Policy (India)
WHO	World Health Organization
WWF	World Wildlife Fund

List of Figures and Tables

Glossary of Terms and Names of Key Figures in Asian Languages*

ahiṃsā	Skt., non-violence
akhlak	Indonesian, Islamic ethics
Arjuna	Skt., a character in the epics
Atharvaveda	Skt., the fourth Veda, largely devoted to magic spells
baray	Khmer [Cambodian], water reservoir
bhikkhu	Pali, monk
dharma/dhamma	Skt./Pali, religious law, Hindu rules of social life and religious observance; Buddha's teaching
diṭṭhi	Pali, (false) views
fiqh	jurisprudence
hadith	Arabic, sayings of prophet Muhammad
Indra	Skt., king of the gods
kaliyuga	Skt., Dark Era
karma/kamma	Skt./Pali, action, or the fruits of action
khalifa	Arabic, vice regent of God on earth, lit. successor
kiai	Indonesian, Islamic teachers and leaders of Islamic boarding schools
kitab kuning	Indonesian, the *Yellow Books*
Kṛṣṇa	Skt., an incarnation of Viṣṇu
madrasah	Arabic, Muslim religious schools
nafsu	Indonesian, desire, yearning, passion
nibbāna	Pali, literally: "extinction (e.g. of a flame)", in Buddhism designation of the state of liberation, release from birth and death
paññā	Pali, wisdom
pesantren	Javanese, Islamic boarding schools
phatthana	Thai, development
phi	Thai, spirit
Phra Khru/Phrakhru/Phra	Thai, ecclesiastic titles for Thai Buddhist monks
Phya	Thai, a title
Pṛthivī	Skt., personified earth, earth goddess
Pṛthu	Skt., a mythical king
purāṇa	Skt., literally: "ancient"

* Cases of multiple mentioning only.

Ṛgveda	Skt., the most ancient sacred text in India
samādhi	Skt./Pali, concentration
santri	Javanese, student of an Islamic boarding school/ a pious person
sīla	Pali, morality
Śiva	Skt., a great Hindu god
sṛṣṭi/visṛṣṭi	Skt., discharge, emission, production
tawhid	Arabic, the oneness of God
thau jam	Thai, mediator between humans and spirits
ulama	Indonesian, religious scholars
Viṣṇu	Skt., a great Hindu god
(*yang*) *pen thammachat*	Thai, natural
yoga	Skt., literally: "the act of yoking", as in to yoke horses to a chariot
yuga	Skt., age of time, era

Notes on Contributors

Arnez, Monika

is Assistant Professor in Austronesian studies at the Department of Southeast Asian Languages and Cultures, University of Hamburg, Germany. Her research interest is on Islam as related to social practices and the environment, youth movements, gender and literature. She has published on the transformation of gender dynamics in Indonesia, the propagation of Islam through narratives, women's empowerment in Muslim mass organisations and religion in the practice of daily life in Indonesia, and edited two volumes: *Repräsentationen von Moral und Sexualität in literarischen Texten der malaiischen Welt*. Hamburg: Hamburger Südostasienstudien, 2011; with Sabine Müller, *Das Theater sind wir: Ausgewählte Theaterstücke von Goenawan Mohamad und Iswadi Pratama*. Hamburg: Hamburger Südostasienstudien, 2013.

Baghel, Ravi

is a geographer and Post-doctoral Research Fellow at the Cluster of Excellence "Asia & Europe in a Global Context" at Heidelberg University, Germany. His research has focused on large river control projects in India, and he is currently working on the epistemology of Himalayan glaciers, which is a part of his broader research interest in the role of knowledge in mediating human–environmental interactions. Selected references: with Marcus Nüsser, "Discussing Large Dams in Asia after the World Commission on Dams: Is a Political Ecology Approach the Way Forward?" *Water Alternatives* 3.2 (2010): 231–248; "Knowledge, Power and the Environment: Epistemologies of the Anthropocene." *Transcience* 3.1 (2012): 1–6.

Batabyal, Somnath

is an anthropologist of the media and lecturer in "Media in Development" at SOAS, University of London, United Kingdom, where he teaches critical media theory with a focus on emerging technology in Asia and Africa. His present research focus is on election processes in India and participatory democracy at the margins. His most important publications include: "Constructing an Audience: News Television Practices in India." *Contemporary South Asia*, 18.4 (2010) 387–399; *Making News in India*. New Delhi, London, Routledge, 2012.

Betz, Joachim

is Professor of Political Science and Senior Research Fellow at GIGA Institute for Asian Studies, Hamburg, Germany. His current research interests lie in the field of political parties in India, regional powers in Africa, Asia, and Latin America, and the social policies of Indian states. Among his recent publications are: "Schwerpunktverschiebung nach Asien? China und Indien als aufstrebende Großmächte." *Peripherie* 28 (2009): 111, 280–300; "India: The Interaction of Internal and External Factors in Foreign Policy." In *Regional Leadership in the Global System*, edited by Daniel Flemes, 237–254. Farnham, Surrey, England: Ashgate 2010.

Darlington, Susan M.

is Professor of Anthropology and Asian Studies at Hampshire College in Amherst (MA), US, where she teaches environmental anthropology and Buddhist studies. Her research focuses on socially engaged Buddhist monks in Thailand who undertake rural development and environmental conservation based on their interpretations of Buddhist teachings. Her important publications include: "The Good Buddha and the Fierce Spirits: Protecting the Northern Thai Forest." *Contemporary Buddhism* 8.2 (2007): 169–185; *The Ordination of a Tree: The Evolution of the Thai Buddhist Environmental Movement*. Albany, NY: SUNY Press, 2012.

Eucker, Dennis

is a political scientist. He holds a PhD degree from Hamburg University, Germany. His research interests include regional and local responses to climate change, environmental resource management and institutional adaptation capacities. Among his recent publications are: with Jonas Hein, *Klimawandel in Südostasien: Die ASEAN als Wegbereiter einer regionalen Klimapolitik? GIGA Focus Asien* 4. Hamburg: GIGA, 2010; *Facing Climate Risks, Variability, and Change – A Capacity Gap Analysis (CGA) of Local Government in Mozambique*. Maputo: Africa Climate Change Resilience Alliance (ACCRA), 2011.

Haum, Rüdiger

is a political scientist. Currently, he is a Research Analyst in Technology and Innovation at the German Advisory Council on Global Change (WBGU), Berlin, Germany and Research Associate with the Environmental Policy Centre at the Free University of Berlin. His main research interests are ecological modernisation and international transfer of sustainable technologies. Among his recent publications are: "Energy Security, Rural

Electrification, and Market-Based Climate Change Intervention by the Global Environment Facility in India: The Limits of Transformation." *International Energy Journal* 11.4 (2010): 115–125; "Project Based Market Transformation in Developing Countries and International Technology Transfer: The Case of the Global Environment Facility and Solar PV." In *Low-Carbon Technology Transfer to Developing Countries: From Rhetoric to Reality*, edited by David Ockwell and Alexandra Mallet, 185–210. London: Earthscan, 2012.

Nugteren, Albertina (Tineke)
was trained an Indologist and is Associate Professor of Religious Studies at Tilburg University, the Netherlands. Her current research interests range from the interaction between nature and religion to ritual dynamics, especially among Hindus in diaspora. Her most important publications in these fields include: *Belief, Bounty, and Beauty: Rituals around Sacred Trees in India.* Leiden, Boston: Brill, 2005; "Hindu Ritual Dynamics: Case Studies from Contemporary Western Europe." *Journal of Religion in Europe* 2.2 (2009), Special Issue.

Nüsser, Marcus
holds a chair for Geography at the South Asia Institute at Heidelberg University, Germany. His research focuses on issues of human–environment interactions, landscape transformation, glacier studies and regional development processes in high mountain regions, especially the Himalayan region. Selected references: "Kilimanjaro and Mount Kenya: Colonized Mountains and their Rediscovery as Symbols of Global Climate Change." *Geographische Rundschau – International Edition* 5.4 (2009): 26–32 and supplement (map); with Susanne Schmidt, "Changes of High Altitude Glaciers from 1969 to 2010 in the Trans-Himalayan Kang Yatze Massif, Ladakh, Northwest India." *Arctic, Antarctic and Alpine Research* 44.1 (2012): 107–121.

Schuler, Barbara
is an Indologist (classical and modern), with a focus on Tamil, History of Religion, and Social Anthropology at Hamburg University, Germany, where she until recently held the position of the Associate Director of the Center for Buddhist Studies at the University. Her current research focuses on emotion studies, rituals and texts of indigenous South Indian traditions, visual anthropology and memory studies. Among her most recent publications are: "The Dynamics of Emotions in the Ritual of a Hot Goddess." *Nidan: An International Journal for the Study of Hinduism* 24 (Special Issue, December 2012): 16–40; "Stifter und Religion." In *Stifter und Mäzene und ihre Rolle in der*

Religion: Von Königen, Mönchen, Vordenkern und Laien in Indien, China und anderen Kulturen, edited by Barbara Schuler, 1–16. Wiesbaden: Harrassowitz, 2013.

Seeger, Martin
is Associate Professor of Thai Studies at the University of Leeds, Great Britain. Much of his research focuses on the life and work of the Thai Buddhist thinkers Buddhadasa Bhikkhu (1906–1993) and Phra Payutto (b. 1939). He has also studied the work of development monks in the northeastern region of Thailand and biographies and the veneration of female saints in modern Thai Buddhism. His publications include: "Phra Payutto and Debates on 'On the Very Idea of the Pali Canon' in Thai Buddhism." *Buddhist Studies Review* 26 (2009): 1–31; "'Against the Stream': The Thai Female Buddhist Saint Mae Chi Kaew Sianglam (1901–1991)." *South East Asia Research* 18.3 (2010): 555–595.

Stargardt, Janice
is a Professorial Research Fellow in Historical Archaeology and Geography of Asia in the Department of Geography, University of Cambridge and Fellow and Director of Studies in Archaeology and Anthropology at Sidney Sussex College in Cambridge, United Kingdom. Two themes in her research in Andhra Pradesh, Sri Lanka, Burma and Thailand have been, firstly, early agricultural intensification and irrigation, and secondly, early maritime trade in the Indian Ocean and the South China Sea. She is the author of: with G. Amable and B. Devereux, "Irrigation is Forever: A Study of the Post-Destruction Movement of Water across the Ancient Site of Sri Ksetra, Central Burma." In *Remote Sensing: A New Tool for Archaeology*, edited by R. Lasaponara, and N. Masini, 247–268. Heidelberg, London, New York: Dordrecht, Springer, 2012; with G. Amable, "Water from the Ancient City: A New Method of Satellite Surveys of Irrigation Water at Sri Ksetra, Burma." In *Proceedings of the SEAMEO-SPAFA First Conference on Southeast Asian Archaeology*, 26pp. Bangkok: SPAFA (in print).

Conventions

The orthographic conventions followed in chapter 1 (Nugteren) and chapter 2 (Seeger) use for Sanskrit and Pali terms the transcription system followed in the Monier Williams's *A Sanskrit-English Dictionary*. Generally, the non-English terms are italicised, and plurals are indicated by the addition of a non-italicised English -s. These chapters dispense with such italics when such words are frequently used (e.g. *pesantren*, *santri*, *kiai*, and *ulama*). Words that have become part of the English language (such as *Brahmin*) have been written without diacritical marks. This is also the case for names of languages (*Sanskrit, Pali*) and well-known geographical names. Any brackets [...] used within citations contain text added by the contributors, if not stated to the contrary.

The international bibliographical convention for Thai authors calls for authors to be listed alphabetically according to given (first) names rather than family names, which were only introduced around 1920. Given and family names are not separated by commas.

Introduction

The Dynamics of Climate and the Dynamics of Culture

Barbara Schuler

> [Nature's] condition, rightly interpreted, reveals a society's culture and traditions as directly as does a novel or a newspaper or a code of laws
>
> (HISTORIAN RODERICK F. NASH)[1]

Smog over Mumbai and Delhi.[2] Who has not seen pictures of it? But how many people know that the air of some megacities contains so great a quantity of serious pollutants that in India's capital alone more than 10,000 deaths per year are caused by them?[3] How this risk is perceived is a question that usually

1 Nash, Roderick Frazier, ed., *American Environmentalism: Readings in Conservation History* (New York: McGraw-Hill, 1990), 1.

2 Regional masses of haze or the so-called atmospheric brown cloud (ABC) to a great extent is caused by black carbon emissions. Black carbon has been identified as a climate forcing source; see Bano, Tarannum et al., "Variation in Aerosol Black Carbon Concentration and its Emission Estimates at the Mega-City Delhi," *International Journal of Remote Sensing* 32.21 (2011): 6749–6764. doi: 10.1080/01431161.2010.512943.

3 Mudur, Ganapati, "Atemlos in der Großstadt [Breathless in the City]," *MaxPlanckForschung* [MaxPlanckResearch] 4 (2010): 24. Models like Ri-MAP (Risk of Mortality-Morbidity due to Air Pollution) allow estimates to be made of how many deaths and hospitalisations can be ascribed to a particular pollutant in the air. Ri-MAP was developed by Bhola R. Gurjar together with Jos Lelieveld, director of the Max-Planck-Institut für Chemie in Mainz, Germany. See also Gurjar, Bhola R. et al., "Human Health Risks in Megacities due to Air Pollution," *Atmospheric Environment* 44.36 (2010): 4606–4613. doi: 10.1016/j.atmosenv.2010.08.011; and Gurjar, Bhola R. et al., "Emission Estimates and Trends (1990–2000) for Megacity Delhi and Implications," *Atmospheric Environment* 38.33 (2004): 5663–5681. – On megacities in India in general, see Ahuja, Ravi, and Christiane Brosius, ed., *Mumbai, Delhi, Kolkata: Annäherungen an die Megastädte Indiens* [Mumbai, Delhi, Kolkata: Approaches to Megacities in India] (Heidelberg: Draupadi-Verlag, 2006); on Mumbai, see additionally the highly instructive essay of Appadurai, Arjun, "Spectral Housing and Urban Cleansing: Notes on Millennial Mumbai," *Public Culture* 12.3 (2000): 627–651.

 | DOI 10.1163/9789004273221_002

comes with the cultural dimension strongly factored in, and so can vary greatly. This introductory essay considers the bidirectionality of climate and culture,[4] and the stark socio-cultural differences that set South and Southeast Asia apart from the West.[5] That it is India that will mainly be serving as an exemplary case is due to its being the country in the region I best know how to read.

---

4 We can no longer ignore the fact that cultures also shape climates and not only vice versa.

5 For an elaboration of this, see Chapman, Graham et al., *Environmentalism and the Mass Media: The North-South Divide* (London, New York: Routledge, 1997), 1–26, offering a concise evaluation of cultural responses to environmentalism in the Indian context. – Differences are revealed in everyday life, quite visibly so in the form of "interdependent" selves – a shared identity with other people, as reflected in language and imprinted emotions (e.g. shame, pride). Markus, Hazel R., and Shinobu Kitayama, "Culture and Self: Implications for Cognition, Emotion, and Motivation," *Psychological Review* 98.2 (1991): 224–253, provide a specific model that characterises the self in Western cultures as "independent" and in Asian cultures – most particularly – as "interdependent". Their model has become an important basis for investigations into concepts of self-conscious emotion. – Differences are further visible in the generation question. Demonstrating respect and acceptance for elders' norms is very high in South and Southeast Asia (though, for instance, in recent Indian films a paradigm shift in this traditional system is detectable). Cf. Assmann, Aleida, *Introduction to Cultural Studies: Topics, Concepts, Issues* (Berlin: Erich Schmidt Verlag, 2012), 119, who discusses the "relations between generations" on the basis of the Lévi-Straussian notion of hot and cold cultures. The obligation to accept the norms and values of the older generations is representative of a cold culture, whereas youth in hot cultures are "constantly finding fault with the older [generation]". – The place of religion and its capacity to exert influence differs, too. Hindu religions, for example, have no centralised clerical structure that could put pressure on the government. There are, however, spiritual leaders who act on the local level as eco-activists (e.g. Shri Shri 1008 Soham Baba, Founder and Chair of the Soham Baba Mission, who participated in an interview at the climate change studio on November 28, 2012 within the framework of the Doha Climate Change Conference, November 2012, accessed December 3, 2012. http://unfccc.int/meetings/doha_nov_2012/items/7285.php?middle=j). The case of Thai Buddhism is different, since Thailand has a centralised Buddhist structure. See Ishii, Yoneo, "Church and State in Thailand," *Asian Survey* 8.10 (1968): 864–871. The Islamic religious establishment in Indonesia, in turn, works in close cooperation with the government (see chapter 3). – For general introductions into the contexts of South Asia, Indonesia, and Thailand, see Europa Publications, ed., *South Asia 2013* (London: Routledge, 2012); Lamoureux, Florence, *Indonesia: A Global Studies Handbook* (Santa Barbara, CA: ABC-Clio, 2003); Terwiel, Barend Jan, *"Siam": Ten Ways to Look at Thailand's Past* (Großheirath: Ostasien Verlag, 2012), Peleggi, Maurizio, *Thailand: The Worldly Kingdom* (London: Reaktion, 2007), Grabowsky, Volker, *Kleine Geschichte Thailands* [A Concise History of Thailand] (München: Beck, 2010), Donner, Wolf, *Thailand: Land zwischen Tradition und Moderne* [Thailand: Country between Tradition and Modernity] (München: Beck, 1996).

1 The Situation Now

In the second decade of the 21st century, the question of climate change – and how to approach it in terms that go beyond the North–South, developed–developing divide – has gained particular urgency in the face of the current crisis of the earth system,[6] especially within the narrative frame of "Asia Rising"[7]. Climate change and how local cultures are coping with it is now the subject of daily media reporting as well as academic research (spanning the fields of political science, economics, sociology, anthropology, and the humanities). This discourse is complicated by the fact that the impressive economic growth in parts of the Asian world, particularly in India, has not been matched with similar progress in dealing with the environmental and climate issues it faces. India and the rising Southeast Asian states are struggling to manage their socio-economic problems (urban–rural, rich–poor, global–local, growing energy demand and population growth[8]). Climate change is a particularly significant stress factor for those areas of the world that already have warm climates.[9] India,

6 Scientists have tried to outline limits that humankind must respect in order to avoid endangering the earth's climate system. But, as we all know, the planetary limits for CO_2 have already been transgressed. See the report of the Intergovernmental Panel on Climate Change (IPCC): Intergovernmental Panel on Climate Change (IPCC), *Climate Change 2007: The Physical Science Basis. Working Group I Contribution to the Fourth Assessment Report* (Cambridge, New York: Cambridge University Press, 2007); Intergovernmental Panel on Climate Change (IPCC), *Climate Change 2007: Impacts, Adaptation and Vulnerability. Working Group II Contribution to the Fourth Assessment Report* (Cambridge, New York: Cambridge University Press, 2007); Schellnhuber, Hans Joachim, "Global Warming: Stop Worrying, Start Panicking?" *Proceedings of the National Academy of Sciences of the United States of America PNAS* 105 (2008): 38.

7 I make use of the book title of Acharya, Amitav, *Asia Rising: Who Is Leading?* (Singapore: World Scientific, 2008).

8 According to Census 2011, India's population reached on 1 March 2011 1,210 million (623.7 million males and 586.5 million females); see Publications Division Ministry of Information and Broadcasting Government of India, ed., *India 2013: A Reference Annual* (New Delhi: Publications Division Ministry of Information and Broadcasting Government of India, 2013), 7.

9 World Wildlife Fund (WWF), "Mega-Stress for Mega-Cities: A Climate Vulnerability Ranking of Major Coastal Cities in Asia," published 2010, accessed September 14, 2013. http://assets.panda.org/downloads/mega_cities_report.pdf; Mitra, A. P., and C. Sharma, ed., *Global Environmental Changes in South Asia: A Regional Perspective* (Dortrecht: Springer Netherlands, 2010); also Nair, K. Shadananan, "An Assessment of the Impact of Climate Change on the Megacities of India and of the Current Policies and Strategies to Meet Associated Challenges" (paper presented at Fifth Urban Research Symposium: Cities and Climate Change: Responding to an Urgent Agenda 2009, Marseille, June 28–30, 2009), accessed September 14, 2013. http://siteresources.worldbank.org/INTURBANDEVELOPMENT/Resources/336387-1256566800920/nair.pdf.

Thailand, and Indonesia, for example, are especially vulnerable to its effects, if only because they themselves have become major emitters. This is precisely what makes it compelling to focus on these three countries.

While in the West the scientific recognition of the importance of environmental and climate change has filtered down to the level of common discourse,[10] in India, Thailand, and Indonesia we should not expect the resonance to be as great. Most people in these countries may not even have heard the expression "global warming", much less have grasped its meaning.[11] This fact and the radical modernisation and high consumer aspirations – at a rapid rate and on a global scale – of the complex societies that form South and Southeast Asia have spurred increasing scholarly interest in how climate matters are being dealt with there. "Drive the car less, take the bus and train more often. Use less energy, cut down on water usage and recycle more"[12] will become the future mantra in South and Southeast Asia as it long since has been in the West.

2 Environmental Histories and Knowledge Transfer[13]

In order to understand the human–environment relationship, it is not sufficient to focus on the present alone. The relationship between individual cultural practices and the environment in the past also needs to be considered, though

10 Over the last 30 years, scientists have successfully developed a high degree of competence in environmental analysis and have put an immense amount of knowledge at our disposal. Our ability today to model complex climate systems is truly amazing, and this achievement has led to the development of a broad spectrum of technical solutions.

11 On India, see Chapman et al., *Environmentalism and the Mass Media*, 259–274; see also chapters 1 and 3 in this volume. – One may note that India's literacy rate is 64.84 per cent, varying in the different states of the union from 90.86 per cent in Kerala to 47 per cent in Bihar; see Publications Division Ministry of Information and Broadcasting Government of India, *India 2013*, 11.

12 Cited from Sutton, Philip W., *The Environment: A Sociological Introduction* (Cambridge, UK: Polity Press, 2007), 142.

13 Environmental history as an academic field emerged in the 1970s in the United States; see Hill, Christopher V., *South Asia: An Environmental History* (Santa Barbara, Oxford: ABC-CLIO, 2008), xii. Nash, *American Environmentalism*, 3, describes the environmental historian as one who "must work not only with the conventional tools of intellectual, political, and social history but also with the biological and physical sciences – particularly ecology and geography".

one difficulty in doing so is that to date we have very little relevant data prior to about 1800 as far as South and Southeast Asia are concerned.[14]

Scholarly approaches to studying how the environment and cultural groups stand in relation to one another have changed substantially over time.[15] Current archaeological work, for one, tends to confirm the bidirectional nature of the relationship.[16] Recognition of such bidirectionality opens up a significant new field to historical inquiry. Works dealing with this double trajectory are rare in the field of South and Southeast Asian environmental history, and even more so in the cultural history of climate change.[17] Two recent volumes are *South Asia: An Environmental History*[18] and *Southeast Asia: An Environmental History*.[19] From these researches the crucial significance of sustained adaptability[20] as a past cultural practice is becoming increasingly clear.

---

14 Grove, R. H., Vinita Damodaran, and Stapal Sangwan, ed., *The Environmental History of South and Southeast Asia: Nature & the Orient* (Oxford: Oxford University Press, 1998), 14.

15 Kealhofer, Lisa, "Human-Environmental Relationships in Prehistory: An Introduction to Current Research in South and Southeast Asia," *Asian Perspectives* 35.2 (1996): 111.

16 Or as Lisa Kealhofer terms it, the relationship is one of "coevolution"; see Kealhofer, "Human-Environmental Relationships," 112.

17 A cultural history of climate change would have to look not only into notions of collective and individual agency, but also into narrative forms of historiography and literary reflections on climate.

18 Hill, *South Asia*. On India, see also Grove et al., *The Environmental History of South and Southeast Asia*; and Dhavalikar, M. K., *Environment and Culture: A Historical Perspective* (Pune: Bhandarkar Oriental Research Institute, 2002). This work mentions the case of Fatehpur Sikri, the capital of the Mughal Empire founded by Akbar in present-day Agra District (U.P.), which had to be abandoned in the 16th century owing to famines (see pp. 266ff), in part as a result of the loss of a source of water caused by a course shift in the Jamuna River (see Hill, *South Asia*, 272). There is also a forthcoming series edited by Markus Nüsser worth noting: "Advances in Asian Human-Environmental Research," see: http://www.springer.com/series/8560 (accessed October 20, 2013).

19 Boomgaard, Peter, ed., *Southeast Asia: An Environmental History* (Santa Barbara, Oxford: ABC-CLIO, 2007).

20 "Adaptation refers to adjustments in ecological, social, or economic systems in response to actual or expected climatic stimuli and their effects or impacts. It refers to changes in processes, practices, and structures to moderate potential damages or to benefit from opportunities associated with climate change." Cited from United Nations, "United Nations Framework Convention on Climate Change," accessed December 3, 2012. http://unfccc.int/focus/adaptation/items/6999.php. The convention goes on to specify the individual stages: "Adaptation activities span five general components: observation; assessment of climate impacts and vulnerability; planning; implementation; and monitoring and evaluation of adaptation actions" (ibid.).

Yet while all adaptive activities, be they of a historical nature or not, are "inherently local" and are "necessarily based on contextual knowledge"[21] or previous experience,[22] it is uncertain that insights from the past could serve other regions in the face of future climate changes.[23] Indeed, this assumption would appear to be true if we consider as crucial factors "the complex interactions of time, geographic scale, the level of environmental data, and our inferences regarding past human and environmental relations".[24] Thus 'knowledge transfer' appears to be limited. However, it is not limited if the transfer of past knowledge happens within the same location, as the archaeologist Janice Stargardt (in this volume) strongly suggests. Her chapter shows vividly what environmental history can teach us, for instance, about adaptation through niche learning.

3 Cultural Dimensions

One may agree that it is hardly a coincidence that differences become particularly conspicuous when a system is in a state of crisis. This is no different in the case of the earth system crisis. People have to cope. But how they cope is a question of culture.

3.1 *Defining Culture*

When talking about culture, we initially need to spell out the special usage accorded to the word "culture" itself in the introductory essay and in the chapters that follow. The term is a much debated issue in sociological and anthropological scholarship, but so far has defied any common definition.[25] In this volume, we conceptualise culture not as a unified system, but rather as a repertoire or "tool-kit"[26] from which actors may choose different components and modules for approaching the world. We privilege culture in the sense of culturally shaped "habits, skills, and styles",[27] and view it as the key factor in many prac-

---

21 Adger, W. Neil et al., "Are There Social Limits to Adaptation to Climate Change?" *Climate Change* 93 (2009): 346. doi: 10.1007/s10584-008-9520-z.

22 Vedwan, Neeraj, and Robert E. Rhoades, "Climate Change in the Western Himalayas of India: A Study of Local Perception and Response," *Climate Research* 19.2 (2001): 109–117.

23 Adger et al., "Are There Social Limits?," 346.

24 Kealhofer, "Human-Environmental Relationships," 114.

25 Swidler, Ann, "Culture in Action: Symbols and Strategies," *American Sociological Review* 51.2 (April 1986): 273.

26 Ibid., 273, 277.

27 Ibid., 273.

tices that to others might appear political, economic, or something else.[28] In the contributions presented here, culture is usefully regarded as heterogeneous, contextual, and elastic, with boundaries that are blurred. It encompasses a broad range of both similarities and contrasts "between all sorts of categories" (classes, roles, groups, or nations).[29] In defining culture in Asia, the volume pointedly distances itself from the historical consciousness of culture deeply rooted in the intellectual tradition of Europe.

3.2 *Doing Culture*[30]

Doing culture in South and Southeast Asia goes hand in hand with multiple-coded practices which may easily overlap. That India, Thailand, and Indonesia cannot be treated in the first place against the background of postmodern experience is evident.[31] The cultural features that define them are too complex and contradictory. On the one hand we have cultural globalisation[32] (Darlington, Seeger in this volume) and the diffusion of Western lifestyle models (see Batabyal), which in local contexts give rise to hybridity[33] and creolisation,[34] thereby simultaneously repositioning the economy, technology, and nature (see Betz, Haum). On the other hand we witness these cultures being embedded in many temporalities at once, in the form, for instance, of archaic traditions that

28 The present volume thus extends – in terms of the *practice turn* (which goes beyond the *cultural turn*) – the object of cultural analysis to potentially all domains of society; see Hörning, Karl H., and Julia Reuter, ed., *Doing Culture: Neue Positionen zum Verhältnis von Kultur und sozialer Praxis* [Doing Culture: New Viewpoints regarding the Relation between Culture and Social Practice] (Bielefeld: Transcript-Verlag, 2004), 13, who further state that culture as practice connects the cultural dimension with the social dimension (ibid., 10). One of the most prominent theorists of practice is Pierre Bourdieu; see his *Outline of a Theory of Practice*, trans. Richard Nice (Cambridge: Cambridge University Press, 1977); and *The Logic of Practice*, trans. Richard Nice (Stanford: Stanford University Press, 1990).

29 Appadurai, Arjun, *Modernity at Large: Cultural Dimensions of Globalization* (Minneapolis: University of Minnesota Press, 1998), 12.

30 I adopt this wording from the book title of Hörning and Reuter, *Doing Culture*.

31 For a definition of postmodernity, see Bauman, Zygmunt, *Intimations of Postmodernity* (London, New York: Routledge, 1992), 31.

32 In passing one should recall that globalisation, strictly speaking, is not a new notion. The Asian history of globalisation goes at least back to the silk roads that connected eastern and western, and northern and southern, parts of the globe.

33 Despite its shortcomings, the concept of hybridity is useful, if used with care. For its usage, see Bhabha, Homi, *The Location of Culture* (New York: Routledge, 1994).

34 "[C]reole cultures come out of multidimensional cultural encounters and can put things together in new ways," see Hannerz, Ulf, *Cultural Complexity: Studies in the Social Organization of Meaning* (New York: Columbia University Press, 1992), 265.

continue to exist in parallel,[35] but also the cultural reappropriation of pre-modern features of identity, such as ethnicity[36] and religion (see Nugteren, Seeger, Arnez). We should, therefore, ask the question: what "habitus" (Bourdieu) has an individual in a country in which singularity does not exist, but where rather, as in India, a density of languages,[37] socio-cultural codes, and religions (Hinduism, Sikhism, Jainism, Buddhism, Christianity, popular religious traditions) exist. That this heterogeneity extends far beyond language and religion alone and can equally be witnessed in the uneven nature of regional economic and social development (a difference detectable from area to area) is all but evident. The same social group which is dominant in one area could be among those dominated in another. If we are intent on looking at such cultural dimensions in all their complexity, the overall socio-cultural development (the *longue durée*) should not be overlooked.

3.2.1 Adopting and Recreating – A Most Typical Cultural Practice

The most instructive feature of action patterns we can detect in this part of the world is the practice of adopting and recreating. Cultures, as in India still to a great extent, tend to practice inclusivism,[38] that is, to integrate new concepts and practices while at the same time keeping them subordinate to old ones, very much in contrast to Western cultural practices (particularly the United States), where constant innovation is the driving force – replacing old concepts and practices with new ones. Adoption and acceptance is thus typical for India. And this cultural practice we find in the Indian approach to modernity in par-

35 This includes indigenous tribes (Hindi, *ādivāsī*); see, regarding the Kanekes in Indonesia, Landmann, Alexandra, "Endangered Guardians of the Sacred Forest," *The Newsletter* 65 (2013): 42, accessed September 24, 2013. http://www.iias.nl/sites/default/files/IIAS_NL65_42.pdf.

36 On ethnic chic, see Tarlo, Emma, *Clothing Matters: Dress and Identity in India* (London: C. Hurst & Co Publishers Ltd., 1996), 284–317.

37 India boasts 122 recognised languages, not including the dialects; see Government of India, Ministry of Home Affairs, Office of the Registrar General & Census Commissioner, India, *Census 2001, Data on Language*, accessed August 20, 2013. http://censusindia.gov.in/Census_Data_2001/Census_Data_Online/Language/data_on_language.html. India is known for its language divide. Though Hindi is the official language throughout the country, along with English, in South India the official and administrative languages are the Dravidian tongues, such as Tamil.

38 On the typical Indian phenomenon of inclusivism, see Hacker, Paul, "Inklusivismus [Inclusivism]," in *Inklusivismus: Eine indische Denkform* [Inclusivism: An Indian Mode of Thought], ed. Gerhard Oberhammer (Wien: Institut für Indologie der Universität Wien, Sammlung De Nobili, 1983), 11–28, who coined the term "Inklusivismus", first and foremost within a religious context.

ticular. Modernity has typically been adopted and recreated.[39] It is not only Western inheritance, but also the circulation of India's own cultural heritage that goes into forming modernity in India. Adoption in India always involves adapting to local conditions, whether the adaptation is of techniques (Haum in this volume) or otherwise. Two outstanding examples are the Ambassador car, a creative adaptation of the Morris Oxford model,[40] and the Indian film industry (Bollywood), another locally adapted form of modernity, and at the same time a profoundly self-assertive alternative to Hollywood.[41] Each such case (and there are many others that could be mentioned) represents a creative adaptation, a transformation within the confines of its own cultural grounding.

Transformation of this kind and to this degree we can witness in the approaches and strategies of certain influential Thai Buddhist thinkers as well (Seeger in this volume). Their creative method of recycling past nature-related concepts and adapting them to current needs is truly amazing, and this achievement has led to the development of indigenous solutions to environmental problems.

3.2.2 Multiple Forms of Knowing

It is self-evident that scientific knowledge occupies a privileged place amongst the available sources of knowledge today (see Nüsser and Baghel, Stargardt, Darlington in this volume), but it is not the only form of knowledge. Cultural and local knowledge is being tapped as well. To this day religious traditions in Asia have performed some of the same functions as natural science, namely providing people with explanations of why things happen the way they do and when they do. Thus religions supply various models for the human–environment relationship (see Nugteren, Seeger, Arnez in this volume), one example being animistic systems (see Seeger), wherein trees, for instance, are social spaces sensitive to human action. Such intimate interwovenness between the

---

39 In the rest of this passage I draw on a talk given by Ashcroft, Bill, "The Multiplicity of Modernity: Post-colonial Studies and Globalization," podcast published July 15, 2009 by Internationales Zentrum für Kultur- und Technikforschung, Universität Stuttgart, lecture series "Kulturtheorien", accessed August 20, 2013. http://www.uni-stuttgart.de/izkt2/2009-07-15-Bill_Ashcroft-The_Multiplicity_of_Modernity_Post-colonial_Studies_and_Globali zation.mp3.

40 The Ambassador is a car manufactured by Hindustan Motors of India. It has been in production since 1958 with few improvements or changes. Accessed August 21, 2013. http://en.wikipedia.org/wiki/Hindustan_Ambassador.

41 See Ashcroft, Bill, "Bollywood, Postcolonial Transformation, and Modernity," in *Travels of Bollywood Cinema: From Bombay to LA*, ed. Anjali Gera Roy and Chua Beng Huat (Oxford: Oxford University Press, 2012), 3.

natural and the human spheres that is so central to the cultural imagination contrasts starkly with the axioms of scientific knowledge, wherein the two are divided.[42] One more notable phenomenon susceptible to cultural framing is natural catastrophes. Many rural people may well view a natural disaster in religious terms, seeing it as a divine form of intervention or punishment.[43] Nonetheless, they may be equally prepared to consult scientists rather than priests should they find themselves in a position of authority.[44] Environmental perceptions may be shaped as well by local knowledge (see Darlington, Nüsser and Baghel, Stargardt). Local practices and oral traditions testify to the dimensions of this knowledge, which may well include empirical knowledge of geophysical change or the like.[45]

Ultimately the multiple forms of knowledge concerning the human–nature relationship appear both contradictory and overlapping.[46] This corresponds to differences in frames.[47] None of the individual knowledge domains by itself captures the full repertoire of knowledge available to people. Thus to try to conceptualise the dimensions of knowledge from a single consistent point of view (valid at all times in all contexts for all people) is to ignore the fact of situatedness. This leads us to the premise that each frame of knowledge (spanning

42 Whether this suffices to explain why scientific arguments relating to the environment seem not to have made many inroads yet within traditional societies (see Seeger in this volume) remains an open question.

43 On local interpretations of natural disasters as divine intervention, see Vedwan and Rhoades, "Climate Change in the Western Himalayas of India," 113; they write: "[The] flood [of 1947] was, interestingly, always recalled in association with religious riots that occurred in the same year. Divine intervention to wash away bloodshed in the valley was given as an explanation of the phenomena."

44 Cf. Sutton, *The Environment*, 13.

45 See Cruikshank, Julie, *Do Glaciers Listen?: Local Knowledge, Colonial Encounters, and Social Imagination* (Vancouver: UBC Press, The University of British Columbia, 2005), 23–124.

46 Perception of the human–nature relationship is reflected in, among other things, astrology (cf. Sutton, *The Environment*, 2), a field of knowledge of great repute in India. Astrological charts, based on an understanding of star constellations and the natural rhythms and processes of their movement, are assiduously consulted as guidelines for managing both planned events in life (e.g. marriage, travel) and what to do in crises and other unforeseen situations.

47 *Frames* in the sense of Gitlin, Todd, *The Whole World Is Watching: Mass Media in the Making & Unmaking of the New Left* (Berkeley et al.: University of California Press, 1980), 7, as "patterns of [...] interpretation, and presentation".

the scientific, cultural, local, temporal, and spatial)[48] needs to be seen in its own right as one among many that may be emphasised in any given situation. That people in South and Southeast Asia can shift with ease from one frame to another at will has been made abundantly clear.

Knowledge has causal force, because it shapes the capacities for developing strategies of action. This is precisely what makes it important for the present volume (see e.g. Nüsser and Baghel who challenge some knowledge practices).

4 Dimensions of Agency

One particularly useful way to approach the volume is by considering the agentic activities described in it. Agency is the key theme woven throughout its chapters. The volume does not set out to find perfect answers to all possible questions of agency, but all the essays do attempt, either directly or indirectly, to come to grips with the basic question of agency in the cultural practices we are studying. The theme of agency addresses a central question: What kind of agency is involved in the development of crafting strategies in regard to climate and environmental change? Who exercises the influence needed for putting climate strategies into place? Is it governmental institutions, single actors, urban entities, or international bodies?[49] Within this context we take on the challenge of analysing the various dimensions of agentic activity, particularly in the case of India, inasmuch as this country is taking ambitious steps to position itself globally as an economically strong modern country, if at the cost of bidding to become the world's third largest pollution emitter by 2035 (see Betz in this volume).

4.1 *Defining Agency*

In this volume, following Mustafa Emirbayer and Ann Mische, we conceptualise agency as a

> temporally embedded process of social engagement, informed by the past (in its "iterational" or habitual aspect) but also oriented toward the future (as a "projective" capacity to imagine alternative possibilities) and

48 The Himalayan glaciers exemplify a context that gives rise to multiple dimensions of knowledge particularly well (see Nüsser and Baghel).

49 It is, of course, evident that each individual domain is part of a larger complex field of discourse in which the different actors position themselves in relation to other groups and diverse frames of meaning, interest, and value.

> toward the present (as a "practical-evaluative" capacity to contextualize past habits and future projects within the contingencies of the moment).[50]

I conceive agency in terms of three basic dimensions, based on the theoretical model of Emirbayer and Mische: the iteration component, the projective component, and the practical-evaluative component. The first dimension of agency refers to the

> selective reactivation by actors of past patterns of thought and action, as routinely incorporated in practical activity, thereby giving stability and order to social universes and helping to sustain [...] institutions over time.[51]

The second dimension of agency comprises

> actors of possible future trajectories of action, in which received structures of thought and action may be creatively reconfigured in relation to actors' hopes, fears, and desires for the future.[52]

And the third dimension of agency involves the

> capacity of actors to make practical and normative judgements among alternative possible trajectories of action, in response to the emerging demands, dilemmas, and ambiguities of presently evolving situations.[53]

These specific views on agency offer opportunities for systematic arguments about the way in which agency shapes action and about which of the agentic dimensions is most forceful.

The underlying view which I put forward will derive from the various agentic activities that go hand in hand with different actor frames described in the volume's studies. According to these latter we can recognise agency (the competence to act upon and within the world) being exercised by religious authorities (Arnez, Nugteren, Seeger), agency being exercised by the state, business, urban entities, and international bodies (Batabyal, Betz, Eucker,

50 Emirbayer, Mustafa, and Ann Mische, "What Is Agency?," *American Journal of Sociology* 103.4 (January 1998): 962.

51 Ibid., 971.

52 Loc. cit.

53 Emirbayer, and Mische, "What Is Agency?," 971.

Haum, Nüsser and Baghel, Stargardt), and, finally, agency being exercised by local rural people (Darlington, Stargardt).

4.2 *The Iteration Dimension of Agency*

Religion seems to provide the best evidence for the iterational dimension of agency. The three chapters (Nugteren, Seeger, Arnez) that examine attempts by religious groups to draw on and affirm religious heritage so as to find a place for their agenda within the context of climate-related discourses offer good examples. Having recourse to past models enables religious actors to picture historical continuity and rootedness while simultaneously responding to an actual challenge. This is particularly well documented in Albertina Nugteren's chapter, which studies the Indian Hindu context. From this chapter it becomes abundantly clear that tapping the Hindu archive for coping strategies seems to happen in a way that confirms non-reflective reproduction of past routine.[54] Yet it would seem crucial for religious agents who are concerned with climate matters to explore past notions of the environment by examining their validity today. The question of religious 'model transfer' is therefore one extremely important consideration, first of all since pre-modern societies cannot of course be considered to have been ecologically aware in the modern sense.[55]

The urban entities that Somnath Batabyal describes are another category that falls under the iterational dimension of agency.[56] Due to the fast growing urban middle class that has profited from economies' recent growth, this category of agency is now the subject of interest in academic debates and discussions.[57] It has not been so for very long. The focus of most analyses of

54 For arguments that social norms prevent individuals from acting environmentally responsibly, see Fransson, Niklas, and Tommy Garling, "Environmental Concern: Conceptual Definitions, Measurement Methods, and Research Findings," *Journal of Environmental Psychology* 19.4 (1999): 373. People in India, and traditional Hindus in particular, have rigid criteria for determining individual purity, but this is offset by their great toleration of environmental impurity around them.

55 I echo here Harris, Ian, "Getting to Grips with Buddhist Environmentalism: A Provisional Typology," *Journal of Buddhist Ethics* 2 (1995): 180.

56 Of course, I assume with Emirbayer, and Mische, "What Is Agency?," 1008, "that it is possible to be (primarily) [nonreflective-]iterational in one frame, projective [i.e. highly future-oriented] in another, and practical-evaluative in yet a third" (my square brackets, BS).

57 For an introduction, see Brosius, Christiane, *India's Middle Class: New Forms of Urban Leisure, Consumption and Prosperity* (New Delhi, London: Routledge), 2010; Brosius, Christiane, "Die Vorstellung vom 'Neuen Indien': Anmerkungen zur unfassbaren Mittelklasse [The Image of 'New India': Annotations on the Incomprehensible Middle Class]," *Der Bürger im Staat* [The Citizen in the State] 59.3–4 (2009): 220–227, accessed August 31, 2013.

environmental discourse has been on rural and forested areas, and on subaltern classes (as opposed to the elite), and only rarely on urban areas and their populations.[58] Yet in terms of environmental footprints there is good reason to focus on the agentic activity of urbanites, whose consumer aspirations and demands are high. A member of the affluent globalised strata in India emits four and a half times more CO_2 than those who are poor.[59] In the capital of India, as Batabyal shows, this category of actor aims at global networks and contrastive lifestyles. Its notion of urban space conforms to political models of economic development. Indeed, the agency of the globalised middle class is geared not only to past routine but also positions itself within a whole complex of processes involving constellations of power and control in the new global imaginary.

4.3 *The Projective Dimension of Agency*

The alternative dimension of agency is considered by Martin Seeger with regard to Thailand. The two leaders of Thai Buddhist eco-religion that he deals with are good examples of the projective, or "creative reconstructive"[60], dimension

URL: http://archiv.ub.uni-heidelberg.de/savifadok/volltexte/2010/1161. For figures regarding the middle class population, see Farrell, Diana, and Eric Beinhocker, "Next Big Spenders: India's Middle Class," *Newsletter International*, May 19, 2007, McKinsey Global Institute, accessed August 31, 2013. http://www.mckinsey.com/insights/mgi/in_the_news/next_big_spenders_indian_middle_class: "The middle class currently numbers some 50 million people, but by 2025 will have expanded dramatically to 583 million people – some 41 percent of the population. [... In 2025 India will have become] the world's fifth largest consumer market." As Brosius, "Die Vorstellung vom 'Neuen Indien'," 221f., states:

"One can understand the significance of the new middle class only against the background of deeply rooted primary loyalties and relationships of a social nature, which constitute a major part of business life, everyday activity, and religious practice in the [...] Hindu [...] society. Very generally put, the 'classical' Indian society is [...] highly hierarchical and segregated. [...] Within a 'macrostructure' that governs social status and job classifications as well as ritual impurity and purity, countless other 'subcastes' (jati) exist as well. [...] Economic liberalisation turned the system upside down, with status coming no longer through birth but through achievement. Therefore, a blend of socially mobile groups developed in the 1990s. [...] Today, it is those groups that significantly shape the new middle class [...]. This wide variety is complicated by [people's] association with language groups, religions, and sects. [...] In many cases, however, India's backbone is still the family or the clan [...]." (Translated from German, BS.)

58 Mawdsley, Emma, "India's Middle Classes and the Environment," *Development and Change* 35.1 (2004): 79–103.

59 See Billett, Simon, "Dividing Climate Change: Global Warming in the Indian Mass Media," *Climatic Change* 99 (2010): 3. doi: 10.1007/s10584-009-9605-3.

60 Emirbayer, and Mische, "What Is Agency?," 984.

of agentic activity. These actors recompose or otherwise innovate practices based upon traditional nature-related concepts so as to meet new needs. Giving rise as it does to new visions, this form of agency may prove helpful and applicable when it comes to getting people to tackle climate concerns, particularly in view of the religious identities that are embedded within it.

In Thailand's changing world, marked by the increased impact of globalisation on society, we find further examples of the projective dimensions of agency (see Darlington). Here the agency rests not with religious authorities but rather with local rural people – the relatively economically weak group that has least contributed to the climate problem. Susan M. Darlington's study offers a case of self-organised groups that by choice have taken on the role of 'catalysts of change' rather than of victims. Here solutions do not come from Thai authorities engaged in implementing a particular model of economic development (for example, industrial agriculture, industrialisation, dam construction, globalised finance), but from single farmers and communities, and from, in particular, Buddhist monks who engage in rural development activism (for example, integrated agriculture) and environmental conservation. The case study offers a good example of how in the face of climatic uncertainties individuals take precautions of their own, and in doing so reveal their perception of risk and controllability,[61] their own decision-making processes, and the intended result of their collective actions. Most importantly, it shows how climate solutions may proceed through the transformative and circulatory dynamic of local rural agency with a clear trajectory to future possibilities.[62]

Green Islam in Indonesia offers another case of agentic capacity (Arnez in this volume); all the more since being guided by Islamic mass organisations. From its attempt to put responses to environmental challenges on a religious footing and integrate eco-theology into Indonesian culture (and particularly the educational system), we can gain some understanding of the dynamics driving this new direction in agentic orientation. Here the actors work hand in hand with state leaders to administer a nation-wide regime of environmental control and sustainable development that is compatible with Islam. The complex agency of the religious bodies is thereby fairly clear. If we see green Islam as being

61 On the perception of risk, see particularly Adger et al., "Are There Social Limits?" See also Lupton, Deborah, ed., *Risk and Sociocultural Theory: New Directions and Perspectives* (Cambridge: Cambridge University Press, 1999).

62 The break with routine, new forms of action, and the creativity of individual actions emerge here as the practice of problem solving. On the creative dimension of action, see Joas, Hans, *Die Kreativität des Handelns* [The Creativity of Action] (Frankfurt a. M.: Suhrkamp, 1996), 15.

more a potential for change than a protector of the status quo, we will have gained a better understanding of its place in this transformative process.

4.4 *The Practical-Evaluative Dimension of Agency*

Joachim Betz, Dennis Eucker, Rüdiger Haum, Marcus Nüsser and Ravi Baghel, and Janice Stargardt bring in another angle on agency. The actors whom they describe focus on evaluation, decision-making, and execution. The agents here are the state, business, and international bodies. Betz, for instance, takes India as a cautionary example of how state actors only after a long, circuitous process arrive at a decision to shift towards a more climate-friendly national and international policy. It is the ranking of problems within the hierarchy of national issues that is of most concern here,[63] above all in terms of the dynamics of different interest groups. But even though the state provides the main framework for cultural organisation in this case, we should not expect results to be inordinately great. Despite common agreement, in practice fractures occur at other stages, emerging most clearly in the wide gap between introduced regulations (e.g. relating to low carbon emissions or energy audits) and their execution. Agentic capacity is only one side of the coin. The mere fact that agents attempt to take action in a regulatory capacity does not guarantee a successful outcome.[64]

This is not to say that the state does not exercise agency successfully, as demonstrated in the case of a rural electrification programme (see Haum), the Solar Photovoltaic Programme, underwritten by the Indian government and meant to phase out domestic carbon dioxide emitters (such as kerosene lanterns or paraffin lamps), while in the process seeking to promote the solar industry. In a country that is rich in sunlight, the governmental agency is here responding perfectly well to the challenges and contingencies of climate change. This agentic activity is at the same time both protective and transformative.

New notable agentic interventions are also coming from influential business associations (see Betz) – interventions, it is natural to think, that are based on seeing climate change in monetary terms (financial benefits through solar

63 As to ranking, Sutton, *The Environment*, 28, asks a very important question: "Are environmental problems ranked in order of their real potential harm or has the ranking been created using some other criteria?" The study carried out by Betz in this volume comes up with some answers to this question.

64 Compare Dembowski, Hans, "Environment and Politics in India," in *Taking the State to Court: Public Interest Litigation and the Public Sphere in Metropolitan India*, ed. Hans Dembowski (Asia House's online version of the book originally published in 2001), 63–82, accessed November 28, 2012. http://www.asienhaus.de/public/archiv/taking_the_state_to_court.pdf. Cf. Emirbayer, and Mische, "What Is Agency?," 1008.

production), while at the same time positioning themselves globally as key players in green growth. These actors are important in climate change terms because they offer a very dynamic example of the transformational and cultural process by which adaptation proceeds.

One interesting nexus of agencies comes to light in Stargardt's study of a revived ancient irrigation system. Her chapter connects the line between climate-coping strategies of ancient rural people, its rediscovery by an international scientific body, and the action this triggers within present Thai governmental policy. It is precisely the 'relational' concept of agency (actors enter into relationship with both past actors and present co-actors) that makes this case study so compelling.

The need to find new approaches to coping with global warming brings new actors from outside to the scene: a global scientific community and acknowledged regional institutions. Eucker, for instance, describes the work of the regional organisation ASEAN towards implementing a climate governance regime among its members, which are marked by stark differences in their economies (for example, Thailand, Malaysia vs. Laos, Cambodia) and highly varying levels of greenhouse gas emissions (Indonesia and Singapore are two of the largest emitters per capita worldwide, whereas other members contribute little). The strength of this regional actor lies in its institutional capacity that allows considered decisions and coordinated advances towards sustainable development.[65]

Another international key player studied in our volume is the scientific community (see Nüsser and Baghel). As the most privileged agent of production of knowledge in the climate debate, it is not only responsible for how environmental knowledge is produced, but also for when, how, and by whom such knowledge is transferred. The special place that science enjoys is dependent on its reliability. And indeed, as Nüsser and Baghel state, "it is in their [i.e. scientists'] power to discredit and distort climate change knowledge", if only by infiltrating 'symbolic' knowledge (e.g. glaciers as icons of global warming) into public discourse. If we take the dynamic geophysical processes as real, but the interpretation given to them as a result of scientific framing (frame-setting and

65 On sustainable development, see the report of the Brundtland Commission in World Commission on Environment and Development (WCED), *Our Common Future, Report of the World Commission on Environment and Development* (Oxford: Oxford University Press, 1987), 8, and German Advisory Council on Global Change (WBGU), *World in Transition*, 395 (sustainable development as inclusive of ecological, economic, and social balancing).

frame-sending),[66] we will gain a better idea of the complex agency of scientific bodies.

Agency shapes action in various ways. As the above analysis shows, projectivity and practical-evaluative activity are the most forceful agentic types in times of crisis and uncertainty.[67] The projective agents are particularly effective in change management, because in the transition to new situations they act creatively, developing alternatives that integrate past and future approaches. Assuming that the projective-transformational and the directive practical-evaluative types of agency do indeed play a significant role in exercising the influence needed for putting climate strategies into place, a further question arises: Are actors in interface positions[68] in times of crisis and uncertainty more inclined to break with routine and to embrace untried possibilities and laboratory-like experimentation than those in more contained settings? In view of the monks, ASEAN, and the global scholarly community's successes as bridgeheads, it would indeed seem so.

5 About the Chapters

The complexity of our subject is apparent from the table of contents. This book brings together scholars from very different disciplinary perspectives and presents material relating to India, Thailand, Indonesia, ASEAN and the Himalayan glaciers. Some of the essays explore ideas and the past, while others examine contemporary practices or prospective courses of action. Included are essays on religion, archaeology, geography, political and media sciences, anthropology, technology, and international affairs. While none of them presupposes the others, their seemingly distinct ambits are interlinked.

---

66 I owe the expressions "frame-setting" and "frame-sending" to the research project "Framing Climate Change" based at the Universität Zürich, International & Comparative Media Research, accessed September 15, 2013. www.ipmz.uzh.ch/Abteilungen/International-ComparativeMediaResearch/Research/Framing.html.

67 On uncertainty, see Hornidge, Anna-Katharina, and Christoph Antweiler, ed., *Environmental Uncertainty and Local Knowledge: Southeast Asia as a Laboratory of Global Ecological Change* (Bielefeld: Transcript-Verlag, 2012); see also Casimir, Michael J., ed., *Culture and the Changing Environment: Uncertainty, Cognition and Risk Management in Cross-cultural Perspective* (New York, Oxford: Berghahn Books, 2008).

68 Emirbayer, and Mische, "What Is Agency?," 1007, call it "bridging positions". On transformer types, cf. Hannerz, *Cultural Complexity*, 257–258.

5.1 *Ideas*

Ideas form the starting point of the present volume. The essays in this part of the volume delve into various religious traditions in different geographical regions: Hinduism in India, Buddhism in Thailand and Islam in Indonesia.[69] The questions raised concerning green Islam, in particular, have long been neglected, even though it is obviously crucial to understand its exact nature. Buddhism and Hinduism are often associated – in both popular and academic discourse – with an attitude of wishing to avoid harming the environment,[70] yet few have investigated the concrete impact these traditions and their adherents have had on the natural world. All three chapters in this section agree that eco-religion is a positive factor in learning how to act in a climate-friendly manner. No chapter, to be sure, paints the situation in utterly stark terms, though each finds clear evidence that might justify doing so. The essays also reflect a serious and ongoing theoretical debate, namely the long-term prospects or future direction of eco-religion in a time of climate change (Nugteren and Seeger).

Albertina Nugteren examines texts – both old and contemporary ones – that promote traditional Hindu notions of environmental awareness, looking penetratingly at the rhetoric practised and fallacies produced within the Indian discourse relating to eco-issues. She reminds the reader of the clear distinction between what is purposefully and what is only incidentally eco-friendly in South Asian traditional worldviews and practices, noting that ideas and day-to-day practices diverge considerably.[71]

Studying a very different context, situated not in the distant past but in the present, Martin Seeger explores empirically the contemporary Thai Buddhist view of nature-related concepts and lived practices. The chapter describes the transitional period through which eco-Thai Buddhism is currently wending its way. It focuses on various new approaches being taken by influential actors

69 Hinduism and Buddhism are each a complex fabric woven out of many strands of tradition. Thus neither term stands for a homogeneous religion. Similarly, no one unified group comprises the totality of Muslims.

70 On Buddhist religions, see Schmithausen, Lambert, "Buddhism and the Ethics of Nature: Some Remarks," *The Eastern Buddhist* 32.2 (2000): 26–78; Sørensen, Henrik H., "Of Eco-Buddhas and Dharma-Roots: Views from the East Asian Buddhist Tradition," in *Nature, Environment and Culture in East Asia: The Challenge of Climate Change*, ed. Carmen Meinert (Leiden: Brill, 2013).

71 For similar findings for Western societies, see German Advisory Council on Global Change (WBGU), *World in Transition*, 77; see also Eckes, Thomas, and Bernd Six, "Fakten und Fiktionen in der Einstellungs-Verhaltens-Forschung: Eine Meta-Analyse [Fact and Fiction in Attitude-Behavior Research: A Meta-Analysis]," *Zeitschrift für Sozialpsychologie* [Journal for Social Psychology] 15.4 (1994): 253–271; Swidler, "Culture in Action," 275.

within the religious community in an effort to find solutions to environmental problems, and concludes that charismatic Buddhists and state authority have been sharing guiding roles, but that scientific arguments relating to climate and the environment seem not to have made much of an impact yet on the general population.

Monika Arnez's paper focuses on the nexus between green Islam, Indonesia, and its government. She examines a religion that thinks of time as being linear and propounds the oneness of God, very much in contrast to the religions that originated in India, which propound rebirth and see time as cyclical. She shows in particular that Muslim mass organisations in Indonesia (a country that is among the largest greenhouse gas emitters worldwide) have been instrumental in addressing environmentalist concerns there, and describes the great influence their projects have had on government policy.

Arnez's chapter makes one thing perfectly clear: While the South Asian religions and Abrahamic religions could hardly be more different in their theology, in practice there is little difference in the environmental policies of predominately Hindu India and predominately Muslim Indonesia. This would suggest that the notion that Hindu and Buddhist religions respect nature more than the monotheistic one is not borne out by the facts.[72]

5.2 *Past*

Changing climatic conditions is the theme of the next part of the volume. The two case studies in this section deal with Thailand and the Himalayan glaciers. A key concept running through both essays is that of knowledge practices.

In her contribution, Janice Stargardt provides a unique and exciting focal point for studying changing human–environment relationships. The chapter introduces the reader to an ancient irrigation system in South Thailand that had been in use for hundreds of years, and thus provides a record of efforts to cope with extreme climatic instability (rains or drought). The paper is based on various historical and archaeological evidences, and on natural sources of data as well. That the old irrigation system has been recently revived not only shows the value of local past knowledge in the face of future climate change, but also offers an example of the archaeological value of reconstructing the human past, while at the same time reconnecting with disconnected past perceptions.

---

72 It has been suggested that values that have grown out of pro-environmental and altruistic orientations can be expected to give rise to long-term sustainable actions; e.g. Fransson and Gärling, "Environmental Concern;" Nilsson, Andreas, Chris von Borgstede, and Anders Biel, "Willingness to Accept Climate Change Strategies: The Effect of Values and Norms," *Journal of Environmental Psychology* 24.3 (2004): 267–277.

In a very different vein, Marcus Nüsser and Ravi Baghel look at the Himalayan glaciers as objects of various forms of knowledge and imagination. Spurred on by the 'vanishing glacier' debate, they engage critically with various epistemological questions relating to glacier research and thereby challenge knowledge practices that go far beyond the geophysical framework. The glacier type in the global warming debate stands for both the real geophysical phenomenon and the contested narrative; the two authors coin the term "cryoscape" for this.

5.3 *Present*

India is one of the very few states where environmental protection is the bound duty of the state and of citizens (Environmental Protection Act of 1986). The two chapters in this section, both on India, raise the basic question of the agenda and politics of, on the one hand, the state, and on the other, the urban elite.

Joachim Betz sets the stage with his analysis of India's paradigm shift in climate policy towards a more open stance in international climate negotiations, a shift to the point where it has committed itself voluntarily to reducing carbon emissions. He unfolds the various reasons for this decision, while making clear that institutional failures continue to be stumbling blocks.

With the final chapter in this section we shift perspective. If we are indeed to explore the sources that are casting ever greater shadows over the climate, then we must go to those locations in which the climate of the future is increasingly being made[73] and determined – locations like urban India.[74] Somnath Batabyal takes up this task. Through a series of notes, queries, and vignettes, he seeks to position the Indian urban middle class and media within the climate debate, while trying to define the internal asymmetric social debate that is going on between the cosmopolitan and urban poor. His study reveals the politics behind environmental policy in India's capital. The role of the media in

73 Notable effects include "[w]orsening air pollution, noise, and solid waste accumulation; enhancing urban heat islands and impacts on health and infrastructure; increasing greenhouse gas emissions," see Association of Academies of Sciences in Asia (AASA), *Towards a Sustainable Asia: Environment and Climate Change* (Berlin, Heidelberg: Science Press Beijing and Springer Verlag, 2011), 31.

74 For a general discussion of urban areas in Asia, see Association of Academies of Sciences in Asia (AASA), *Towards a Sustainable Asia*, 53ff.

reporting on climate change has been well documented in recent years.[75] However, such research with respect to India has been rare.[76] Batabyal fills this gap.

5.4 *Prospects*

The articles that follow approach the object of research at the local, national, and indeed international levels. In their essays on Thailand, India, and ASEAN, the authors all address, under the heading "Prospects", the different types of climate coping strategies.

Susan M. Darlington's paper interlinks climate change, cultural/religious views, and local eco-friendly strategies in contemporary Thailand. She refers to environmental justice and makes clear that disadvantaged rural people are willing and able, if given the chance, to decisively manage their own situation.

Another angle on finding solutions is brought in by Rüdiger Haum, who looks at climate protection policy and economic development in India. In highlighting a major effort on the part of India to reduce carbon dioxide emissions by supporting solar technology, he challenges all win-win political rhetoric that leads one to believe in a causal connection between pro-climate innovation and economic benefit.

In the final essay, Dennis Eucker focuses on the regional organisation ASEAN and its potential role in promoting sustainable development and climate adaptation. The chapter expresses clear advocacy of institutional efficiency in the drive for sustainability.

6 Concluding Remarks

There is still a general perception that addressing climate change is not the responsibility of the newly industrialising countries.[77] The unbending attitude of the United States as the world's highest emitter only invites such a reaction.[78] However, in the view of an increasing number of key players in the countries

75 See Burgess, Jacquelin, "The Production and Consumption of Environmental Meanings in the Mass Media: A Research Agenda for the 1990s," *Transactions of the Institute of British Geographers* 15.2 (1990): 139–161, accessed December 22, 2012. http://www.jstor.org/stable/622861.

76 Among the rare examples is Billett, "Dividing Climate Change." See also Chapman et al., *Environmentalism and the Mass Media.*

77 See the analysis and evaluation of media in Billett, "Dividing Climate Change."

78 Though the article King, David A., "Climate Change Science: Adapt, Mitigate, or Ignore?" *Science* 303 (2004): 176–177, accessed December 8, 2012, http://www.sciencemag.org, was written eight years ago, the points it makes are still valid.

with emerging economies, including scholars, business and regional bodies, religious organisations, and environmentalists, this reaction is untenable. These countries have themselves become opponents of a human-friendly climate.

Indisputably, the dominant discourse relating to climate change reflects the view of Western cultures. Adding indigenous views and practices from India and Southeast Asia, of course, complicates the matter. While the pressure coming from the climate is incontestable, cultural responses are much more complex and diffuse. This fact reminds us that the climate is global *and* plural.

If we ask: Is sustainability something that works in South and Southeast Asia?, then, as far as ecology is concerned, the answer is yes, and if only because of the cultural practice, or operating principle, of inclusivism. Obviously, South and Southeast Asian cultural practices do not exclude alternatives; on the contrary, if needed its peoples carry out new experiments and adapt themselves to new situations both from the bottom up and from the top down following their own logic. The ability to adapt is, as Adger et al. state, related to knowledge (past and present), social practices, cultural stability, economic wealth, societal and political structures, and (on the individual level) status and age.[79]

Change in the interest of climate solutions is achieved in many ways, but first and foremost through engagement. Despite common assumptions, the know-how and technologies needed for pro-climate coping strategies already exist (see Betz, Haum).[80] What is needed is for them to be financed and tailored to the broad range of specific social and cultural settings.[81] Indeed, all other factors are dependent on cultural strategies of action and new ways of thinking. How

79 See Adger et al., "Are There Social Limits?"

80 Including the knowhow for carbon capture and storage (CCS technology); see Wuppertal Institute for Climate, Environment and Energy, "CCS Global, Prospects of Carbon Capture and Storage Technologies (CCS) in Emerging Economies, Final Report to the German Federal Ministry for the Environment, Nature Conservation and Nuclear Safety (BMU), Part II: Country Study India," published June 30, 2012, accessed September 21, 2013. http://epub.wupperinst.org/frontdoor/index/index/docId/4582.

81 As leading experts in climate matters rightly claim, a central challenge lies in understanding how cultures interlink techniques in intelligent ways and how technology can be introduced at the societal level, be it for the creation of new infrastructure or in order to change lifestyles; see Schneidewind, Uwe, "Technik allein bringt's nicht: Ohne soziale Innovationen wird der Klimawandel nicht zu beherrschen sein [Technology Alone Won't Do: Climate Change Cannot Be Managed without Social Innovations]," *Die Zeit* [The Times] 2, 2012, 25; also "Transition Research," published by Wuppertal Institute for Climate, Environment and Energy, accessed October 14, 2012. http://wupperinst.org/en/our-research/transition-research/; on cities as laboratories, see Schneidewind, Uwe, and Hanna Scheck, "Cities as 'Real World Laboratories' for System Innovations: Theories, Models and Empirical Designs," *Proceedings of the Third International Conference on*

these latter come about is less clear, and it is precisely this phenomenon that the chapters that follow illuminate.

Green religions, the middle class, media, knowledge practices, climate injustice, and state and regional policies – our collection of essays examines all of this. Yet even this is still only part of the picture. My hope is that the present volume will be a further step in a continuing discussion of climate and culture in South and Southeast Asian settings.

Sustainability Transitions, August 29–31, 2012 (Lyngby: Technical University of Denmark, 2012), 361–379.

PART 1

Ideas

CHAPTER 1

Cosmos, Commodity, and Care: Three Layers in Hindu Environmental Awareness

Albertina Nugteren

Abstract

In this article I trace some of the major points within the nexus culture–religion–nature in India. Many of the specifically Hindu elements in India's environmental lobby represent an appeal to the past. Mining religious heritage with the objective of finding inspiration for dealing with today's issues invariably results in an uncritical, selective, non-contextual and a-historic reproduction of inspiring imagery. Yet the cultural continuum, in a modest way, proves to be one potential resource for raising people's consciousness about environmental issues. Case studies from all over the globe show that cultural-spiritual values may induce sustainable use of resources, but often the relation is merely indirect. In matters of global anthropogenic climate change the scope may simply be too overwhelming to come up with culture-specific responses. A focus on place-based local initiatives reveals that, indeed, addressing religious sentiments may work in India, but what currency do these have in the global marketplace?

1 Introduction

The apocalyptic urgency in matters of the environment, and more recently climate change, sparked a torrent of studies on the complicated and often ambiguous relation between culture, religion, worldviews, actual behaviour, and nature. Ever since the historian Lynn White jr., in his famous 1967 article, spoke the ominous words "[...] the remedy must also be religious", we notice a selective reading of Asian religious texts, especially, and a romanticising view on indigenous traditions, in a desperate need for alternatives and a persisting belief in change *through* culture and religion.[1]

Such a selective, de-contextualised and a-historic search proved to be contagious. Not only Westerners fell into the trap, also in India many scholars and activists enthusiastically began to demonstrate, both to themselves and to the

1 White, Lynn jr., "The Historical Roots of Our Ecologic Crisis," *Science* 155 (1967): 1203–1207.

 | DOI 10.1163/9789004273221_003

Western world, that the Vedic-Hindu continuum was replete with environmental awareness: inspiring cosmologies, deep ecological insights, and rich expressions of interconnectedness and sacredness. Poul Pedersen, in 1995, critically termed this the 'Religious Environmentalist Paradigm' (REP).[2]

In various ways the study of the religion–nature nexus has now matured as an academic field. It is widely recognised that cultural and spiritual values may have determined much of the human behaviour towards nature in the past, alongside other factors such as social and economic goals. In order to categorise the various situations in this religion–nature nexus we need to clearly and systematically distinguish the ways in which South Asian worldviews either intentionally or accidentally worked out on the ecological environment. We summarise the debate by indicating at least five empirically diverse situations: (1) spiritual values inform a particular community's practices in such a way that its overall habitus is environmentally friendly, either intentionally or accidentally; (2) overall religious values inform a particular community's practices in such a way that the effects on a sustainable environment are ambiguous or even opposed to it; (3) a particular community may foster some worldviews that could be considered eco-friendly in theory but the day-to-day practices of both individuals and management institutions are not; (4) there is a disconnection between local-individual values and supra-local management institutions involved in resource management; and (5) there is a trend towards reconnection between communities' inherited cultural-spiritual values and their local, regional, national and even international implementation.

Case studies from all over the globe show that cultural-spiritual values may induce conservation and sustainable use of resources, but the relation is often merely indirect. In countries like India and Nepal belief systems may simultaneously be eco-friendly and environmentally destructive. At present it may be concluded that the REP still holds sway over many especially since it is hoped that, in addition to scientific prognoses and governmental regulations, an appeal to inherited cultural-religious values may turn the tide of the environmental crisis as manifesting in the subcontinent today. In modest ways the potential of an activist appeal to this heritage has been acknowledged as a serious factor also in the formal (i.e., scientific) discourse on sustainable management.

This seems to be less so in matters of climate change. Apart from the fact that many local people may have never even heard of global warming, some who are familiar with the term may dismiss it as irrelevant to their own life since it

2 Pedersen, Poul, "Nature, Religion, and Cultural Identity: The Religious Environmentalist Paradigm," in *Asian Perceptions of Nature: A Critical Approach*, ed. Ole Bruun and Arne Kalland (Richmond: Curzon Press, 1995), 265 ff.

is seen as a problem caused by the rich countries elsewhere in the world. Those in South Asia who are in a state of alarm about any local effects that may or may not be attributed to global warming processes may tend to blame others and insist on the industrialised nations to take full responsibility. But both in formal science and in politics climate change is beginning to be taken as a serious issue. It is gradually acknowledged that the subcontinent may not merely be a passive victim of the greed, carelessness and unlimited expansion going on in other parts of the world. Instead, it is admitted that through its own population pressure, economic boom, and emission rates it has become a major player itself. Whether and how this translates into actual regulations is discussed in other chapters within this volume.

In civilisations as ancient and continuous as in South Asia there is a marked tendency to rhetorically and romantically refer to their rich cultural heritage as the cause of their glory, on the one hand, and as a repository of remedies for any contemporary doom scenario, on the other. In this contribution I discuss three major layers in Hindu responses to the present environmental crisis from the perspective of their traditional worldviews. I look critically at the rhetoric and fallacies of the environmental discourse in Hindu South Asia, but also indicate its creativity and its ability to raise environmental activism in religiously motivated people by engaging their myths and rituals. This situates my presentation at the intersection of several historical, theoretical, practical and ideological discourses.

2 Cosmos

2.1 *Creation*

First of all it may be required to characterise the so-called Hindu tradition as a conglomerate of organically grown worldviews, i.e., a naturally diverse heritage that was fed by innumerable individual, local, and regional expressions concerning the overarching questions of life and death, of individual and collectivity, and of humans wondering about the surrounding cosmos. The resulting narratives underwent diversification over time and space. Some versions were streamlined by brahminical editorial engagement. Oral versions always persisted alongside written texts, however, and stories or inspired visions kept being reproduced in other art forms as well. This makes South Asian classical heritage a many-splendored thing. For any creation myth I would relate here I could present at least a dozen others, and for any version of the cosmos presented here someone else could voice a considerable number of alternatives. This is especially true when the indigenous traditions are included as well. How to give a more or less coherent characterisation of South Asian cosmology in the face of such enormous imagination, creativity and diversity?

I will limit my sketch to those views that have actually entered the environmental discourse and thus represent argumentative currency in the debate on climate change. The first cluster of views I introduce here circles around the topic 'creation' in its wider sense. The Sanskrit term often used here is *sṛṣṭi* or *visṛṣṭi*, discharge, emission, production. Like all ancient world views, the various creation stories present a mythical-religious pattern that has become the basis of South Asian images of the cosmos, and the source of the civilisation's morals and aspirations. Some of the most delightful passages in the *Ṛgveda* are those hymns in which the poets simply ponder such questions of origins without supplying definitive answers, so much that at some point the very question became the answer: a nebulous divinity Ka, meaning 'Who?', was produced. Probing the depths of their imaginations the poet-singers "went around, with mouths stuttering, frustrated".[3] How to pierce the mystery of that first creative moment when non-existence became existence? One of the often-quoted phrases goes:

> In the beginning, darkness was obscured by darkness;
> All was water, indiscriminate.
> Then, stirring that which was hidden in the void,
> The One emerged through *tapas* (heat).
> Then Desire arose in that One, in the beginning:
> That was the first seed of thought.[4]

We thus see that carefully, probingly, words were formed and names were given to those powers that may have started the processes of evolution. We find imaginings of primordial waters, and of heat or a fiery glow that may have caused the first stirring in those dark unformed waters. Once things were set into motion, tremendous powers were unleashed. Some hymns say that this fiery explosion shattered whatever pre-material may have been there, and that the dispersed bits and pieces of this very first life-body (*puruṣa*) formed land and heaven and all entities in between.[5]

From one to two to multiplicity: this becomes *virāj*, the widespread. The cosmic drama has begun and is being played out on – and with – the original material. The world is thus god's body, not only as a *causa efficiens* but also as a

3 *Ṛgveda* 10.82.7. The *Ṛgveda* is the oldest of the four *Vedas*.

4 *Ṛgveda* 10.129, 3–4.

5 *Taittirīya Brāhmaṇa* 2.2.9, 1–10.

causa materialis. It was through this creative explosion, this generative act of self-sacrifice, that order arose from chaos, and being from nonbeing. But once diversification had started, tensions and struggles were always there. Many myths relate such cosmic threats in terms of the elements and ancient antagonisms between primeval water and primeval fire, land and heaven, mountains and sea, forces of obstruction and forces of release, and, of course, life and death. This is where contemporary climate change scenarios sound almost tame.

In the rich heritage of cosmogonic myths and their concurrent cosmic upheavals many South Asians today are not overly impressed by climate change statistics and prognoses. Some may actually still live in a Purāṇic world, as we will also see in the next trope, that of the *kaliyuga* setting against which contemporary environmental degradation and climate change are made sense of.[6] Forces of nature, even anthropogenic climate change, may be threatening and involving great risks for those living in deltas, by endless stretches of beach, and in the merciless path of melting glaciers, mentally and visually such occurrences may even appear familiar as they were 'foretold' in their traditions. Especially in times of challenge for many it becomes part of their culture to live out Purāṇic scenarios, and it is in their chain of memory to mentally or actually re-live what was there in the beginning. In a cyclic universe nothing is really completely new and unprecedented.[7]

Stories of great floods abound. In Hindu mythology the deluge is connected with the god Viṣṇu's incarnation as a fish. It is said that a small fish swam into

6 *Purāṇas* are 'ancient' scriptures. In this particular context they are referred to because some of them speak of a system of four eras (*yuga*). It is often maintained that our present era is the darkest of all, 'foretold' as the disastrous *kaliyuga* or Dark Era.

7 That this narrative heritage remains a resource for everyday religion has been aptly demonstrated by studies on the Ganges (see Alley, Kelly D., "Idioms of Degeneracy: Assessing Gaṅgā's Purity," in *Purifying the Earthly Body of God: Religion and Ecology in Hindu India*, ed. Lance Nelson (Albany: SUNY Press, 1998), 306–315), the Yamuna (see Haberman, David L., *River of Love in an Age of Pollution: The Yamuna River of Northern India* (Berkeley: University of California Press, 2006) and recently on the Gangotri-Gaumukh Glacier (see Drew, Georgina, "A Retreating Goddess? Conflicting Perceptions of Ecological Change near the Gangotri-Gaumukh Glacier," *Journal for the Study of Religion, Nature and Culture* 6.3 (2012): 344–362). See also the title of a two-day international conference organised by the India China Institute, The New School, New York, March 7–8, 2013: *Everyday Religion and Sustainable Environments in the Himalayas*, cf. http://www.indiachinainstitute.org, accessed August 27, 2013). In general, see also Tomalin, Emma, *Bio-divinity and Biodiversity: The Limits of Religious Environmentalism* (Aldershot: Ashgate, 2009).

Manu's hands when he was taking his morning bath in a pond.[8] The fish asked to be saved. Manu, out of compassion, took it and kept it in a jar; when it outgrew the jar he put it into a pond; and when it outgrew the pond, he brought it back to the sea. As a sign of gratitude the mature fish warned him about the coming flood and advised him to build a boat and await the appointed time. Manu did so, and when the flood came the fish re-appeared and pulled Manu's boat to the northern mountains. After the water had subsided, Manu disembarked, and became both the saviour and the progenitor of humanity in a new cycle of earthly existence.[9]

Whereas some may take recourse in such mythical imagery, and trust in divine intervention since gods have been known to mercifully descend to earth in a new incarnation whenever a crisis arose, for others the threats are shockingly real.

2.2 *Kaliyuga*

One of the most widespread culture-specific responses to environment-related loss or the threats of climate change has to do with the present cosmic cycle in which humanity supposedly lives today, the so-called Dark Era (Sanskrit: *kaliyuga*). *Yuga*s in Purāṇic literature are calculated as long cycles. In traditional presentations of the four major cycles that the world is imagined to pass through in endless repetition, the present era is the darkest, full of gloom, degeneration, sin and decay. Virtue is at its lowest. Man himself is coarse, and as he has lost any sense of moral refinement and responsibility he exploits the world around him to his own ends. He has forgotten the divine order (*dharma*) almost completely and lives a selfish, careless, and disconnected life. There is a proliferation of disease, exhaustion, anger, hunger, fear, and doubt. Men live without a proper orientation and ignore the ancient injunctions about proper behaviour.[10]

I will give some examples of how people express this sense of gloom when correlating environmental degradation and the present Dark Era. Some contemporary authors make extensive use of the model of the Dark Age – the ancient kaliyuga model (according to which we are now living in the darkest of four eras) – in contextualising ecological degradation, either as a theoretical

8 Manu is often seen as the progenitor of mankind, the first law-giver, and our common ancestor who saved mankind from the universal flood.

9 This story from the *Matsyapurāṇa* is retold in: Klostermaier, Klaus K., *A Survey of Hinduism* (Albany: SUNY Press, 2007), 97. In popular etymology the associations between Manu (the first man), *manuṣya* (coming from Manu, i.e. humanity) and even English terms like man, mankind and humanity are well-established.

10 *Dharma* in Hinduism designates those behaviours considered necessary for the maintenance of the natural order of things.

construct or as a view from the field encountered in interviews. In the context of the environmental crisis, Kapila Vatsyayan makes explicit use of doomsday scenarios found in epics and *Purāṇas* to show how moral and environmental degradation were foretold in the past. Apparently without inhibition, she speaks of "desecration of the bowels of the earth through excessive quarrying", that "Pṛthivī, the eternal mother, has been desecrated", "the rape of tree- and river goddesses", "destruction of the gods of the woods, the *vandevata*", "dangerous play with the mythical centre – Sumeru, the world axis, the Himalayas", "the pollution of the holy space – the air, *vayu*", "*asuric* [demonic] chimneys of black *tamasic* [obstructing] forces", and that "we have polluted holy sound, the primeval *nada*".[11]

Vasudha Narayan, in her contribution to the Summer 1997 issue on environmental ethics of the *Journal of the American Academy of Religion*, evokes vivid epic and Purāṇic images of what the world would look like at the lowest end of the kaliyuga:

> At the end of the Eon *the population increases* – and odor becomes stench, and flavors putrid. When the Eon perishes, women will have too many children. [...] The cows will yield little milk, and the trees, teeming with crows, will yield few flowers and fruits. [...] Brahmins will plunder the land bare for alms. Householders, out of fear of the burden of taxes, will become thieves.[12]

Ann Grodzins Gold, during her field research within a farming community in Rajasthan, noted down various references made to the present kaliyuga as the cause of environmental degradation. In general it is mostly the high status Brahmins who use the kaliyuga label, but she heard it evoked by uneducated farmers as well. She concludes,

> In textual depictions of Kali Yuga from the Sanskrit Puranas, as in the social and environmental change narratives we elicited from

11 Vatsyayan, Kapila, "Ecology and Indian Myth," *India International Centre Quarterly* 19.1–2 (1992): 157–180.

12 Narayan, Vasudha, "'One Tree is Equal to Ten Sons': Hindu Responses to the Problems of Ecology, Population, and Consumption," *Journal of the American Academy of Religion* 65.2 (1997): 291–331.

> Ghatiyaliyans,[13] ecological breakdown and moral laxness have a thoroughly interpenetrating logic.[14]

In textual sources like the *Purāṇas* and epics it is assumed to be a characteristic of the degenerate kaliyuga that many individuals neglect the divinely ordained prescriptions for their stage of life and status in society. In such laments there is a close correlation between the degeneration of virtue and the ravaging of the Earth.

It is evident that the designation kaliyuga in popular parlance has become a metaphor, a manner of speech, a common denominator for bad times, and is simultaneously seen as the overall cause of all kinds of degradation. Such a defeatist attitude could be extremely de-motivating in environmental matters. The kaliyuga argument, as used in current environmental discourse, may adequately account for deterioration in morals and living conditions, but may also function as a legitimation of social and economic-ecological evils: the times are to blame, not we.

2.3 *Divine Embodiment*

Whereas such culture-specific coping mechanisms like myths, visions of cyclic time, and trust in periodic divine interventions are obviously ambiguous, in the sense that they may act as buffers against the apocalyptic urgency of scientific predictions and consequently keep people from activist concern and an individual–collective change of behaviour, there may be a third cultural factor that actually works although it has pitfalls of its own. In order to mobilise people the strictly scientific and apocalyptic rhetoric are combined with appeals to religious sensitivities. People may thus be addressed in the ethical or moral dimension by being referred to dharma texts and myths in which their favourite gods themselves once acted to restore order and cleanliness. The idea is: become like Śiva who swallowed the Kālakuṭa poison,[15] or offered his long matted hair for breaking the earth-shattering force when the river Gaṅgā rushed down from heaven. Become like Kṛṣṇa, who fought the poisonous demon Kāliya in

---

13 The author here refers to the villagers among whom she did her fieldwork.

14 Grodzins Gold, Ann, "Sin and Rain: Moral Ecology in Rural North India," in *Purifying the Earthly Body of God: Religion and Ecology in Hindu India*, ed. Lance Nelson (Albany: SUNY Press, 1998), 168.

15 Śiva is one of the main Hindu gods. The poison referred to here was produced during the cosmogonic churning of the ocean. It was so potent that it might have destroyed all creation. Śiva agreed to drink it, and this is why his neck became bluish. A small quantity dropped from his hands, and it is often said that it is because of this that there are poisonous snakes, scorpions, toxic plants, et cetera.

his beloved river Yamunā,[16] or who saved the cowherds from raging storms, flashes of lightning, and day after day of torrential rain, by giving them shelter, defeating Indra's hegemony, and bringing a new order of devotion and care. The message is: do as they did.

Not only gods act as models for environmental action; historic figures are also taken as providing examples in the defence of nature. References to gods, goddesses, historical personages and living environmental heroes singled out as inspiring models for environmental awareness may often be at the forefront, as they appeal to a widespread religious sentiment. But another more basic factor stands out. In Hindu affection the earth itself is often seen as a divine body. Creation myths in which the world's origin is found in a cosmogonic self-sacrifice establish a culture-specific equation between the body of this first being and all resulting entities. Some of those myths even make direct associations between the various body parts blown apart in that generative explosion and features of the natural world: his hair became the trees, his bones became the mountains, his skin became the earth's surface, etc. All life-forms in the universe together constitute this divine body. This conception of the material world as god's body not only makes all life forms interconnected, it also lays open the world as a meaningful and sacred place.

Some activists have used this imagery proficiently. Scientific language about environmental degradation and scenarios of doom connected with climate change may reach and mobilise only some, whereas the factor of recognition may be so much higher when a small dose of science is combined with a good dose of references to deeply reverberating mythic narratives and an appeal to culture-driven individual and collective dharma. Beneath the retelling of such myths there is a widespread preference for immanence. When this world is indeed god's body, divine presence becomes palpable. For instance, in the evocative language of faith-based environmentalism references are often made to the time-honoured way – still used by a few pious Hindus when they utter an apology directed to the Earth before getting up in the morning – to start the day by greeting the Earth respectfully and even ritually apologising that one will tread the sacred ground – god's body! – with one's footsteps.

What a far cry from industrial and domestic reality today! How could humanity tread so carelessly? Such questions are ubiquitous. Some attempts at answers, such as discussed by Vatsyaya Narayan and Lance Nelson, have indicated the in-built ambivalence. So even if the Vedic-Hindu continuum may be a treasure trove of tender expressions of connectedness, noble injunctions of

16 Kṛṣṇa is another of the great Hindu gods, often presented as one of the incarnations of Viṣṇu come to restore dharma on earth.

respectful behaviour, and of gods acting as environmental saviours, there are as many, or clearly even more, tendencies towards carelessness and commodification.

3 Commodity

Although Indian civilisation may have produced an amazing wealth of artefacts in which natural abundance is expressed with artistic delight, such as in religious texts, architecture, sculpture, drama, poetry, music and textiles, this does not necessarily mean that nature was sacrosanct. The Vedic hymns extolling the beauty and power of the elementary forces, of night and dawn, of the alternating seasons, of sun, moon and planets, of silvery rivers and dark-green forests, are not merely wonderstruck by this cosmos the poet-singers found themselves in. On closer investigation such hymns are prayers in which praising words are offered as currency against which the cosmic powers may be willing to continue for the benefit of mankind. The passages in which trees are lauded often appear to be the prelude to the ritual of cutting them in the process of being carved into sacred objects. The song in which cattle are lovingly described may well be the introduction to a massive sacrifice. Passages delighting in one of India's great rivers and mountains serve merely as backdrops to a particular human drama. And the natural settings against which many of the favourite epic stories were played out were mostly places of forced exile.[17]

Humans shape the world, mostly for their own purposes and sometimes by accident. As humans work their will upon the land, they fashion the things around them into instruments of survival and progress. In turn, human lives are given direction and meaning by natural forces. Places of beauty may capture their imagination and affection, but they primarily need to fulfil certain practical criteria. Environmental history can offer valuable perspectives.[18] It is the study of the reciprocal relationship of nature and humankind in a particular period in a particular place. South Asia's environmental history has been 'written' primarily by its peasant society. The early hunters and gatherers may have

---

17 For a more detailed discussion see my book: Nugteren, Albertina, *Belief, Bounty and Beauty: Rituals Around Sacred Trees in India* (Leiden, Boston: Brill, 2005); as well as my article: Nugteren, Albertina, "From Cosmos to Commodity... and Back: A Critique of Hindu Environmental Rhetoric in Educational Programs," in *Religion and Sustainable Development: Opportunities and Challenges for Higher Education*, ed. Cathrien de Pater and Irene Dankelman (Berlin: LIT Verlag, 2009), 159–168.

18 See, for instance, Hill, Christopher V., *South Asia: An Environmental History* (Santa Barbara, Oxford: ABC-CLIO, 2008).

had relatively little long-term effect. Earliest agriculture, in the form of slash-and-burn cultivation, may have had more impact, but its effect may have been relatively light. This changed with sedentary agriculture. Human settlements began to press on their surroundings, over time commerce and even supra-local trade may have set in, and the original subsistence relation with the surroundings may have shifted to aspects of increasing privatisation, commodification, and exploitation. For this transformation of nature by human use we find visual evidence stemming from periods long before the era of Europeanisation. The greatest impact, however, came from the changes brought by intercontinental trade, industrial revolution and, in its wake, monoculture. Commodification became an established policy with British colonisation, and this has continued after independence. Deforestation, hydro-electric engineering, and massive pollution have become almost ubiquitous.

Civilisations are in part formed by their relationship with and attitudes towards nature. Many Indians today pride themselves on their inbuilt reverence for nature by selectively appropriating relevant passages and practices from the past. But in that very search they may stumble upon unwelcome and embarrassing surprises, such as in the following: In a relatively unknown passage it is told that Kṛṣṇa and Arjuna once witnessed a raging forest fire.[19] As the forest burns, frightened creatures start to flee the raging flames. Instead of coming to their rescue it is described how the two princes race around the burning forest in their chariot, catching the escaping animals, and wantonly hurling them back into the flames, all the while laughing loudly and having a great time together. Those who want to present the Hindu tradition as a model of non-violence (*ahiṃsā*) and ecological propriety must feel embarrassed and awkward about such a passage. In general, the episode has been interpreted either as a great sacrifice to placate the Vedic god Agni,[20] or, more realistically, as a reference to the clearance of forests to provide agricultural land. In that case, the fleeing creatures may represent not only animals but also forest tribes.[21]

19 Arjuna, being Kṛṣṇa's relative, is best known as the main archer in the great battle of Kurukṣetra in which Kṛṣṇa acted as his charioteer. For the Khāṇḍava Forest episode, see Nugteren, *Belief, Bounty, and Beauty*, 394–395. Ruth Cecily Katz explains this event as no more than the pranks of two adolescents in her study of Arjuna in the Mahabharata: Katz, Ruth Cecily, *Arjuna in the Mahabharata: Where Krishna is, there is Victory* (Columbia: University of South Carolina Press, 1989).

20 Agni, one of the major Vedic deities, is connected with fire, and thus with the sacrificial cult.

21 Nugteren, *Belief, Bounty, and Beauty*, 394. Cf. also Gosling, David, *Religion and Ecology in India and Southeast Asia* (London: Routledge, 2001), 11, 16.

Fire can be considered to be, and has traditionally been seen as, the pioneering cultivator's prerogative: burning the forest so that the ashes on the forest floor enable the newly arrived farmer to grow his crops. Swidden cultivation has come under attack ever since colonial times as it was often seen as extremely wasteful and primitive, and this particular story causes additional embarrassment because of the cruel behaviour of the two cousins. It presents us with a Kṛṣṇa quite different from the nature-loving adolescent and the ethically refined teacher of *yoga* we normally encounter. Those who insist upon reading respect for nature into India's ancient scriptures may feel bewildered here. However, the ethical message of the epics takes into account men's and even gods' ambiguity and ambivalence, and this narrative about the two princes enjoying the death agonies of a vast number of animals trying to escape is certainly one such instance. Just as in other passages lies, deceit, cunning, and cowardice are exhibited in the behaviour of epic heroes, so is unnecessary cruelty in this text. Such a story undermines the presupposed ecological and ethical ahiṃsā message of the Indian scriptures, at least as seen with contemporary eyes.

What can we learn from this? At the very least that the religious message is full of contradictions? That this passage should have been erased in a later editorial round but was overlooked? Or that in hindsight, and from today's preoccupations, we tend to look at such issues anachronistically? Or even that the two-week conflagration, instigated by King Pṛthu, reportedly the first royal to clear the forests and establish the first agricultural settlements and townships, can be seen as an act of creation, just as animal sacrifice was seen as a generative event?[22] In the latter case, the royally instigated fire cleared the ground, literally and figuratively, for the foundation of their realm. No ascetic, dharmic or brahminic qualms about violence; instead the ruler's prerogative over the land. The inner conflict of existence is thus rationalised and legitimated by the long-established social stratification and the special place assigned to kingship, in myth but also in history, as we see below.

In the era of the Mauryan empire (ca. 326–184 BCE), especially during the reign of Chandragupta, assisted by his minister Kauṭilya, not only agriculture provided wealth to the royal treasure; every form of commodity and service began to be taxed. It was in this era that the commercialisation of nature was established, and maintained by a large network of collectors, controllers and

22 King Pṛthu, a mythical being referred to as the first king ever, and seen as the prototype of a benign ruler, was invested with the power to rule the earth in the name of the gods. He is popularly seen as the first to level the earth, to build villages and cities, and develop agriculture, trade and cattle-breeding.

spies. Revenues were exacted in the name of and for the welfare of the state. That this 'common good' was not mere make-believe is illustrated by the reign of King Aśoka around 250 BCE whose role as an ecologist *avant la lettre* has recently been indicated. His concern for the environment seems genuine. In his rock and pillar edicts (a considerable number of which survive) he emphasised righteousness by and for all. He forbade animal sacrifice, propagated public works such as the planting of trees for shade and the digging of wells as watering places for both men and animals, established eateries along the major routes in his realm, and encouraged animal welfare even to the point that animal hospitals were set up. His program was a combination of Buddhist ethics and a pragmatic attitude. Although the Mauryan Empire began to crumble after Aśoka's death, a lasting administrative infrastructure had been established. When new dynasties arose, public works were continued. Especially irrigation, notably in South India and Sri Lanka, became one of the focal points. Indian settlements elsewhere, rising first by trade and then by the establishment of royalty, such as in Thailand and Cambodia as well as in Java, Sumatra, and the Malay Peninsula, carried this practice and expertise further afield.

What we have thus seen is a gradual commodification and exploitation of natural resources, the flourishing of engineering and mechanical control, and royal prestige connected with ambitious public works in the heydays of Hindu dynasties. Whether one sees the following millennium as predominantly an era of Muslim rule, European merchants, and British colonisation or as a merely lightly ruffled continuation of internal dynamics is a matter of perspective. There is some truth in the statement that successive waves of invasion, infiltration or influence from culturally 'others' have not greatly altered the texture of native society. What is clear, however, is that the commodification of nature had set in with Mauryan tax administration, in an internal and peculiarly South Asian way. Especially trade and royal taxes had been driving factors in this. Since ancient days products from India had evoked yearnings for luxury in far-away places. Commodities like timber and cotton but also exotic animals and their skins or feathers, as well as the fabled jewels and spices, had to be taken from nature. With trade relations increasing, the demand for such goods sometimes irretrievably changed local infrastructures, local landscapes and local power relations.

4 Care

Climate change: who cares? Whereas locally experienced degradation, pollution and overexploitation may lead to place-based activism, and whereas the sorry state of a sacred river may incite devotees to a collective clean-up around 'their' part of the river, the hazy notion of anthropogenic climate change often meets with unconcern. The idea of global warming is simply too big to grasp. Moreover, isn't change the natural corollary of existence?

Indian civilisation had been able to develop and spread thanks to the subcontinent's natural bounty and beauty. This implied tremendous changes over the millennia. Human habitation pressed on forests and rivers, on mountains and oceans, on soil and air, on flora and fauna. And it continues to do so, probably at an unprecedented rate. Those outsiders who have personally had the opportunity to compare the state of human settlements, rubbish heaps, quarries, landscapes, cities, villages, roads, waterways and coastlines over the last few decades have an inkling of the rapid changes taking place under their very eyes. Changes, however, are multidirectional. Whose perspective do we take when looking at those changes? That of the essentialising tourist on a search for the authentic India? That of the business partner who becomes impatient when changes are too slow? That of the alarmed environmentalist? Moreover, people with disastrous droughts imprinted on their national history may not be overly alarmed by prospects of glaciers melting and rivers swelling. Instead they may be inclined to give water a warm welcome. But ever since the 2004 tsunami hit South Asia with disastrous consequences the image of rising tides has touched a nerve.

Apart from climate change scepticism and unfamiliarity with the anthropogenic climate change discourse on the one hand, and strictly scientific engagement with facts and prognoses, on the other, there is a reflex to take recourse to tradition's golden heritage in such matters. Authoritative texts from the past and imagined indigenous wisdom are thus mined with the idea that they may offer time-honoured clues to right behaviour. Indian culture provides various options. Some retreat into myth to make sense of changing times. Others devotedly take action in an imitation of their beloved gods. Some may have kings and heroes from the past as their model. Others may feel bewildered, weighed down by so many things beyond their control and comprehension. Sometimes various coping styles are combined, and Indian citizens are addressed on various levels and in various modalities simultaneously. One example of this is to be found in a governmental publication from the Ministry of Environment and Forests in 2002, titled "India: Climate-friendly Development". On some pages, alternating scientific texts, graphs, and policy statements we find, located in a visually differentiated column and a distinct letter type, quotations from the *Atharvaveda*,[23]

23 The *Atharvaveda*, often called the fourth *Veda*, incorporates much of the early traditions of healing and magic.

such as: "What of thee I dig out, Let it quickly grow over. Let me not hit thy vitals, Or thy heart", and: "O Mother Earth! You are the world for us and we are your children; let us speak in one accord; let us come together so that we live in peace and harmony!", and: "O Mother Earth! Sacred are the hills, snowing mountains, and deep forests, be kind to us and bestow upon us happiness. May you continue supporting people of all races and nations!".[24]

Without being precise it is often loosely stated in such reports that "a strong environmental ethic is embedded in India's culture. This remains unchanged despite increased prosperity".[25] It is this ethic that is appealed to even in otherwise dry documents. This proud reference to India's traditional culture has been taken up by various Hindu nationalists as well: "This is an opportunity to show how our tradition can help with modern concerns and issues."[26] Gopal-Lila Patel, project manager for the Bhumi Project,[27] commented: "We're not asking people to do anything new. We're simply encouraging them to re-learn the teachings of the past."[28]

What such appeals have in common is that they mobilise people by using Hindu perspectives and ancient culture in order to encourage them to take a more active role in the care and protection of the planet. The blending of science and religion appears to work, in India, especially when it addresses place-based devotional sentiments, such as concern over the Ganges and Yamuna Rivers, evocatively referred to as India's dying goddesses.

On the international level India has become one of the major vectors in global climate summits, such as recently in Durban. In what has been referred to as 'the age of Asia' or a process of 'Asianisation' of the world, India may hold

24 Quoted on pages 7, 13, and 39, respectively, in this recent publication: Ministry of Environment and Forests, "India: Climate-friendly Development," published October 2002, accessed November 28, 2011. http://www.envfor.nic.in/cc/cop8/moefbk/moefbk.pdf.

25 Ghosh, Prodipto, "Climate Change: Is India a Solution to the Problem or a Problem to the Solution?" in *Climate Change Perspectives from India*, published by United Nations Development Programme, India (New Delhi: UNDP India, 2009), 17–36, http://www.in.undp.org/content/dam/india/docs/undp_climate_change.pdf.

26 Oxford Centre for Hindu Studies, "Bhumi Project: Hindu Groups Address Climate Change," published January 6, 2011, accessed November 28, 2011. http://ochs.org.uk/news/bhumi-project-hindu-groups-address-climate-change. Here, Sudhir Garg, of Patanjali Yogpeeth Trust, is quoted on the occasion of a meeting in New Delhi (January 2011), convened by the Bhumi Project. The Bhumi Project is an international Hindu response to environmental matters such as climate change, and is facilitated by the Oxford Centre for Hindu Studies, in partnership with the worldwide Alliance of Religions and Conservation, and backed by the UNDP.

27 *Bhūmī* means "Mother Earth."

28 Oxford Centre for Hindu Studies, "Bhumi Project."

one of the keys to the future. To what extent its past, its heritage, its cultural continuum, and the choices made in the projection of its contemporary self-image may be or even should be a decisive factor in handling this key wisely is a matter beyond the scope of this article.

CHAPTER 2

Ideas and Images of Nature in Thai Buddhism: Continuity and Change

Martin Seeger

Abstract

In 2005, during the course of a research project on the involvement of Buddhist monks in development projects in the north-eastern region of Thailand, our research team was engaged in discussions with a Thai 'development monk' about the motivations of local people for environmental protection. The monk drew attention to significant shifts in the attitudes of rural Thai people to nature and also argued that nowadays, the combination of a variety of motivational factors is necessary to understand the reasons for the desire to protect the environment. The following sources of motivation have to be taken into account: respect for charismatic figures, fear of state authority, sense of communal ownership, respect and awe for the supernatural character of nature, and scientific understanding of the importance of environmentalism. Based on these empirical observations, I examine concepts of Buddhist environmental ethics as proposed by one of the most influential Thai Buddhist thinkers, the monk Phra Payutto, in connection with shifting meanings and ambiguities in Thai conceptions of nature. I also consider conceptions of the forest as a spiritual 'training ground' and of karma as potential motivational factors for environmental protection. The focus of this article is to consider how a number of Thai Buddhists have sought to explain the causes of environmental degradation and how they have responded to it by proposing a variety of different strategies informed by religious beliefs, practices and teachings.

* For comments on previous drafts of this paper, I would like to thank Victor King, Michael Parnwell, Sue Darlington, Justin McDaniel, Baas Terwiel, Adcharawan Seeger and Khun Naris Charaschanyawong.

 | DOI 10.1163/9789004273221_004

•••

"Cut down the forest but do not cut down the trees."
(*DHAMMAPADA*, MAGGAVAGGA, VERSE 283)[1]

∴

1 Introduction

During the first three months of 2011, at one of Bangkok's major roads, Phetkasem Road, a four-metre tall banana tree was the centre of attention and focus of veneration by a large number of Thai people. Surrounded by the dense and noisy concrete jungle of Bangkok and in the middle of a pile of debris, the banana tree grew on an untended site where once a factory had stood. The crowd visiting the tree was sometimes so large that there were traffic tailbacks of up to three kilometres. It was reported that in the evening, between 9 and 10 pm, the number of people who wanted to get close to this tree increased significantly. In particular, during the time close to lottery draws, the number of people "multiplied and to establish order was extremely difficult."[2] The banana tree has been regarded as being very unusual in that its trunk was rather large and strong. But what was perceived as being even more astonishing (*mahatsajan kwa nan*)[3] was that the tree's blossom, unlike that of a 'normal' banana tree, was not only green instead of red and had a jade-coloured bud. The bud's form also resembled that of a lotus flower. The lotus flower is, of course, one of the most significant symbols in Buddhism,[4] signifying

---

1 This is my own translation from Pali, having consulted various Thai translations of this passage. When referring to sources in Pali, I have used the *Chaṭṭha Saṅgāyana* CD-ROM, Version 3, published by the Vipassana Research Institute. The references to the Pali texts are to the volume and pages (or verse number) of the text editions by the Pali Text Society, as given on the *Chaṭṭha Saṅgāyana* CD-ROM.

2 Http://www.thailuckynumber.com/mum_inside.php?mum_lucky_number_id=70; accessed December 16, 2011.

3 I have used a standardised system for the romanisation of Thai script in this article except in cases where proper names have an established transliteration of their own. Throughout this paper Thai words are not visually differentiated from Pali words but their origin can be looked up in the glossary. All the translations from Thai are my own, unless stated otherwise.

4 When I refer to Buddhism in this article, I am focusing exclusively on Thai Buddhism and/or Theravada Buddhism and not taking into account other forms of Buddhism, such as Mahayana, Chinese Buddhism, etc. I differentiate between Thai Buddhism and Theravada Buddhism

mental purity and spiritual beauty. Thai people who flocked to the tree believed that the benevolent Goddess Mae Tani, here called by the name 'Sri Nuan', inhabited (*sing sathit*) the tree.[5] Thais venerated Mae Tani by making offerings such as water and flower garlands, and igniting incense sticks and candles. In front of a statue representing Mae Tani, a sign is attached, on which visitors are instructed as to how to invoke her. First, 'Lord Buddha' is to be praised in Pali,[6] a sacred language of Theravada Buddhism, followed by a stanza asking for Mae Tani's benevolence (*metta*) in a linguistic hybrid of Thai and Pali. Also, sets of female silk clothes in 'seven colours' were offered to Mae Tani. Mae Tani expressed her benevolence by revealing lucky lottery numbers in numerous cases, which of course led only to the number of visitors increasing. Various techniques were used to make the

---

when I deem this to be relevant for the understanding of the specific context. I deliberately avoid using 'Theravada Buddhism' as a general term in this paper as most Thai Buddhists would answer the question 'which religion they believe in/adhere to' by saying that they are Buddhist. 'I am a Theravada Buddhist' would sound rather unnatural. Using 'Theravada Buddhism' to designate religious beliefs and practices in Thailand may be problematic, as Prapod Assavavirulhakarn has masterly shown in his recently published book: Prapod Assavavirulhakarn, *The Ascendancy of Theravāda Buddhism in Southeast Asia* (Chiang Mai: Silkworm Books, 2010). See also footnote 21.

5 This description of the happenings around the banana tree in Phetkasem Road is based on YouTube videos, Thai newspaper articles and my interviews with people who visited the site or work in proximity to it. I also visited the site on April 10, 2011 after the tree had been felled, talked to people who sold incense sticks and flowers at the site and undertook a detailed photographic recording of the site. Some of the major online sources I have used can be found here: "Chao ban tuen kho huai jau mae ชาวบ้านตื่นขอหวยเจ้าแม่ [Excited Folk Pray for Lotto Numbers from Mother Goddess]," published by YouTube, accessed September 25, 2012. http://www.youtube.com/watch?v=9vcMXrMYf3U; "Hae kho huai mae tani rot tit lai kilomet แห่ขอหวยแม่ตานี รถติดหลายกม [People Pray for Lotto Numbers – Traffic Tailbacks of Many Kilometres]," published by Thairat Newspaper Online, originally published in ไทยรัฐฉบับพิมพ์ [Thairat Newspaper], January 15, 2011, accessed May 20, 2011. http://www.thairath.co.th/today/view/141536; "Klai huai ok! Hae kho tani phetkasem naen ใกล้หวยออก! แห่ขอตานี เพชรเกษมแน่น [Shortly before the Lottery Draw! People Flock to Tani and Pray for Numbers – Phetkasem Overcrowded]," published by Thairat Newspaper Online, January 14, 2011, accessed September 25, 2012. http://www.thairath.co.th/content/region/141354.; "Hae kho lek det jau mae tani naen soi phetkasem แห่ขอเลขเด็ดเจ้าแม่ตานี แน่นซอยเพชรเกษม [People Flock to Mae Tani to Pray for Lucky Numbers, Phetkasem Soi Overcrowded]," published by Webportal mthai.com, accessed September 25, 2012. http://lotto.mthai.com/lotto-news/result-787.html; "Chuan pai du mahat-sajan kluai tani pli dang dokbua ชวนไปดู มหัศจรรย์ กล้วยตานี ปลีดั่งดอกบัว [Be Amazed by the Tani Banana Tree Whose Blossom is Like a Lotus Flower]," Oknation Blog posted by Khasemakhung คาเซมะคุง, 18 January 2011, accessed December, 16, 2011. http://www.oknation.net/blog/kase-makung/2011/01/18/entry-1.

6 Chanting "*namo tassa bhagavato arahato sammāsambuddhassa*" (Hail to the Blessed One, the Worthy One, the Perfectly Awakened One) three times (*tang namo sam jop*).

lucky numbers visible. People rubbed powder on the tree trunk or used torchlight to investigate very thoroughly the trunk's surface in order to decipher numbers. Thais believe that the numbers perceived may vary, depending on the individual's accumulated storage of meritorious karma (*bunwatsana*).[7] Also, people reported that after having taken photographs of the tree, they later discovered luminous shapes on the pictures, which they interpreted as supernatural manifestations of Mae Tani.

FIGURE 2.1 *The banana tree in Phetkasem Road. Thais showing respect to Mae Tani and trying to decipher lucky numbers*[8]

One Thai dreamed of Mae Tani as a young, beautiful looking, white-skinned and smiling girl and, afterwards, won three million Thai Baht in the lottery. In another instance, a woman who was in the process of extracting numbers from the tree was believed to have suddenly become possessed by Mae Tani. Through this woman, Mae Tani asked for restraint on the part of the visitors, saying: "Do

7 See e.g. "Sahasakha sueksa phuea phatthana chonnabot สหสาขาศึกษาเพื่อพัฒนาชนบท [Interdisciplinary Studies for the Development of Rural Areas]," published by Chulalongkorn University Website, accessed June 1, 2011. http://pioneer.chula.ac.th/~vsuntare/report/korn2.html.

8 Photo: © Naris Charaschanyawong.

not light incense and candles too close to me; I feel hot. Do not rub cool powder against me, as it burns." This request was not heeded for long, though, and soon afterwards people were scraping (*khut*) the tree again in the same way as they had done previously. It was also reported that in a dream Mae Tani asked for help to protect her by having a fence erected around her. However, this request was also not complied with as the fencing off was perceived as a way to prevent people gaining access to lottery numbers. It was also reported that an illegal lottery operator (*jau-mue*) was so incensed about the frequent revelations of successful numbers that he planned to get someone to penetrate the tree with an elephant tusk, believing this would be a trick (*khlet*) to drive Mae Tani away. After 8 March 2011, however, the number of people visiting the tree fell abruptly and sharply. The reason for this was that on the morning of that day an evil person (*khon janrai*) cut down the tree. It has been presumed that an illegal lottery operator (*jau-mue*) was behind this act.

2 Thai Beliefs in the Supernaturality of Nature

In the following analysis of the phenomenon of the sacred banana tree in Phetkasem Road, I want to identify significant and relevant themes; the assumption is that the phenomenon of the banana tree in Phetkasem Road serves as an entry point to unravel and examine a number of Thai beliefs, attitudes and practices in connection with nature. Having identified these themes, I discuss them by juxtaposing them with empirical observations made by a development monk (*phra nak phatthana*) involved in environmental protection. This is complemented by linking these observations to the ideas and concepts of one of the most influential contemporary Thai Buddhist thinkers, Phra Payutto. The concepts proposed by Phra Payutto are then further investigated by examining conceptions of forest[9] as a spiritual 'training ground' and of karma as potential motivational factors for environmental protection. The purpose is not only to identify and investigate recently observed shifts in the meaning and conceptions of nature and their ambiguities, but also to discuss models of

9 In this article I will consider Thai attitudes primarily towards trees and forest. I am of course aware that forest does not equate to environment or nature *per se*, but is only an element of it. Often though, forest may be understood as symbolically representing nature and environment. There may however be specifics that solely concern the forest and are not applicable to other aspects of the environment or nature. I have chosen the forest as one major concern when investigating nature and environment in Thailand because the activists and thinkers whose ideas and observations are discussed here are also often concerned with the forest as the representation of nature or environment more generally.

environmentalist ethics that have been proposed by Thai Buddhist intellectuals in connection with these shifting meanings and ambiguities. I must emphasise though that due to the enormous complexity of the issues involved I do not attempt to discuss them systematically or comprehensively but rather intend to provide some insights into relevant ideas and activities in the Thai religious sphere. Thus, I do not consider the impacts of governmental policies, capitalist development, globalisation and (de-)localisation. Nor do I claim that the positions, initiatives and ideas discussed here are representative of larger sections of Thai society. Often they are not but have to be understood to be rather idiosyncratic. I also do not want to pursue a critical comparative investigation of the positions, ideas and actions I propose here. This would be too ambitious for the purposes of this paper. Thus the main aim is to address how a number of Thai Buddhists have sought to explain the causes of environmental degradation and how they have responded to it by proposing a variety of different strategies, which are informed by religious beliefs, practices and teachings. As will be shown, Thai religion is perceived to have enormous potential to address environmental problems effectively and sustainably.

Even though the scale of attention and media coverage of the banana tree in Phetkasem Road was quite unusual, similar numerological practices in connection with sacred trees (*khut lek*) have been in evidence elsewhere in Thailand. A careful observer can notice, often in Buddhist temples, that the practice of rubbing white powder against 'unusual' or sacred trees in order to extract lucky numbers is widespread. Also, there is an elaborate 'tree science' that not only classifies trees into those which are auspicious and inauspicious but also prescribes where specific trees are to be grown and where not.[10] In the case of banana trees believed to be inhabited by Mae Tani, Thai people 'do not dare' plant the tree near a dwelling house. According to the Encyclopaedia of the Thai Royal Institute, there is the Thai belief that if leaves are cut from the Tani banana tree, specific techniques to sever the leaves from the tree have to be applied:

10 See e.g. Anuman Rajadhon, Phya, *Essays on Thai Folklore* (Bangkok: Thai Inter-Religious Commission for Development and Sathirakoses Nagapradipa Foundation, 1988), 337–343; Sponsel, Les, and Poranee Natadecha-Sponsel, "Why a Tree is More than a Tree: Reflections on the Spiritual Ecology of Sacred Trees in Thailand," in *Santi Pracha Dhamma: Essays in Honour of the Late Puey Ungphakorn*, ed. Pipob Udomittipong (Bangkok: Santi Pracha Dhamma Institute, 2001), 364–373; and Suwanna Kriengkraipetch, "Thai Folk Beliefs about Animals and Plants and Attitudes toward Nature," in *Culture and Environment in Thailand: A Symposium of the Siam Society*, ed. Siam Society (Bangkok: Siam Society, 1989), 195–211.

> If a whole leaf is cut from the tree and taken into a house, this is regarded as a bad omen [*lang rai*] that indicates that someone in that house will die soon. This [belief] may derive from the traditional practice of laying out three Tani banana leaves at the bottom of a coffin.[11]

At the same time, when the tree is blossoming, an elaborate ceremony, sometimes with the help of Buddhist monks, is performed in order to ask Mae Tani to protect people living nearby and bestow wealth upon them. The blossom of the Tani banana tree has also been used as an ingredient in the practice of love magic.[12]

Often, its auspiciousness or inauspiciousness depends on the name or shape of the plant;[13] or whether the tree has played a significant role in Buddhist narratives.[14] There are many more examples in the Thai religious belief system that show that in the perception of Thais, nature is not just a purely physical world, but also one that is inhabited by supernatural beings or which possesses supernatural efficacies.[15]

---

11 *Saranukromthai Ratchabanditayasathan* สารานุกรมไทย ฉบับราชบัณฑิตยสถาน [The Thai Encyclopaedia of the Royal Institute], vol. 13 (Bangkok: The Royal Institute, 1981), 7912–7913.

12 Ibid., 7913.

13 This may be due to the natural, common shape of a plant or to what is perceived to be a rather unusual shape in individual cases, as in the case of the Phetkasem Road banana tree. Two further examples for this have also been reported on in major Thai newspapers. Thus, the Thai newspaper Khausot reported on December 19, 2009 that villagers in Ayutthaya Province showed respect to a sugar palm as its top had the shape of a mythical snake ("Taek tuen 'tontan' ok yot phayanak แตกตื่น "ต้นตาล" ออกยอดพญานาค [Excitement About the Sugar Palm that Sprouts into a Naga]," Khaosod Online Newspaper, December 19, 2009, accessed December 23, 2011. http://www.khaosod.co.th/view_news.php?newsid=TUROd2NtOHhOREU1TVRJMU1nPT0). Another example is when Thai people in the southern Thai province of Prajuapkhirikhan paid respect to (*yok mue wai*) and took photographs of three sugar palm trees that looked like Somdet To (a widely and highly revered Thai saint who lived in the 19th century) sitting in meditation posture (see: "Taek tuen du tontan – muean luang pho to แตกตื่นดูต้นตาล-เหมือนรูปหลวงพ่อโต [People Get Excited About A Palm Tree that Looks like Luang Pho To]," published by Webportal mthai.com, accessed December 23, 2011. http://news.mthai.com/hot-news/14484.html). See also: McDaniel, Justin, *The Lovelorn Ghost and the Magical Monk: Practicing Buddhism in Modern Thailand* (New York: Columbia University Press, 2011), 69–70.

14 Thus, Bodhi trees (*ficus religiosa*), under one of which the Buddha is believed to have attained awakening, are highly revered in Thailand, often beautifully decorated, wrapped in saffron robes and surrounded by Buddha statues.

15 See also e.g. Darlington, Susan M., "The Spirit(s) of Conservation in Buddhist Thailand," in *Nature Across Cultures: Views of Nature and the Environment in Non-Western Cultures*, ed. Helaine Selin (Dordrecht: Kluwer Academic Publishers, 2003), 131–132.

FIGURE 2.2 *The bodhi tree in front of the famous Phra Pathom Chedi in Nakhon Pathom*[16]

Other famous representative examples for these beliefs are that in the Rice Goddess Mae Phosop who is venerated in various rituals throughout the farming year,[17] and the Earth Goddess Mae Thorani, who is depicted in one of the most famous iconographies of Thai Buddhism, showing her supporting the Buddha in gaining awakening.[18]

In Thailand, banana trees have been perceived as symbols of "prosperity and success", since they "can be cultivated with little effort [...]"[19] In the case of the banana tree in Phetkasem Road, however, this meaning has been enhanced in that the tree is not only regarded as a symbol of prosperity but also believed to be the potential source of it. This belief is in line with traditional practices to ask for protection and wealth and is therefore certainly not a new element in Thai religion.

16 Photo: © Martin Seeger. This tree is believed to be inhabited by the tree deities Hiranyaket and Porametthawichit.

17 Anuman Rajadhon, *Essays on Thai Folklore*, 351–405.

18 Guthrie, Elizabeth, "A Study of the History and Cult of the Buddhist Earth Deity in Mainland Southeast Asia" (PhD diss., University of Canterbury, 2004).

19 Suwanna, "Thai Folk Beliefs," 207.

FIGURE 2.3 *The takian tree in Wat Pa Yang, near the Sri Satchanalai National Historical Park in Sukhothai Province*[20]

The banana tree in Phetkasem Road, as perhaps a more radical example, can be seen as an exemplification of a complex amalgam of different attitudes towards and perceptions of nature. The banana tree phenomenon in Phetkasem Road is linked to complex cosmological conceptions combined with animistic beliefs and Buddhist elements and motifs.[21] However, it has also become an

20 Photo: © Martin Seeger. This tree is believed to reveal lucky numbers. With the money donated by people who won the lottery with numbers from the tree, a new ordination hall was built.

21 This does not imply that these three realms can, or should, be clearly separated. In fact, Buddhist canonical and post-canonical scriptures contain numerous stories about (tree) deities. See e.g. SN.III.249–250, where the Buddha talks about tree deities. See also SN.IV.302. The commentary text of the canonical book *Dhammapada* (DhA.I.313–316) describes how tree deities (*vanasaṇḍe adhivatthā devatā*) drove 500 monks away from a forest grove they entered in order to practice meditation, by terrorising them. The Buddha then gives to the monks a weapon (*āvudhaṃ*) in the form of a text, dealing with friendliness (*mettā*), and instructs them to recite the text. Once they have returned to the forest and start to recite the text the deities "residing throughout the forest-grove conceived friendly feelings in their hearts for the monks, came forth to meet them, asked the monks to let them take their bowls and robes, offered to rub their hands and feet, posted strong guards on all sides, and sat down together with them." Cf. Burlingame, Eugene Watson, *Buddhist Legends*, trans. from the original Pali text of the *Dhammapada* Commentary, ed. Charles Rockwell Lanman, Part II (Delhi: Motilal Banarsidass Publishers, 2005), 17–19. See

exemplification of the reckless exploitation of nature, despite the repeated requests of the nature deity Mae Tani. In fact, what the banana tree phenomenon in Phetkasem Road seems to reflect is people engaging the spirit world for the sake of their material advancement. Materialistic acquisitiveness is interwoven with beliefs in the supernaturalness of nature. Thus, materialistic pursuit does not deny the actuality of supernaturalness but is rather based on it, and has subordinated it in order to exploit it. Even though trees have also been regarded not only as sources of protection and wealth, but also as potentially dangerous, in the case of the banana tree in Phetkasem Road, materialistic pursuit seems to have been a stronger motivational force than fear and awe of (the supernaturalness of) nature. Sadly, the belief in the supernatural efficacy of the Phetkasem Road banana tree also finally led to its destruction.

3 Environmental Protection in Thailand's Northeast, Past and Present

Many of these interpretations and elements of the banana tree phenomenon in Phetkasem Road seem to be, at least to some extent, in line with, but also sometimes even in opposition to, the observations of a number of monks engaged in environmental protection.[22] In 2005, during the course of a research project on the involvement of monks in development projects in the northeastern region of Thailand,[23] our research team was discussing local people's motivations for environmental protection with Phra Khru Suphajarawat Subhācāro (b. 1940), a locally highly respected abbot of a small temple (Wat Thalat, Amphoe Kutchum) in Yasothon Province. Phra Khru Suphajarawat, who had been involved in or been the driving force of various local development projects for more than two decades, drew attention to significant shifts in rural Thai people's attitudes towards nature. For him, these attitudinal

also: Darlington, "The Spirit(s) of Conservation," 130–134; and DeCaroli, Robert, *Haunting the Buddha: Indian Popular Religions and the Formation of Buddhism* (Oxford, New York: Oxford University Press, 2004).

22 See also: Darlington, "The Spirit(s) of Conservation," 136–137; and Darlington, Susan M., "The Good Buddha and the Fierce Spirits: Protecting the Northern Thai Forest," in *Contemporary Buddhism* 8.2 (2007): 169–185.

23 We conducted several in-depth interviews with Phra Khru Suphajarawat over the course of three weeks during the summer of 2005. For further details on this research project, see: Parnwell, Michael, and Martin Seeger, "The Relocalization of Buddhism in Thailand," in *Journal of Buddhist Ethics* 29 (2008): 78–176. See also: Parnwell, Michael, "Neolocalism and Renascent Social Capital in Northeast Thailand," in *Environment and Planning D: Society and Space* 25 (2007): 990–1014.

changes became apparent in the decreasing belief in, and following from this, diminishing respect and awe for the supernatural qualities of nature. Phra Khru Suphajarawat recollected that in former times villagers did not dare to cut trees in forests believed to belong to local guardian/ancestral spirits (*puta*).[24] However, with the help of a mediator (*thau jam*)[25] who represents the villagers to the spirits, permission from the spirits had to be obtained before trees could be felled.[26] Offerings in the form of alcohol and boiled chickens were made and trees replanted. If trees were cut without the permission from local spirits (*jau thi jau thang*), people could be punished for this by becoming sick due to supernatural intervention and ghosts may have manifested themselves or possessed villagers. Nowadays, Phra Khru Suphajarawat explained, there are some places where this belief is still alive, albeit to a lesser degree. He argued that the decrease of people's respect for numinous (*khlang*) aspects of nature is reflected in the enormous extent to which the forest has been cut down. Furthermore, he pointed out that whereas formerly the motivation for environmental conservation was to a large extent based on the belief that nature is inhabited by powerful supernatural beings, only a combination of a variety of motivational factors is necessary nowadays for the protection of the environment. During our conversations with him, we discussed a number of motivational factors, such as respect for charismatic figures, fear of state authority, sense of communal ownership, respect and awe for the supernatural character of nature, and scientific understanding of the importance of environmentalism. When asked to rank the degree of influence of motivational factors for environmental protection, he explained that belief in and fear (*klua*) of nature's sacredness are still active and it is believed that exploitation of the forest may entail physical punishment by supernatural forces in the form of illnesses. He also mentioned that there is "fear of other villagers" as forest is regarded as communal property. He summarised as follows: "Where the community spirit is strong or [the belief in] the sacredness [of forest] is still active, there you will find forest remaining."[27] He also argued that fear of the state plays a role as the state imposes environmental protection requirements on Thai people in the form of laws and national parks. In addition, he also mentioned the Thai King's and Queen's initiatives for environmental protection which have been translated into policies implemented by the various levels of the Thai administrative system, with the result that

24 See Tambiah, S. J., *Buddhism and the Spirit Cults in North-East Thailand* (Cambridge: Cambridge University Press, 1970), 263–284.

25 Ibid., 274 280.

26 See also: Hayashi, Yukio, *Practical Buddhism Among the Thai-Lao: Religion in the Making of a Region* (Kyoto: Kyoto University Press, 2003), 84–85.

27 Interview with Phra Khru Suphajarawat, July 19, 2005.

plots of land are being demarcated and put under protection against encroachment.[28] Quite interestingly, it seems that for him scientific understanding of the importance of environmentalism did not play an important, if any, role, here at all; at least he did not mention it in this context. At the same time, however, he argued that the belief in the sacredness of the forest was vanishing due to the arrival (*khau ma*) of modern science and technology:

> It has been recounted that through spirit possession [*sing khon*] people learned that spirits [*phi*] are afraid of the big, red eyes of tractors. [Spirits complained] "The tractors are huge; their wheels are trampling on my head which is unbearable!"[29]

But it is not only the disappearance of the belief in local spirits and the supernaturality of nature that Phra Khru Suphajarawat perceives as having been undermined by modern science. He also argued that the belief in traditional cosmology, according to which unwholesome deeds have karmic consequence beyond one's current lifetime has been eroded due to a scientific understanding of the world:

> In the mind of people of the new generation, do hells and heavens have an existence? Are there future lives for them? Nowadays, people do not believe in what they do not see. Thus, they are not as deeply grounded in moral behaviour [*sinlatham*] as they should be. It is not really engrained in their minds and hearts. There is no longer a spirituality that they use as a strong motivation as it was the case in former times. What was factual in the belief of people has become abstract and is therefore not believed in as strongly as in the past.[30]

Over the last three decades, Phra Khru Suphajarawat has been involved in and central to many local development projects, including organic farming, self-sufficient agriculture,[31] preservation and distribution of traditional knowledge

---

28 See also: Parnwell, and Seeger, "The Relocalization of Buddhism," 118.

29 Interview with Phra Khru Suphajarawat, July 19, 2005.

30 Interview with Phra Khru Suphajarawat, July 19, 2005. Compare with Visalo, Bhikkhu, "Buddhism for the Next Century: Toward Renewing a Moral Thai Society," in *Socially Engaged Buddhism for the New Millennium: Essays in Honor of the Ven. Phra Dhammapitaka (Bhikkhu P. A. Payutto) on His 60th Birthday Anniversary*, ed. Pipob Udomittipong and Chris Walker (Bangkok: Sathirakoses Nagapradipa Foundation and Foundation for Children, 1999), 238. See also below.

31 The objective is to produce for local consumption (with surplus being sold) in order to become independent from the market as far as possible. (Prida Rueangwichathon ปริดา เรืองวิชาธร, *Phlik fuen khuen chiwit chiwit lae ngan khong phra khru suphajarawat* พลิกฟื้นคืน

of herbal medicine, the establishment of a health centre in the grounds of his monastery and a community rice mill, the introduction of a community currency (*bia kutchum*), the establishment of a local group of development monks (*klum sangkha asa phatthana*) and forest rehabilitation.[32] These many projects have aimed not only at establishing sustainable development of local communities but also to preserve nature. According to Phra Khru Suphajarawat, due to their charisma and central role in communities, monks play a very important part in forest conservation and reforestation projects:

> Monks are the spiritual centre of and refuge for communities [...] and Buddhism is traditional capital that has, for a long time, been deeply engrained in Thai society and thus has a [strong] influence on Thai culture and the way Thais lead their lives.[33]

In this way, he asserted the centrality of monks in forest protection projects:

> If monks get involved in reforestation and forest protection projects, villagers will follow suit. They will not only preserve existing forest but also plant additional trees, too. Sometimes, villagers may even donate [*thawai*] huge plots of land to the monastery [for the purpose of reforestation or conservation].[34]

Phra Khru Suphajarawat states that in order to be sustainable, development has to be based on Buddhist principles that aim to 'build' and develop people (*sang khon/phatthana*)[35] and

> a very important thing is that there must not be greed [*lobha*]; natural agriculture means to avoid destroying our environment. All things, earth, plants, animals and humans support and are dependent on each other [*uea kan*]. None of them can be absent. With natural agriculture, both the environment and human health will be fine.[36]

ชีวิต ชีวิตและงานของพระครูสุภาจารวัฒน์ [Revitalizing Life: Life and Work of Phra Khru Suphajarawat] (Bangkok: Religious Council for Development, 1992), 34.

32 For details see: Parnwell, and Seeger, "The Relocalization of Buddhism," 108–123.

33 Suphajarawat, Phra Khru พระครูสุภาจารวัฒน์, *Botbat phra song nai chonnabot thamklang phawa wikrit* บทบาทพระสงฆ์ในชนบทท่ามกลาง ภาวะวิกฤติ [The Role of the *Saṅgha* in Rural Areas in the Middle of the Crisis] (Bangkok: Komonkhimthong Foundation, 1999 CE/2542 BE), 9.

34 Suphajarawat, *Botbat phra song*, 12–14. See also Parnwell, and Seeger, "The Relocalization of Buddhism," 157, footnote 16.

35 Suphajarawat, *Botbat phra song*, 10.

36 Ibid., 14.

As with other environmentalist monks in Thailand,[37] Phra Khru Suphajarat has used traditional forms of merit making in order to motivate villagers not only to help conserve nature but also to educate villagers about the environment. The traditional robe-offering (*pha pa*) ceremony is a case in point. *Pha pa* literally means 'forest robe' (it probably derives from the Pali *paṁsukūla-cīvara*: [rag robes]) and refers to cloth that laypeople donate to monks (by discarding it so that a monk can pick it up himself); in Thailand during the *pha pa* ceremony donations are made for the maintenance and development of monasteries. In 1989, Phra Khru Suphajarawat organised the *pha pa* ceremony to distribute to villagers seedlings that had been donated to the monastery during the ceremony (altogether some 4,000 plants).[38] He wanted the villagers to recognise the value of plants, and develop a caring attitude towards them.

In addition, reforestation projects are undertaken as a ceremony with the participation of monks and villagers. This not only creates in villagers a feeling of ownership over the forest but also encourages them not to fell trees due to their respect for monks.[39] Phra Khru Suphajarawat's projects, which pursue local sustainable development and conservation of nature, have been developed on the basis of the Buddha's teachings (*dhammavinaya*) and excel in their holistic, and at the same time diversified and creative, approaches. Thus, the case study of Phra Khru Suphajarawat's social engagement and environmental protection projects demonstrates how multiple strategies and religious motifs have been integrated into endeavours to respond to shifting worldviews and value systems in order to motivate people to participate in Buddhist development (*phutthaphatthana*) projects and conservation of nature. In fact, both things are perceived as one and the same.

4 Phra Payutto: Nature and Buddhist Development

Based on these empirical observations, I would now like to discuss concepts of nature and (environmentalist) ethics proposed by one of the most famous and influential contemporary (Thai) Theravada scholar monks, Phra Payutto (b. 1939). Phra Payutto is a widely acclaimed, enormously prolific writer, thinker and lecturer.[40] In many of his works, he investigates the problems and

37 See e.g. Darlington, "The Spirit(s) of Conservation," and Darlington, "The Good Buddha."

38 Prida, *Phlik fuen khuen chiwit*, 31–32. Compare with Darlington, "The Spirit(s) of Conservation," 138.

39 Prida, *Phlik fuen khuen chiwit*, 43.

40 On the website of his monastery Wat Yanawetsakawan, there is a list of his output containing more than 550 publications in English and Thai (http://www.watnyanaves.net/th/

challenges of modernity from a Buddhist perspective. Thus, he has also given several talks on Buddhist perspectives on nature and development, some of which have been published in book form. In the following discussion I refer to these publications, but also use as important sources a talk that he gave on "Ecology According to Buddhist Principles"[41] some twenty years ago. In addition, I will rely on his explanations on these topics, prompted by my questions I sent to him as an email attachment.[42]

Before I discuss the ideas and concepts raised by Phra Payutto, I should point out that, in his discussion of environmental degradation in Thailand and its causes, Phra Payutto describes what he perceives as relevant trends and shifting values and attitudes in Thai society, rather than attempting to provide an academic and comprehensive investigation of environmental problems. It should also be kept in mind that many of his texts referred to here are often only slightly revised transcriptions of talks which he has given. Thus, in my view, Phra Payutto's comments should not be seen as generalizations of complex changes and issues but rather as an identification of what he perceives to be important factors in these developments. While he acknowledges positive initiatives for environmental protection and caring attitudes towards nature in Thailand,

---

book_list, accessed December 27, 2012). In addition, there is also a list of a huge number of 'Dhamma Talks' published as MP3 files (http://www.watnyanaves.net/th/sound_album_list. Accessed December 27, 2012). For more details on his work and biography, see e.g. Olson, Grant, "A Person-Centered Ethnography of Thai Buddhism: The Life of Phra Rajavaramuni (Prayudh Payutto)" (PhD diss., Cornell University, 1989); Olson, Grant, "Introduction," in *Buddhadhamma: Natural Laws and Values for Life*, by Phra Prayudh Payutto, trans. Grant A. Olson (Albany: State Univ. of New York Press, 1995), 1–33; Seeger, Martin, "Phra Dhammapiṭaka und die Pāli-Kanon-Debatte in Thailand: Ein Beitrag zur Untersuchung des modernen Buddhismus" (PhD diss., Universität Hamburg, 2005).

41 Payutto, Phra, "Ecology According to Buddhist Principles," unpublished paper presented in 1993.

42 I would like to thank Phra Payutto very much for his time and willingness to answer these questions in detail and for giving me references to other relevant ideas in his published works (email from Phra Payutto May 23, 2011). I also wish to thank his monastery for providing me with digital recordings of the long, public talk (lasting more than five hours) which Phra Payutto gave on "Ecology according to Buddhist Principles" in 1993 (2536 BE). These two sources have been invaluable in addition to Phra Payutto's published work on forest, development and environmental protection (see sources referred to below). Needless to mention that due to the enormous breadth of the concepts and ideas that Phra Payutto has been suggesting on these themes (and other related topics) over the years, it is impossible to present and discuss comprehensively all of them in this short article: I will only refer to those that I regard on the one hand as most central to his teachings and, at the same time, most relevant for the purposes of this article (but even with these premises I am unable to avoid serious omissions).

given the immensity of environmental problems, in the sources referred to below, he has chosen to focus on those factors and trends which are detrimental to developing 'right' attitudes towards nature.

In line with Phra Khru Suphajarawat's observations and comments, Phra Payutto states that as a consequence of the emergence of modern education, respect and awe for the supernatural in nature have dwindled. At the same time, however, he also argues that knowledge of scientific rationales for the preservation of nature has so far not entailed the adequate creation of effective environmentalist practice:[43] He maintains that "profit thinking and greed [*lobha*] have defeated understanding and knowledge" of the disastrous ecological consequences of overexploitation of natural resources.[44] Thus, there is the dilemma that respect for, awe and fear of the supernatural have been eroded by scientific knowledge and modern education and, as a consequence of this, has to a large extent become ineffective as a motivation for environmental protection. At the same time, Phra Payutto argues, the diminishing of respect and awe in the supernatural character of nature has not been compensated for by adequate creation of scientific understanding of an environmentalist protection imperative. For him, this discontinuity is a failure of development (*khwamphitphlat khong kanphatthana*).[45]

In his analysis of the Thais' perceptions of and attitudes towards nature, concepts of 'development' play a central role. In this way, Phra Payutto notes that in the 1960s, with the beginning of Thailand's development era (*yuk phatthana*), Thais' perceptions towards nature changed radically. Delcore has described this time period as follows:

> [...] the Thai state has pursued a resource intensive drive for national development, based on the extraction of natural resources, promotion of cash crops for export, and export-oriented industrialization.[46]

43 Payutto, Phra พระพรหมคุณาภรณ์ (ป. อ. ปยุตฺโต), *Khon rak pa pa rak khon* คนรักษ์ป่า ป่ารักษ์คน [If Humans Protect the Forest, the Forest will Protect Humans] (Bangkok: Panyaprathip Foundation, 2011), 21, 43, 152–154; see also: [Payutto, Phra] Phra Rajavaramuni (Prayudh Payutto), *Looking to America to Solve Thailand's Problems*, trans. Grant A. Olson (Bangkok: Sathirakoses Nagapradipa Foundation, 1987), 79–83.

44 Payutto, *Khon rak pa pa rak khon*, 153; personal email from Phra Payutto (May 23, 2011).

45 Payutto, *Khon rak pa pa rak khon*, 43.

46 Delcore, Henry D., "Symbolic Politics or Generification? The Ambivalent Implications of Tree Ordination in the Thai Environmental Movement," in *Journal of Political Ecology* 11 (2004): 3.

For Phra Payutto, the way development was pursued in the 1960s was lopsided in that it focused on materialistic development. This entailed significant changes in Thai attitudes to and perceptions of nature. He argues that

> During that time period, forest was seen as a symbol of backwardness and lack of development, which is the reason why Thais destroyed forest to a large extent. The aim was to create progress [*jaroen*], which was called development [*phatthana*].[47]

Even Buddhist monasteries were not immune to these new views and values, and they cut down trees on a large scale.[48] He states that these attitudes are in stark contrast to how nature was perceived in early Buddhism. According to him, there are three fundamental Buddhist perspectives on nature.[49] Firstly, Buddhist literature contains numerous texts that describe nature as something that is beautiful and that can make humans happy.[50] This happiness can be an important motivation for environmental protection: "In order to preserve nature we have to love nature; in order to love nature we have to experience happiness when with nature."[51] Secondly, since all sentient beings are equally subject to the universally valid natural laws (*kot thammachat*) of impermanence (*anicca*) and unsatisfactoriness (*dukkha*), according to which they all have to experience not only happiness but also the unpleasantness of birth, aging, pain and dying, Buddhism teaches that humans should have a benevolent attitude (*mettā*) and a sense of friendship (*maitri*) toward all sentient beings. Finally, nature is perceived as very valuable in that it offers the amicable

---

47 Payutto, *Khon rak pa pa rak khon*, 47. Compare with Rigg, Jonathan et al., "Understanding Languages of Modernization: A Southeast Asian View," in *Modern Asian Studies* 33.3 (1999): 584–586; see also Darlington, "The Spirit(s) of Conservation," 130–131.

48 Payutto, *Khon rak pa pa rak khon*, 46–47; and Payutto, Phra พระธรรมปิฎก (ป. อ. ปยุตฺโต), *Kan phatthana thi yangyuen* การพัฒนาที่ยั่งยืน [Sustainable Development], second edition (Bangkok: Buddhadhamma Foundation, 1996), 180.

49 Payutto, *Kan phatthana thi yangyuen,* 188–191; Swearer, Donald, "The Hermeneutics of Buddhist Ecology in Contemporary Thailand: Buddhadāsa and Dhammapiṭaka," in *Buddhism and Ecology: The Interconnection of Dharma and Deeds*, ed. Mary Evelyn Tucker and Duncan Ryūken Williams (Cambridge: Harvard University Press, 1997), 32–33. Here, Swearer gives a slightly different summary of these attitudes.

50 Payutto, *Kan phatthana thi yangyuen*, 188–191. Compare with Schmithausen who comes to the conclusion that "early Buddhism was impressed not so much by the beauty of nature as by its sombre aspects" (cf. Schmithausen, Lambert, "The Early Buddhist Tradition and Ecological Ethics," *Journal of Buddhist Ethics* 4 (1997): 2).

51 Payutto, *Kan phatthana thi yangyuen*, 190.

environment of secluded (*kāya-viveka*) training for spiritual progress, which should ultimately lead to the attainment of *nibbāna* (*upadhi-viveka*).[52]

In this way, Phra Payutto maintains that the development of positive attitudes toward nature as a consequence of these three perceptions can be immensely conducive to the creation of motivation for environmental protection, but in itself it is not what Buddhism is ultimately aiming for.[53] The reason for this is that the values that are attributed to nature may be based on an anthropocentric view that accords values to nature based on selfish (*henkaetua*) motivations. In other words, according to Buddhist principles, anthropocentric and instrumentalist values of nature are ultimately to be transcended.[54] For Phra Payutto, solely to solve the physical problems of preserving an ecological balance is insufficient.[55] He argues that ultimately, protection of nature should be linked intrinsically to or be a natural ([*yang*] *pen thammachat*)[56] consequence of human, inner development that culminates in the attainment of genuine freedom (*kan khau thueng itsaraphap thi thae jing nai thi sut*).[57] Or, put differently, "through this kind of human development, environmental protection happens automatically and instantly, without the need for any intention to protect it."[58]

In many of his books, Phra Payutto outlines in detail how this human development should be pursued. It entails a holistic, integrated approach with three dimensions, which are interrelated and mutually supportive of each other. These dimensions, specifically applied to the protection of nature, are, firstly, 'behaviour', which includes legislation, social rules and regulations and behavioural patterns (*phruettikam khoei-chin*), anchored in culture and traditions, which aim to protect nature and create a positive relationship with it; secondly, 'mind-development', which comprises the development of positive, happy and responsible attitudes toward nature; and, thirdly, the creation of knowledge and insight (*paññā*) in order to realise the 'causal interrelatedness of all things'.

52 Compare with Prayun, Phra พระเมธีธรรมาภรณ์ (ประยูร ธมฺมจิตฺโต), *Thamma lae kan anurak singwaetlom* ธรรมะและการอนุรักษ์สิ่งแวดล้อม [The Dhamma and Environmental Conservation] (Bangkok: Buddhadhamma Foundation, 1995), 20.

53 Payutto, *Kan phatthana thi yangyuen*, 187.

54 Payutto, *Khon rak pa pa rak khon*, 100–103. Compare with Swearer, "The Hermeneutics of Buddhist Ecology," 33.

55 Payutto, *Khon rak pa pa rak khon*, 76.

56 Ibid., 114.

57 Ibid., 76.

58 Ibid., 113; a similar argument was put forward by Buddhadasa Bhikkhu พุทธทาสภิกขุ, *Phutthasatsanik kap kan-anurak thammachat* พุทธศาสนิกกับการอนุรักษ์ธรรมชาติ [Buddhists and the Conservation of Nature] (Bangkok: Komonkhimthong Foundation, 1990), 6–7, 12.

These three dimensions correspond to the Buddhist principle of the Threefold Training of morality (*sīla*), concentration (*samādhi*) and wisdom (*paññā*), into which the aspects of Buddhist practice of the Noble Eightfold Path can be divided. Phra Payutto explains that this approach constitutes a systematic approach to human development that takes into account the different levels of inner development. Human development in accordance with these principles, he remarks, allows the creation of an 'ethics of happiness' instead of an 'ethics of fear, reluctance or compromise'. Perceiving significant differences between Western understandings of ethics and Buddhist ethics,[59] Phra Payutto explains that the Buddhist 'ethics of happiness' implies that environmentalist action and restraint in consumption (*bhojane mattaññutā*)[60] are to be practised automatically, naturally and wholeheartedly without the fear of having to face the dire consequences of the exploitation and destruction of nature or the fear of punishment.[61] It also implies that humans do not perceive environmentalist action as a necessity that entails the compromise of one's personal needs or freedom. In this way, he argues:

> If humans develop themselves in the right way, they will be able to defeat their inner nature [*au chana thammachat phainai tua eng*]. [That means humans will be able to defeat their mental defilements. Defeating inner defilements] is more feasible than defeating the external nature, which is something the West has been trying to achieve and which has caused problems for humans.[62]

Elsewhere, Phra Payutto expresses a similar idea by referring to a seemingly paradoxical saying, which is ascribed to the Buddha and transmitted in the Pali canonical book Dhammapada. The saying goes as follows: "Cut down the forest but do not cut down the trees." He discusses this 'Dhamma riddle' or 'Koan' (*pritsana-tham*) by explaining that depending on context, both in Thai and Pali language, the forest (*pa*) can have a number of connotations with displeasing emotional connections. Thus, the word 'forest' may be associated with emotions not only of fear, disorder, and lack of principles and cleanliness but also of backwardness, remoteness, lack of development and ignorance. Phra Payutto argues that

59 Payutto, *Kan phatthana thi yangyuen*, 90.

60 Compare with A.I.113.

61 Payutto, *Khon rak pa pa rak khon*, 72–76.

62 Payutto, *Kan phatthana thi yangyuen*, 90.

> Humans presently face huge problems as they still have "forest" [*pa*] in their mind. This forest has to be eradicated by developing humans [through inner development]. We cannot eliminate this forest by felling trees for the sake of developing the country.[63]

As shown, even though always informed by and directed towards Buddhist soteriology the concepts regarding nature conservation proposed by Phra Payutto pursue an integrative, holistic, flexible and continuous approach. Awe of supernaturality and fear of undesirable consequences may be a necessary means of motivation to some extent but are, in a process of inner development, to be transcended as motivation for environmentalist action.[64] Also, even though nature is initially perceived in a more instrumentalist way, protection of nature is ultimately to happen as an automatic and inherent consequence of a practice that pursues the realization of 'natural law'. In this process of realization humans learn that they are an inseparable part of nature, entirely subject to its laws and ultimately unable to control it (*anattā*).

5 Nature and the Thai Forest Tradition

Another Thai perspective on forest, which seems to exemplify nicely some of the aforementioned concepts proposed by Phra Payutto, can be identified in biographies of Thai forest monks. For a large number of Thais the practice to attain spiritual perfection (*nibbāna*) and Buddhist sainthood (*arahattaphala*) are epitomised by the lives of monks belonging to the Thai forest tradition. Many of these monks, most of whom lived and became famous during the twentieth century, are believed to have achieved *nibbāna*, the summum bonum in Buddhism. These monks spent a significant amount of time in the forest in order to develop their minds according to Buddhist canonical principles, or put differently in the words of Kamala, 'deep forest and the forest cemetery' were the 'training grounds' for the spiritual practice of these monks towards

---

63 Payutto, *Khon rak pa pa rak khon*, 118–119. See also Ivarsson, Søren, "Man, Nature and Environmentalism in Thailand: The Role of Buddhism," in *Forest in Culture – Culture in Forest: Perspectives from Northern Thailand*, ed. Ebbe Poulson et al. (Tjele: Research Centre on Forest and People in Thailand, 2001), 38–39.

64 Email from Phra Payutto. See also Payutto, *Khon rak pa pa rak khon*, 43–44 and Löhr, Sabine, "Zwischen Ost und West: Ansätze einer Ökologischen Ethik im Modernen Buddhismus" (PhD diss., Universität Hamburg, 2004), 166–167.

spiritual liberation.[65] Here, the wilderness of the forest allows re-establishing authenticity through the re-enactment of crucial parts of the Buddha's biography. By testing determination on an extreme level, the forest allows monks to realise and authenticate Buddhist truths. These monks thereby have become themselves the centre of veneration as an instantiation of Buddhist truths and ideals, as a consequence of which they have built up enormous charisma. In particular, the icon and 'father' of the Thai forest tradition, Ajarn Man Bhuridatto (1871–1949) exemplifies this idea very well. Expressed in the words of the Thai scholar Thanwa Jaithiang,

> Ajarn Man was walking on foot in the wilderness of the forest and [facing] things that a normal person is probably not capable of bearing. Thus, he was sometimes sleeping in a forest replete with tigers, or in caves with malevolent demons [...]. These things are something which is most unbelievable for ordinary beings [*puthujjana*]:[66] How could someone be able to live all alone in isolation in the middle of the forest? This is only possible when one has strong ideals and confidence [*saddhā*] in the principles of the *dhamma* and the avoidance of harming others.[67]

Thanwa Jaithiang has also argued that prior to Ajarn Man's meditative accomplishments it was believed that spiritual perfection was no longer possible,[68] as the monks' life had approximated that of the villagers too much. This approximation of monastic life and worldly life, however, is in stark contrast to the

---

65 Kamala Tiyavanich, *Forest Recollections: Wandering Monks in Twentieth-century Thailand* (Chiang Mai: Silkworm Books, 1997), 79. See also Taylor, J. L., *Forest Monks and the Nation-State: An Anthropological and Historical Study in Northeastern Thailand* (Singapore: Institute of Southeast Asian Studies, 1993), 239; and Schedneck, Brooke, "Forest as Healer: Reinterpretations and Hybridity within the Forest Tradition of Thailand," unpublished paper. In his biography of his teacher Ajarn Man Bhuridatto, Phra Mahabua describes the forest environment as "the most appropriate battlefield to choose in one's struggle to attain all levels of Dhamma" (Boowa Ñāṇasampanno, Ācariya Mahā, *Venerable Ācariya Mun Bhūridatta Thera*, trans. Bhikkhu Sīlaratano (Baan Taad: Wat Pa Baan Taad, 2003), 65).

66 *Puthujjanas* are those who have not yet achieved any levels of Nobility, that is they are not Noble Persons (*ariyapuggala*).

67 Thanwa Jaithiang ธันวา ใจเที่ยง, *Phra ajan man phurithatto phu nam haeng kan-sang thammikasangkhom chao na lum nakhong* พระอาจารย์มั่น ภูริทัตโต ผู้นำแห่งการสร้างธรรมิกสังคมชาวนาลุ่มน้ำโขง [Venerable Ajarn Man Bhuridatto, the Leader of a Dhammic Peasant Society in the Lowlands of the Mae Khong River], published 2006, accessed December 28, 2011. http://61.47.2.69/~midnight/midnight2545/document95155.html.

68 Cf. also Seeger, Martin, "'Against the Stream': The Thai Female Buddhist Saint Mae Chi Kaew Sianglam (1901–1991)," *South East Asia Research* 18.3 (September 2010): 574.

spiritual path of the Buddha who "by going into the forest left society behind and relinquished an advanced social position [*sakdina*] in order to be equal with the people and to live peacefully and in a non-harming way with nature".[69] This shows that here the forest is perceived as a crucial factor in allowing determined practitioners to follow Buddha's footsteps in order to attain ultimate freedom through mind development. Thus, the forest allows verification and authentification of Buddhist soteriological truths. Based on these perceptions and concepts, monks of the forest tradition, together with lay-people who want to gain merit by supporting the preservation of an authentic Buddhist environment, have been able to preserve sometimes rather large areas of forest by incorporating them into the compounds of their monasteries.[70] In this way, over the course of the last 50 years or so, and as a consequence of extreme deforestation in Thailand, these monasteries have become forest sanctuaries, forested islands in vastly deforested areas, such as e.g. the northeastern monastery Wat Pa Nakhamnoi, of which some two million square metres are covered with forest. Forest monasteries are however not only perceived as physical islands created as a result of deforestation, but also as spiritual islands or an 'oasis in the desert' amidst "a society where materialism is epidemic".[71]

The well-known Thai intellectual and activist monk Phra Phaisan (Visālo) argued in 1990 that in modern Thai society, which according to him "is increasingly influenced by degraded values and obsessed by material growth",[72] forest monasteries are potentially healing places for society where people can recover mentally by going into retreat and making themselves aware/reminding themselves of the importance and benefits of spiritual development and non-materialistic values. Thus, forest monasteries are perceived here as strongholds that "question the validity of popular values"[73] whilst preserving "traditional wisdom so badly needed by Thailand and the modern world": the monastic community of "the forest monastery is a society aiming for human development

69 Thanwa Jaithiang, *Phra ajan man phurithatto*.

70 Isager, Lotte, "History and People of North Thailand," in *Forest in Culture – Culture in Forest: Perspectives from Northern Thailand*, ed. Ebbe Poulson et al. (Tjele: Research Centre on Forest and People in Thailand, 2001), 110.

71 Phaisan Visālo, Phra, "The Forest Monastery and its Relevance to Modern Thai Society," in *Radical Conservatism: Buddhism in the Contemporary World: Articles in Honour of Bhikkhu Buddhadasa's 84th Birthday Anniversary*, ed. Sulak Sivaraksa, et al. (Bangkok: Thai Inter-Religious Commission for Development/International Network of Engaged Buddhists, 1990), 295.

72 Phaisan Visālo, Phra, "The Forest Monastery and its Relevance to Modern Thai Society," 294.

73 Loc. cit.

amidst the natural environment. In this system of relationships we can see the wisdom which stresses the interrelatedness and interdependence of persons, society and nature".[74]

At the same time, it should be noted here that, as Darlington argues, Thai Buddhist monks have also to some extent facilitated the cutting down of trees in various ways:

> Forest monks brought elements of the civilized world to the forest, blurring the distinctions between the two. Their presence helped quell villagers' fears of the forest spirits, leading them to move into the forest, settling previously wild, dangerous and unlivable areas.[75]

Also, by participating in two important programmes supported by the government (both launched in the 1960s) and other development programmes, monks "inadvertently were a factor in [...] environmental degradation" as "elements of fear of and respect for spirits that limited encroachment into forested land in the past were negated by monks' engagement in and support of national economic development".[76] Darlington also reports on monks who, responding to villagers' requests, subdued spirits so that trees could be felled without having to fear punishment by the spirits.[77]

6 Individual and Social Karma and Environmental Protection

Phra Phaisan has argued quite interestingly on the role of Thai beliefs in the supernatural in connection with environmental protection. According to him, Buddhism in the past was important in the creation of social awareness (*sang samnuek to suanruam*). He argues that one of the reasons for this was that

> [...] despite Buddhism's strong emphasis on dhammic principles [*lak tham*] on a personal level, working for the common good was present in the communities [in the past], as "spirits" [*phi*] had [a significant role] in this respect [...]. *Phi* did not tolerate the unsanctioned felling of forest

74 Ibid., 297. See also Sumedho, Bhikkhu, "Forest Tradition as a Challenge to the Modern World," in *Socially Engaged Buddhism for the New Millennium: Essays in Honor of the Ven. Phra Dhammapitaka*, ed. Pipob Udomittipong and Chris Walker (Bangkok: Sathirakoses Nagapradipa Foundation and Foundation for Children, 1999), 470–482; and Schedneck, "Forest as Challenge."

75 Darlington, "The Spirit(s) of Conservation," 134.

76 Ibid., 135.

77 Loc. cit.

> trees and the destruction of rivers and rivulets; this would have been a violation of the rules of the guardian spirits or ancestors (which [at the same time] would have been [understood as] a violation of the regulations of the community). The violation of moral rules of Buddhism [*sīla*] was regarded as a demeritorious deed for which karmic consequences [*vipāka*] were entailed, such as having to go to hell after death, having a short lifetime for one's future life or having to face hardship in one's future life. The violation of spirits' rules [*phit phi*], however, did not only imply immediate consequences for the culprit, such as incurring an illness; the community may also have had to suffer from drought and so forth.[78]

Thus, Phra Phaisan observes a former complementarity[79] in Thai Buddhism and beliefs in and practices of supernaturalism, in which Buddhism was more concerned with the remote and individual whereas 'spirits' were invoked to deal with the immediate and the social. Similarly, the famous Thai historian and social commentator, Nidhi Eoseewong argued that

> Since ancient times, the social dimension of Buddhism has not really been emphasized in Thai society, as the beliefs in spirits already had the function of ensuring that social regulations were adhered to. Buddhism has given more importance to personal spiritual development.[80]

Both Phra Phaisan and Nidhi stress however, that Theravada Buddhism itself does not lack a social dimension (*miti thang sangkhom*) at all but it is the particular form of Thai Buddhism in which the teaching of these social aspects has been neglected.[81] Both point out that in particular the monks Buddhadasa Bhik-

78 Phaisan Visālo, Phra พระไพศาล วิสาโล, *Phutthasatsana thai nai anakhot naenom lae thangok jak wikrit* พุทธศาสนาไทยในอนาคต แนวโน้มและทางออกจากวิกฤติ [Thai Buddhism in the Future: Trends and Solutions to the Crisis] (Bangkok: Sotsri-Saritwong Foundation, 2003), 228–229.

79 Ibid., 230.

80 Nidhi Eoseewong นิธิ เอียวศรีวงศ์, *Phutthasatsana nai khwamplianplaeng khong sangkhom* พุทธศาสนาในความเปลี่ยนแปลงของสังคม [Buddhism in the Context of Social Change] (Bangkok: Komonkhimthong Foundation, 2000), 28.

81 Phaisan Visālo, *Phutthasatsana thai nai anakhot*, 228–229; and Nidhi, *Phutthasatsana nai khwamplianplaeng khong sangkhom*, 29. Since I do not have the space for this here, I will not further pursue the questions as to how far early or canonical Buddhism had a social dimension. Gombrich, for one, states that the Buddha's "concern was to reform individuals and help them to leave society forever, not to reform the world" – cf. Gombrich, Richard, *Theravāda Buddhism: A Social History from Ancient Benares to Modern Colombo*,

khu and Phra Payutto have demonstrated in their works that in fact canonical Theravada Buddhism possesses a strong social dimension.[82]

Not only do the arguments of Phra Phaisan and Nidhi recapture some of the above-mentioned empirical observations of Phra Khru Suphajarawat and arguments of Phra Payutto, they also point to another religious concept often referred to in discussions about motivational factors for environmental protection or degradation. In this respect, Kraft has stated that: "In all Buddhist cultures the primary arena of karmically significant action has been the individual." However, given the '[c]urrent nuclear and environmental problems', Kraft argues that "[w]here Buddhism has focused on individual karma, now we also need better ethical analysis of collective behaviour. How might notions of group karma be rendered in modern terms?"[83]

In line with Kraft and the arguments of Phra Phaisan and Nidhi, the Thai scholar Suwanna Satha-anand argued, specifically with regard to the Thai context, that

> [o]ne of the major concerns relating to the understanding of karma in Thai society is the fact that karma is understood, not only primarily, but perhaps exclusively, in terms of the individual dimension.[84]

---

second edition (London, New York: Routledge, 2006), 30. See also Parnwell, and Seeger, "The Relocalization of Buddhism," 92–100. Compare with Schmithausen, "The Early Buddhist," 11.

82 See also: Phaisan, Phra พระไพศาล วิสาโล, *Phra thammapidok kap anakhot khong phutthasatsana* พระธรรมปิฎกกับอนาคตของพุทธศาสนา [Phra Payutto and the Future of Buddhism] (Bangkok: Wutthitham, 1999), 24–30, 101–103; and Phaisan, *Phutthasatsana Thai nai anakhot*, 235–240. In connection with this line of argument, Phra Phaisan also reflects on the reason(s) why "the majority of Thai people regard social problems as a result of doings by individuals or personal factors (e.g. selfishness or laziness), neglecting structural causes. Is this mainly because of the influence of Buddhism or is it a consequence of the influence of textbooks and nationalistic ideas during the last decades, according to which the family or the village is the model of the nation? These [latter] ideas caused an understanding that suggests that social problems are not different from problems within the family or the village. That means that these problems emerge as a consequence of the behaviour of selfish individuals" (cf. Phaisan, *Phutthasatsana thai nai anakhot*, 233–234, footnote 6).

83 Kraft, Kenneth, "Nuclear Ecology and Engaged Buddhism," in *Buddhism and Ecology: The Interconnection of Dharma and Deeds*, ed. Mary Evelyn Tucker and Duncan Ryūken Williams (Cambridge: Harvard University Press, 1997), 276.

84 Suwanna Satha-anand, "Karma as Moral Justice in Thai Buddhism," *Manusya: Journal of Humanities*, Special Issue No. 4 (2002): 80.

Here, it is telling to look at comments made in connection with severe natural disasters that have occurred in Thailand in recent years. In the context of the heavy flooding in Thailand during the monsoon season of 2011, the famous Thai Buddhist writer, Dangtrin,[85] reported that many people perceived a connection between past collective karma and the fact that a large number of people died together and still die as a consequence of natural disasters.[86] In his analysis of Thai beliefs with regard to karma in connection with the Tsunami of 2004, Terwiel states that "[w]hen a disaster suddenly occurs in Thailand, [Thai] Buddhists have the feeling that they somehow have brought this upon themselves [as a result of their own doing]."[87] In his analysis, Terwiel quite interestingly alludes to a 'perception of a collective karmically-conjoined society'.[88] Thus,

85 'Dangtrin' is the penname of Saran Maitreewech.

86 Greenista Society, accessed January 6, 2012. http://greenistasociety.com/read/read_12.php.

87 My translation from German. Cf. Terwiel, Barend Jan, "Die Naturkatastrophe aus der Sicht des thailändischen Buddhismus," in *Jahrbuch der Akademie der Wissenschaften zu Göttingen* (Göttingen: Vandenhoeck & Ruprecht, 2005), 65. See also: Merli, Claudia, "Religious Interpretations of Tsunami in Satun Province, Southern Thailand: Reflections on Ethnographic and Visual Materials," *Svensk Religionshistorik Årsskrift* 14 (2005): 154–181.

88 Cf. Terwiel, "Die Naturkatastrophe," 67. Here it is quite interesting to note that, as Schmithausen observed, "already in the Pāli canon we encounter the idea that the external world is decisively influenced by the *moral* and *spiritual* behaviour of living beings or at least that of human beings, and that this influence has repercussions for the situation of human beings; e.g. moral misbehaviour of people may cause lack of rain, which in its turn causes famine" (Schmithausen, Lambert, "Buddhism and the Ethics of Nature: Some Remarks," *The Eastern Buddhist* 32.2 (2000): 65). See also Schmithausen, "The Early Buddhist Tradition," 13, 24. Schmithausen is referring to the *Palokasutta* (A.I.159–160) and the *Adhammikasutta* (A.II.74–76) and also to the famous *Aggaññasutta* (esp. D.III.85–92). A well-known example from Thai history for the belief in a connection between ethical behaviour and the external world is the cholera epidemic, which reached Bangkok in 1820. Terwiel describes the severity of this disaster and the attempts to deal with it thus: "The death toll in Bangkok was great. The monasteries could not keep up with the cremations; many corpses were thrown in the river. During the emergency the king [Rama II] issued a decree to his people explaining how the sickness had come from abroad and how ordinary medicine had no effect. Therefore a special royal ceremony for the expulsion of evil was held: sacred stanzas were chanted, all around the city large guns were fired, and a procession was held with the nation's most sacred objects. The king temporarily relieved all people of their duties and exhorted all to make merit, chant sacred mantras, and practice munificence [...] The populace was asked not to kill any living being, thus avoiding harmful acts" (Terwiel, Barend Jan, *Thailand's Political History: From Fall of Ayutthaya to Recent Times* (Bangkok: River Books, 2005), 94–95). See also Ivarsson, "Man, Nature and Environmentalism," 36–38, where Ivarsson discusses the interrelatedness of human's

Dangtrin's and Terwiel's observations show that karma in Thai Buddhism is not solely perceived to be operating on an individual level, but may also be understood as operating on a communal or social level. At the same time, and also to some extent demonstrated by Dangtrin's and Terwiel's observations, karma in Thai Buddhism is also very often understood in negative terms and invokes adverse feelings or is sometimes even perceived, in the words of Phra Payutto, as a "mysterious, secretive force that ambushes people through punishment".[89]

In many of his works, Phra Payutto seeks to rectify 'narrowed' Thai perceptions of karma that focuses solely on negative, past karma.[90] The fixation on old karma and its fruition in the present may cause not only indifference and lack of a sense of responsibility but also inertia due to the absence of motivation to improve one's situation. Thus, this perception may be harmful. Phra Payutto also argues that past karma should not be understood as a "supernatural force to become attached to or to which one entrusts one's fate" but rather as a chance to learn from and improve one's situation.[91] He argues that a correct engagement with karma aims at fostering self-reliance, a sense of responsibility, active

morality and nature as depicted in one of the most significant cultural texts in Thailand: the Trai Phum Phra Ruang, a cosmological treatise, believed to have been composed in the 14th century by Phya Lithai, a king-to-be of the Thai kingdom Sukhothai. In the Trai Phum Phra Ruang's section describing 'cosmic destruction' it is said that "When the fire consumes the *kappa* [cosmic age], it consumes the people who do evil deeds with their bodies; those who do evil deeds with their mouths; those who do evil deeds with their minds; those who do not know any merit or Dhamma at all; those who do not respect their fathers and mothers; those who do not respect monks, *brahmins*, and others who follow the Dhamma; those who do not treat their brothers, sisters, relatives, and friends in a proper way; and those who treat one another like deer or hog-deer, like ducks or chickens, like pigs or dogs, like elephants or horses, and chase, hurt, and kill one another. Because of the effect of such evil deeds that they have done, disasters occur everywhere throughout the *cakkavāḷa* [world system]" (cf. Reynolds, Frank E., and Mani B. Reynolds, trans., *Three Worlds According to King Ruang: A Thai Buddhist Cosmology* (Berkeley: Asian Humanities Press, 1982), 307). Compare also with Prayun, *Thamma lae kan anurak singwaetlom*, 21–23.

89 Payutto, Phra พระธรรมปิฎก (ป. อ. ปยุตฺโต), *Narok sawan samrap khon run mai* นรก – สวรรค์ สำหรับคนรุ่นใหม่ [Hells and Heavens for the People of the New Generation] (Bangkok: Buddhadhamma Foundation, 1995), 76.

90 Payutto, Phra พระธรรมปิฎก (ป. อ. ปยุตฺโต), *Phutthatham* พุทธธรรม [Buddhadhamma] (Bangkok: Mahachulalongkorn University, 1995 CE/2538 BE), 9.

91 Ibid., 207; and Payutto, *Narok sawan samrap khon run mai*, 3–12.

exertion and a rational understanding of causal processes (instead of, for example, the blind belief in the supernatural).[92]

As already mentioned above, against the background of Thailand's intellectual history, Phra Payutto excels (amongst other things) in that he is regarded as having, in the words of Phra Phaisan, significantly 'revived the social dimension'[93] of Buddhism: in his teachings of Buddhism he "has arguably been more successful than any other Thai monk in pursuing to widen the scope of Buddhism so that it not only comprises the level of the [individual] mind but also the social sphere".[94] Thus, Phra Payutto maintains that Theravada Buddhist teachings on karma also contain both an individual and social level.[95] He explains that the coming together of karmic action of individuals amongst and towards each other is karma as the way society operates (*withi khong sangkhom*). In his teachings on karma on a social level, Phra Payutto very much emphasises the importance of the three mental defilements (*kilesa*) craving (*taṇhā*), conceit (*māna*) and (false) views (*diṭṭhi*):[96] "Human history is largely the story of these defilements."[97] These three defilements, Phra Payutto has repeatedly argued, are also the main obstacles in solving environmental problems.[98] In particular, *diṭṭhi* is karmically very significant. It comprises beliefs, theories, ways of thinking, ideologies and value systems, and "directs the fate

92 Payutto, *Phutthatham*, 213–214.

93 Phaisan, *Phra thammapidok kap anakhot khong phutthasatsana*, 24–30.

94 Ibid., 29.

95 The same, so Phra Payutto proposes, applies to another core, very closely related teaching: the Dependent Origination (*paṭiccasamuppāda*) has been taught by the Buddha as applicable not only for the individual mind but also on a social level (cf. Payutto, *Phutthatham*, 137–140). Thus, the *paṭiccasamuppāda* may be used to explain the arising of environmental problems (email from Phra Payutto). Even though the *paṭiccasamuppāda* is probably the most often invoked teaching in arguments for Buddhist environmental ethics, and would certainly be a most interesting topic, in particular when connected to Western scholarships (e.g. Schmithausen, "The Early Buddhist Tradition"), due to limitations of space I am not able to pursue a detailed examination as to whether and how far Dependent Origination, both as understood in early Buddhism and taught in modern Thai Buddhism, is concerned with social or other aspects of life other than the process of individual rebirth.

96 These three defilements, grouped together, are also known as the three *papañca* (mental diffusions, burdens causing delay).

97 Payutto, P. A., *Good, Evil and Beyond: Kamma in the Buddha's Teaching*, trans. Bruce Evans, third Impression (Bangkok: Buddhadhamma Foundation, 1996), 62.

98 Payutto, *Kan phatthana thi yangyuen*, 87–88.

of society and the world and shapes [human] history".[99] Phra Payutto identifies the following interrelated, mistaken views as major problems in the relationship between humans and nature: (1) That humans are separate from nature; (2) That real happiness is the result of materialistic consumption (as a consequence of which nature is shaped and exploited in order to satisfy the demands of humans); and (3) That humans must conquer and manage nature (*phichit/ jatkan thammachat*). Phra Payutto points out that Western thinkers have studied the historical development of these premises in the West and how these ideas underlie Western religions, philosophy, science, and economy.[100] These ideas may also have affected the current understanding and attitudes of Thais towards nature.[101] Phra Payutto argues that

> [i]n order to resolve environmental problems, the idea [*diṭṭhi*] to have to conquer nature, which clearly is in stark contrast to Buddhist ideas, has to be resolved. This [however] will necessitate an enormous effort in developing human beings [*tong chai kansueksa thi mi kamlang sung*].[102]

7 Discussion

The purpose of this article has not been to investigate whether and how early Buddhism, whose texts are often referred to as an important source for arguments for 'green Buddhism', was concerned with positive or caring attitudes towards nature; much has been written on this elsewhere and there is no unanimous view.[103] Neither have I attempted to examine pre-modern Thai views of and attitudes towards nature. Here it should be mentioned, though, that the impact that Thai people in pre-modern times have had on the environment has been quite small (which is hardly surprising given the rather low population and the technological possibilities available). Terwiel has argued that this

99 Payutto, Phra พระธรรมปิฎก (ป. อ. ปยุตฺโต), *Phutthatham chabap doem* พุทธธรรม (ฉบับเดิม) [Buddhadhamma (Original Version)], 23rd edition (n.p., 2003), 208.

100 Phra Payutto states, however, that there is now a significant rethinking in the Western world in this respect (cf. Payutto, *Khon rak pa pa rak khon*, 49–54).

101 Payutto, *Khon rak pa pa rak khon*, 44.

102 Email from Phra Payutto.

103 A good overview with regard to these different positions is provided by Swearer, Donald, "An Assessment of Buddhist Eco-Philosophy," *Harvard Theological Review* 99.2 (2006): 123–137.

> [...] may have led to the hypothesis that in premodern times the Thais, possibly inspired by their Buddhist ideology, handled their environment gently. However attractive such an approach to early modern times may be to present-day environmentalists, it does not fit with an objective look at the available data [...]. The case of Thailand demonstrates that the premise that nature is mild, in equilibrium, and inherently good (until modern times, when mankind lost its true place in nature) is a false one. Nature was dangerous, uneasy, and threatening [...].[104]

Rather than engaging in historical issues or problems of analysing Buddhist canonical texts to identify ecological ethics, the main purpose of this paper has been to examine contemporary Thai views and strategies that conceptualise nature as a response to or in connection with environmental degradation. This necessarily brief discussion of a number of observations and concepts in the Thai religious sphere has revealed that there is a wide range not only of complex ambiguities and shifts in the meaning of nature but also of different approaches and hermeneutical strategies, both on a practical and a theoretical level, responding to these ambiguities and meanings in order to address huge ecological problems. As shown above, very often seemingly inextricably linked to and complementing each other, both Buddhist and animistic elements have been influential lenses through which nature has been perceived. For this reason, both these belief systems are regarded, often in a harmonious marriage, as having significant potential to deal effectively with environmental problems. Another, very important and, despite initial criticism, successful example for this is the so-called 'ordination of trees' ritual (since the late 1980s) that incorporates both Buddhist and animistic ideas to sanctify trees and, by so doing, protect them from being felled.[105] As I have demonstrated above, in addition to

104 Terwiel, Baas, "The Physical Transformation of the Central Thai Region in Premodern Times," in *A History of Natural Resources in Asia: The Wealth of Nature*, ed. Greg Bankoff and Peter Boomgaard (Basingstoke: Palgrave Macmillan, 2007), 55–56.

105 As 'tree ordinations' in Thailand have received wide attention in the scholarly world, I did not consider them in this article. For tree ordinations in Thailand, see e.g.: Taylor, Jim, "Social Activism and Resistance on the Thai Frontier: The Case of Phra Prajak Khuttajitto," *Bulletin of Concerned Asian Scholars* 25 (1993): 3–16; Darlington, Susan M. (1998), "The Ordination of a Tree: The Buddhist Ecology Movement in Thailand," *Ethnology* 37.1 (1998): 1–15; Darlington, Susan M., "Buddhism and Development: The Ecology Monks of Thailand," in *Action Dharma: New Studies in Engaged Buddhism*, ed. Christopher Queen, Charles Prebish and Damien Keown (London: RoutledgeCurzon, 2003), 96–109. Darlington, "The Good Buddha;" Tannenbaum, Nicola, "Protest, Tree Ordination, and the Changing Context of Political Ritual," *Ethnology* 39.2 (2000): 109–127; Isager, Lotte, and Søren Ivarsson, "Contesting Landscapes in Thailand: Tree Ordination as Counter-

initiatives on a practical level, there are also influential Thai Buddhist thinkers who have proposed diverse concepts of environmentalism basing their arguments on the Pali canonical scriptures.[106] Thus, it can be argued that there is a large number of diverse, and often complexly intertwined, shifting and ambiguous factors that have not only generated changes in Thai views of nature, but also informed and motivated environmental action and perspectives. In summary, these factors are the charisma of (individual) Buddhist monks and the Thai monarchy, intervention by the state, Buddhist-animistic rituals, (decreasing) belief and awe in the supernaturality of nature, teachings of Buddhism as a 'green religion', Buddhist soteriology (through the re-enactment of the spiritual quest towards awakening), increasingly widespread scientific understanding of the ecological system, Western-influenced environmentalism, views of forest as a place for recreation and leisure (especially among the urban middle class)[107] and also, unfortunately, the too obvious dire results of the over-exploitations of nature.

We have also seen that monks who are concerned about environmental degradation have combined a variety of these elements in order to motivate their local communities to help actively to protect the environment. Here, what Darlington calls a "creative use of [...] Thai traditions and beliefs"[108] can be observed. In fact, what I find remarkable is the way in which a variety of cultural elements have been creatively re-arranged, combined, re-interpreted and, possibly also invented, sometimes despite strong criticism.[109] This has been done with the aim to find indigenous and ingenious solutions to environmental

---

Territorialization," *Critical Asian Studies* 34.3 (2002): 395 417; Delcore, "Symbolic Politics or Generification?" 8–22; Brown, Kevin, "Spectacle as Resistance: Performing Tree Ordination in Thailand," *The Journal of Religion and Theatre* 5.2 (2006): 91–103. For a very interesting discussion on why it is problematic to use the term 'ordination' in 'tree ordination', see Gabaude, Louis, "Note sur l''ordination' sans ordre des abres et des forêts," *Aséanie* 25 (2010): 91–125.

106 In addition to Phra Payutto (on whom I was focusing in this respect in this article) Buddhadasa Bhikkhu, Phra Payun, and Chatsumarn Kabilsingh (now increasingly known under her ordination name Dhammananda), all of whom are also influential thinkers of modern Thai Buddhism, have also been elaborating on environmental protection from a Buddhist perspective. See e.g.: Buddhadasa, *Phutthasatsanik kap kan-anurak thammachat*; Prayun, *Thamma lae kan anurak singwaetlom*; and Chatsumarn, Kabilsingh et al., *Buddhism and Nature Conservation* (Phra Nakorn: Thai Tibet Center, 2010).

107 Stott, Philip, "*Mu'ang* and *Pa*: Elite Views of Nature in a Changing Thailand," in *Thai Constructions of Knowledge*, ed. Manas Chitakasem and Andrew Turton (London: SOAS, 1991), 152.

108 See also Darlington, "The Spirit(s) of Conservation," 141.

109 Darlington, "The Good Buddha," 177.

problems,[110] whilst often being critical or suspicious of Western approaches and ideas. This approach was expressed in the vernacular with the following words: “what is old is not forgotten; what is new one is not intoxicated by” (*mai luem khongkau mai mau khongmai*).[111] Here, (indigenous) spiritual approaches to and understanding of nature are not perceived as obsolescent or naive. In fact, these rituals, together with their symbolism and motivational potential, are often much more valued than modern scientific knowledge which, so it has been argued, may lead to the hubris that humans are able to defeat nature.

The initiatives and concepts introduced in this article point to a Thai Buddhism that is far from being a static, passive entity, but is rather creative, flexible, inclusive, innovative, dynamic, self-critical and socially-engaged (all this while still being concerned with continuity and the preservation of traditional values). It must be noted however that even though these initiatives of alternative development and environmental protection may be locally successful (and to some extent translocally through Buddhist networking groups such as Sekhiyadhamma), similarly to the above introduced concepts of environmental ethics and (Buddhist) development, it seems that on a national level these ideas are far less powerful and so far not representative of lasting national trends and policies.[112]

110 Another very interesting case in point is the so-called ‘river life prolonging’ ritual (*suep chata maenam*) as performed in the northern Thai province of Nan: here, to some extent similar to the ‘tree ordinations’, rearranged cultural elements and ritualistic patterns together with deliberate inventions are used to resacralise nature in order to protect it from exploitation and environmental pollution. This ritual, which is based on an astonishingly elaborate and profound symbolism, also aims at the creation of community spirit and pride. It is also ensured that younger generations are involved in these rituals in order to achieve long-term results (cf. Banjit Sairokham, and Thanat Baiya บานจิตร สายรอคำ และ ถนัด ใบยา, *Buat pa suep chata maenam kan jatkan pa lae nam choeng watthanatham* บวชป่า สืบชะตาแม่น้ำ การจัดการป่าและน้ำเชิงวัฒนธรรม [Ordaining Trees and River Life Prolonging (Ceremony): Cultural Management of Forest and Water] (Nan: Hak Mueang Nan Foundation, 2006).

111 Banjit, and Thanat, *Buat pa suep chata maenam*, 88.

112 The ‘tree ordinations’ may, to some extent at least, be an exception to this as ‘tree ordinations’, which started as grassroots activities, have been rather successful on a national level, too. Thus, in 1996 a programme to ordain 50 million trees in honour of the King of Thailand was initiated by non-governmental and governmental groups. Also, as Darlington remarks, “[t]hrough their preaching, educational programs, and conservation activities, the ecology monks have influenced Thai society’s view of Buddhism and, to some degree, its practice. They have raised the nation’s consciousness regarding its environmental responsibilities as their activities have drawn significant attention and media coverage” (cf. Darlington, “The Ordination of a Tree,” 5).

CHAPTER 3

Shifting Notions of Nature and Environmentalism in Indonesian Islam

Monika Arnez

Abstract

This article traces global ideas of nature in Islam and analyses local developments in environmentalism in Indonesia, the most populous Muslim country. Based on theological considerations of cosmos, earth and humans, it investigates how religious scholars (*ulama*) of the Muslim world argue for environmentalism. Moreover, since Muslim actors in environmental debates have a tendency to differentiate from the West and resort to Islamic norms and values, this essay addresses the question of how Muslim authorities, from international as well as national Indonesian backgrounds, articulate cultural and religious identities. In this context, the issue addressed is how classical literary sources and traditional cultural elements are used in controversial debates on environmentalism. Furthermore, this paper analyses how the largest Indonesian Muslim welfare organisation, Nahdlatul Ulama, and its affiliated Islamic boarding schools (*pesantren*), have been responding to environmental challenges.

1 Introduction

Across the globe, climate change has a variety of impacts on human life. The frequent occurrence of natural disasters like forest fires, droughts and floods has plunged the livelihoods of many people into crisis. Although often affected by these negative impacts, many countries with a predominantly Muslim population face economic and political problems, and as such climate change is not their first priority. This is exacerbated by the fact that the local population is often unaware of global warming so it cannot personally relate to it.[1]

* The author would like to thank Nicole Lin and Liam Finnie for editing and proofreading this article. Furthermore, she thanks the European Commission for financing part of her fieldwork for the project "Integration in Southeast Asia: Trajectories of Inclusion, Dynamics of Inclusion (SEATIDE)". She would also like to thank Isti Zusrianah for her support and hospitality.

1 Nugteren (see her article in this volume) has also observed this for South Asia.

 | DOI 10.1163/9789004273221_005

Foltz observed that Islam "has not figured prominently in emerging discussions on religion and the environment",[2] but environmental awareness of Muslim authorities has been increasing in the last few years. This is reflected, for instance, in measures taken by institutions, Civil Society Organisations (CSOs), Muslim actors and governments. Since the new millennium, especially the World Summit on Sustainable Development of 2002, the idea of Islamic sustainable development has taken hold in the Muslim world. From that time onwards, ulama (religious scholars) in Muslim countries have united efforts in addressing climate change, hoping that an appeal to the Islamic *ummah* (community) can impact positively on Muslims' behaviour towards nature and thus mitigate the effects of the environmental crisis. In this context, it is also worth mentioning that predominantly Muslim countries have organised an increasing number of environmental workshops and conferences in the last few years. For instance, the Muslim Seven Year Action Plan (M7YAP) was drawn up at the 2008 conference in Kuwait city.[3] The M7YAP proposed ideas on introducing new concepts for green mosques and green *hajj* (pilgrimages to Mecca), distributing books on Islamic environmentalism and training of local authorities on climate change issues and the re-introduction of Islamic rituals from an environmental perspective.[4] Complementing this cross-border initiative, local Muslim leaders have brought in their own ideas on how to address climate change.

As far as international governments are concerned, they acknowledge that climate change is a serious issue, and many of them have taken concrete steps to mitigate the effects of global warming by introducing new regulations and establishing environmental institutions. The current Indonesian government, aware of its high emission rates, prides itself on ensuring sustainable development. At the 2009 United Nations Climate Change Conference in Copenhagen,

2 Foltz, Richard C., "Islam," in *The Oxford Handbook of Religion and Ecology*, ed. Roger S. Gottlieb (Oxford, New York: Oxford University Press, 2006), 212.

3 Saniotis, Arthur, "Muslims and Ecology: Fostering Islamic Environmental Ethics," *Contemporary Islam* 6 (2012): 163. M7YAP was declared in June 2009 in Istanbul, and the Muslim Association for Climate Change Action (MACCA) was established to implement it. For a summary of the project see Earth-Mates Dialogue Center (EDMC) with the United Nations Development Program (UNDP) and the Alliance of Religions and Conservation (ARC). "The Muslim 7-Year Action Plan (M7YAP) To Deal with Global Climate Change," last modified September 20, 2009, accessed December 7, 2012. http://www.arcworld.org/downloads/Muslim-7YP.pdf.

4 For a summary of the project see: "The Muslim Seven Year Action Plan on Climate Change 2010–2017 – Summary," last modified September 20, 2009, accessed December 7, 2012. http://www.bcca.org/ief/fl/WindsorARCMuslim_summary091020.pdf.

 | DOI 10.1163/9789004271494_005

it formulated the objective to reduce greenhouse emissions in Indonesia by 26 per cent by 2020.[5]

However, the general acceptance of the fact that environmental issues deserve more attention does not mean that ideas related to these topics are uncontested. In Indonesia, for instance, the exploitation of resources, in particular timber, is a major issue, and influential actors remain engaged in illegal logging practices.[6] A recent example demonstrating this is the crackdown on Greenpeace Indonesia. In August 2011, the influential Indonesian Ulama Council (Majelis Ulama Indonesia, MUI) accused this organisation of having received money from gambling operations in the Netherlands,[7] and the Betawi Brotherhood Forum (Forum Betawi Rempug, FBR) wanted to dissolve it, alleging that it had not been legally registered.[8] However, the reproaches rather seem to have been a pretext for keeping Greenpeace from meddling in unfair practices of palm oil production.[9] For prior to FBR's protests, Greenpeace had publicly criticised logging practices, and, as a result, the export of crude palm oil (CPO) to several European countries had been prohibited.[10]

Although power political interests are not the focus of this article, the example of Greenpeace points to tension between 'western ideas' and a wish for indigenous authenticity through religion and/or culture. The growing influence of transnational Islamist groups in Indonesia, such as Hizbut Tahrir or the

5 REDD Indonesia, "Pidato Presiden Republik Indonesia, Susilo Bambang Yudhoyono di CIFOR tentang <kesetaraan pertumbuhan pembangunan berkelanjutan> [Speech by the President of the Republic of Indonesia, Susilo Bambang Yudhoyono in CIFOR about <Equal Growth for Sustainable Development>]," last modified July 10, 2012, accessed October 1, 2012. http://blog.cifor.org/10156/pidato-presiden-republik-indonesia-susilo-bambang-yudhoyono-di-cifor-tentang-kesetaraan-pertumbuhan-berkelanjutan/#.UMrz1Kx9QfY.

6 Cribb has shown that the New Order took some significant steps in environmental protection; however, illegal logging had already been a problem in this period. Cribb, Robert, "Environmentalism in Indonesian Politics," in *Towards Integrated Environmental Law in Indonesia?*," ed. Adriaan Bedner and Nicole Niessen (Leiden: CNWS), 48.

7 Puji, Siwi Tri, "MUI nyatakan dana lotere Greenpeace haram [MUI Declares Lottery Funds of Greenpeace Forbidden by Islam]," published by *Republika.co.id*, August 4, 2011, accessed August 12, 2013. http://www.republika.co.id/berita/nasional/umum/11/08/04/lpeec7-mui-nyatakan-dana-lotere-greenpeace-haram.

8 "5 Pemicu Kontroversi RUU Ormas [5 Initiators of the Controversy about the Mass Organisation Bill]," published by Tempo.co, July 3, 2013, accessed August 12, 2013. http://www.tempo.co/read/news/2013/07/03/078493040/5-Pemicu-Kontroversi-RUU-Ormas.

9 The attempt at relocating the organisation was successful; the office of Greenpeace Indonesia is now in Tebet, South Jakarta.

10 Puji, Siwi Tri, "MUI nyatakan dana lotere Greenpeace haram."

Muslim Brotherhood, has a negative effect on the perception of Western institutions.[11] However, Muslim minorities, for example, Shiite circles[12], also criticise a loss of values in the context of environmental problems. For example, Seyyed Mohsen Miri, the former rector of the Shiite Islamic College for Advanced Studies (ICAS), directs his criticism at the West and indigenous Muslims when he states that the 'Western' preoccupation with inventing new technologies and striving for modernising human life has resulted in a neglect of fundamental Islamic values. In this context Kula's statement, "the current Islamic environmental concern is often more related to the West than to the environment itself",[13] may be exaggerated, but it is undeniable that many Muslim scholars identify Western secularism, claims to power, and capitalism as causes of environmental problems.

There is consensus in the Muslim world that humans' failure to behave in an appropriate ethical manner is a result of having lost Islamic values and ethics; however, there are divergent opinions on the causes of the loss. As reviving traditional cultural elements seems to be one strategy of counteracting the degradation of nature, this article analyses the respective transnational ideas of different Muslim actors. Within the Indonesian context, it looks at the ways in which the Muslim mass organisation Nahdlatul Ulama and its affiliated Islamic boarding schools (*pesantren*)[14] understand environmentalism. Which local strategies do they apply to achieve an increasing awareness of environmental issues in Muslim communities? Which role do cultural elements play here? Before analysing these issues in more detail, this contribution starts by describing central elements from Islamic sources many Muslim scholars refer to, thus characterising the relationship between cosmos, earth and humans. Therefore, the following paragraph provides the basis for the subsequent discussion on

---

11 Some recent studies reveal that this influence also has its limits. For example, van Bruinessen has shown for Cirebon in North Java that smaller Islamic NGOs in cooperation with Muslim mass organisations and pesantren can serve as a partner by reviving cultural traditions such as arts as a resource. Cf. Bruinessen, Martin van, "Ghazwul fikri or Arabisation? Indonesian Muslim Responses to Globalization," in *Dynamics of Southeast Asian Muslims in the Era of Globalization*, ed. Ken Miichi and Omar Farouk (Tokyo: Japan International Cooperation Agency), 30.

12 Although Sunnis are the majority in Indonesia, the number of Shiites has been growing since democratisation. Although sometimes associated with liberal Islam, they do not necessarily agree with its practical interpretations. See: Marcinkowski, Christoph, "Aspects of Shi'ism in Contemporary Southeast Asia," *The Muslim World* 98.1 (2008): 50.

13 Kula, E., "Islam and Environmental Conservation." *Environmental Conservation* 28 (2001): 2, accessed October 10, 2012. doi 10.1017/S0376892901000017.

14 Another term used for these Islamic boarding schools is *pondok pesantren*.

different ideas related to environmentalism. How do Islamic scholars argue for environmentalism, based on ideas of nature found in the Qur'an and *hadith* (sayings of prophet Muhammad)?

2 Theological Considerations of Cosmos, Earth and Humans

2.1 *The Oneness of God, and the Balanced Cosmos*

Although religions provide a variety of answers on the intended relationship between the cosmos, humans and the earth, those answers are guided by theological concepts on how the universe is organised. Islam is based on the concept of the oneness of God (*tawhid*),[15] the essence of Islamic monotheism. *Tawhid* refers to the belief in God's singularity. As the fundamental principle of Islam, *tawhid* is expressed through the first words of the testimony of faith: "There is no God but God." Islam strongly rejects polytheism, and, in addition to the testimony of faith, several verses of the Qur'an indicate that there is only one God.[16] According to the Qur'an, God's oneness can neither be broken down into smaller units nor can it be divided. All forms in the universe other than God are unique variants of His creation. By worshipping Him, they can come closer to God, although they can never reach divinity. As far as humans are concerned, *tawhid* is at the heart of human action and social life. Thus, the term also serves as a guide to contemporary Muslims on how to behave adequately: to serve God, one must do his or her best for the benefit of all creatures. Therefore, it can be argued that *tawhid* is an Islamic doctrine with an ethical dimension.

According to Muslim scholars, unity, balance and harmony are important aspects of the universe. Several verses of the Qur'an describe the universe as a beautiful and harmonious place, a result of divine creation. The first line of argument, which relates to the issue of cosmic symbiosis (*takāful*),[17] is referred to here because Islamic scholars use it to argue for the necessity to preserve nature and counteract climate change. God has created the universe, the environment, humankind and other creatures, and all of them are imbued with their own value. Each creature and everything that God created points to His greatness and wisdom, as well as the enormous diversity of life. They praise God in their own way, even though humankind might not be able to understand this:

15 Spelling variants Tauhid, tawhid, Ar. *tawhīd*.

16 For instance: 'there is no God but God' (*Surat* 47:19) or He is the "creator of all things" (*Surat* 6:102).

17 This term is also often used to refer to joint guarantee in a business undertaking, a principle accepted in Islam.

> Prophet, do you not see that all those who are in the heavens and earth praise God, as do the birds with wings outstretched? Each knows its [own way] of prayer and glorification. God has full knowledge of what they do.[18]

The Qur'an highly values natural resources, especially trees and water. A good world is compared to a fruit-bearing tree with strong roots and branches reaching high into the sky.[19] God rewards humans who believe in this world, but punishes those who don't for their wrongdoings, in this life and the hereafter. Islamic tradition rejects cutting down trees without an apparent legitimate reason because such behaviour is in contradiction with God's creation. It is noticeable that in Islamic tradition the *himā*, areas the local population protects, contribute to the health of the environment.

Qur'anic verses mention that God has created water, which ensures humans' survival on earth. When performing the five prayers, water is a necessary, purifying resource for ablution, and in paradise believers are promised rivers of pure water.[20] Islam, based on the concept of goodness (*khayr*), will protect nature if the protection of nature is considered positive in itself. However, the only creatures able to preserve the environment are humans, since God entrusted them with the task to look after the earth, to be his representatives on earth.

2.2 *Distinguished by God: Humans' Cosmic Duty*

In the context of environmentalism, the cosmic balance between God, humans and nature is often addressed to highlight their responsibility. To maintain this balance, humankind has to worship the creator, obey His rules, act responsibly towards nature, preserve life and avoid abuse and wastefulness. Humans are capable to do so because God created a perfect world in which man, as the caliph or vice regent of God on earth (*khalifa*, lit. successor), was given a unique position: only to humans did God confer the right to shape nature according to his own wishes. In the environmental discourse, especially on climate change, Islamic scholars frequently draw on verses of the Qur'an to highlight man's role as *khalifa*, such as "I am putting a successer on earth";[21] and, "It is he who made you successors on the earth and raises some of you above others by rank, to test

18 Haleem, M. A. S. Abdel, *The Qur'an: A New Translation of M. A. S. Abdel* (Oxford: Oxford University Press, 2005), 223–224 (Surat 24:41).

19 Haleem, *The Qur'an*, 160 (Surat 14:24).

20 Haleem, *The Qur'an*, 332 (Surat 47:12).

21 Haleem, *The Qur'an*, 7 (Surat 2:30).

you through what He gives you".[22] These verses, especially the latter, do not only indicate that the Almighty has created everything on earth and in the sky for man, but they also appeal to the faithful to maintain the balance on earth by behaving friendly towards other creatures and preserving nature. Hence, based on the concept of *khalifa*, several Muslim scholars such as *kiai*[23] Muhammad indicate that nature is not man's property:

> All was water, indiscriminate. [...] Nature is not man's true property. Man's property is only left for trust, a deposit or a loan which, when time comes, must be returned in its original condition.[24]

If they are able to do so, humans are supposed to return this loan in a better condition, more beautiful than before. To achieve this aim, the Qur'an reminds them of their duty to refrain from overexploitation of natural resources:

> It is He who produces both trellised and untrellised gardens, date palms, crops of diverse flavours, the olive, the pomegranate, alike yet different. So when they bear fruit, eat some of it, paying what is due on the day of harvest, but do not be wasteful: God does not like wasteful people.[25]

Other Islamic sources also reject the excessive use of commodities, as the following example reveals. A *hadith* narrates that people would not receive God's grace on the Day of Resurrection if they had not shared abundant water with others. Withholding Grace would be the punishment for those having withheld resources they had not created.[26] Another *hadith* says that Prophet Muhammad, when being passed by a Bedouin called Sa'd, asked him about his wastefulness since he used a lot of water for his ablution. When Sa'd asked the Prophet if there could be any squandering in ablution, he responded: "Yes, even if you

22 Haleem, *The Qur'an*, 93 (Surat 6:165).

23 *Kiai* is a title given to religious teachers and leaders of Islamic boarding schools (*pesantren*) on Java. A kiai carries out various roles as he owns and manages a pesantren, but he also teaches and leads prayers.

24 Muhammad, Husein, "Manusia dan tugas kosmiknya menurut Islam [Humans and their Cosmic Duty According to Islam]," in *Menanam sebelum kiamat: Islam, ekologi, dan gerakan lingkungan hidup* [Planting before Doomsday: Islam, Ecology, and the Environmental Movement], ed. Fachruddin M. Mangunjaya, Husain Heriyanto and Reza Gholami (Jakarta: Conservation International 2007), 4.

25 Haleem, *The Qur'an*, 91 (Surat 6:141).

26 Sahih Al Bukhari, chapter no. 41, *hadith* no. 553, accessed March 3, 2013. http://ahadith.co.uk/chapter.php?cid=127.

are on the bank of a flowing river."[27] Wastefulness towards nature in any form is clearly discouraged here because excessive use of any natural resources contradicts with Islamic teachings.

However, humans, as God's vice regents on earth, do not necessarily perform their task well. In contrast to other creatures, humans are inclined to violate God's rules since they are able to choose freely and exert power over others. In this regard it is frequently stated that man's thoughts and actions must be based on Islamic ethics (*akhlak*)[28] to preserve nature. Consequently, human freedom is limited by the ethical codes of *akhlak*. Only those Muslims who have adhered to the moral codes in this world will be rewarded with Paradise, whereas those who have not will suffer in Hell. When caring for nature in this world, the faithful act in a manner that is consistent with *akhlak*, and at the same time they prepare for the afterlife, where they will live in a peaceful and verdant environment, in communion with God and the angels. According to Muslim faith, all devotees who have been rewarded for their good deeds with a blessed afterlife will enjoy the pleasures of paradise together. Doing something good for nature, be it in terms of planting trees or behaving in an environmentally friendly way, is tantamount to serving God through stewardship. Muslims regard a harmonious communal life with nature as a necessary prerequisite for reviving spiritual and cultural values that have been lost in modern society. However, many Muslim actors taking part in environmental debates have a tendency to differentiate from the West, resorting to Islamic norms and values, setting a religious basis for environmentalism. After providing an insight into this debate, this paper will outline specific Indonesian responses to traditional environmental discussions. Which ideas do Indonesian Muslims, arguing within the framework of Islam and local culture, propose for the purpose of environmental protection?

3 Articulating Cultural and Religious Identities

3.1 *Islamist Environmentalism and Eco-Mysticism*

Islamists, who have gained importance in the Muslim world due to the revival of Islam since the Iranian revolution in 1979, reject materialistic and capitalistic approaches as they are 'not in accordance' with Islamic values. Relating to environmental problems, Ignatow argues that nearly all Islamists claim for

27 Sunan Ibn Majah, chapter no. 2, *hadith* no. 425, accessed March 3, 2013. http://ahadith.co.uk/chapter.php?cid=147&page=16.

28 The term refers to individual behaviour and the code of ideal behaviour for all Muslims, which is based on the commands and prohibitions of the Qur'an and Sunna.

cultural authenticity, viewing Islam as "the authentic source of environmental consciousness, preceding Western environmentalist ideas",[29] whereas modernism is rejected. Environmental Islamists believe that the West has caused the present environmental problems, including global warming, by overproducing resources and leading an excessive lifestyle.[30] Ignatow gives Ali Bulaç as an example, who claims in his book *Religion and Modernism* (*Din ve Modernizm*, 1995) that Western welfare and civilisation have produced too many consumables and possessions, thus falling into the trap of materialism.

However, it is not only Islamist environmentalists who argue for the need of cultural authenticity, but also Muslim intellectuals from other backgrounds, for instance mysticism. The following examples reveal how selected Sufis refer to and interpret classical Islamic literature and Islamic philosophical considerations on the relationship between man and nature, as well as indigenous concepts of environmental protection, to argue for a revival of tradition and authentic Islamic values. Seyyed Hossein Nasr, a well-known Islamic scholar born and raised in Iran, has written extensively about the relationship of God, nature and humans.[31] After writing his PhD thesis about cosmology at Harvard University, he has taught Islamic studies in the US since 1984. He propagates the argument that the loss of values in Muslim countries can be attributed to the influence developed countries have exercised on the Muslim world:

> In its effort to look for power within the intrigues of the political world and the world's present economic situation, in spite of its best efforts, the Muslim community is always behind in creating ever changing technologies, and has to borrow them constantly from the Western world [...].[32]

As far as the stances of Muslims towards environmentalism are concerned, this illustrates that the border between Islamism and Sufism may become blurred.

---

29 Ignatow, Gabriel, "Globalizations and the Transformation of Environmental Activism: Turkey since the 1980s," *Globalizations* 5.3 (2008): 438.

30 Elites in developing countries are also sometimes mentioned in this context. See Foltz, "Islam," 212.

31 Among his well-known books are *The Encounter of Man and Nature: The Spiritual Crisis of Modern Man* (London: Allen & Unwin, 1968) and *An Introduction to Islamic Cosmological Doctrines. Conceptions of Nature and Methods used for its Study by the Ikhwan Al-Safa, Al-Biruni, and Ibn Sina* (Cambridge: Belknap Press, 1964).

32 Nasr, Seyyed Hossein, "Masalah lingkungan di dunia Islam kontemporer [The Problem of the Environment in the Contemporary Muslim World]," in *Menanam sebelum kiamat* [Planting before Doomsday], ed. Fachruddin M. Mangunjaya, Husain Heriyanto and Reza Gholami (Jakarta: Conservation International 2007), 47.

It is interesting to observe here that Nasr argues along the same line when he depicts the century-long dominance and colonisation of the Muslim world by Western countries as the major cause for deteriorating values. This, he asserts, has had the consequence that Muslim countries could not compete with Western countries in technological development. According to him, a significant challenge for Muslim governments and groups now is how to exercise influence on vibrant discourses without losing their Islamic principles. Another problem he addresses is that man has become the measure of all things since the Renaissance era, and a God-centered worldview was substituted by an anthropomorphic one.[33] When God is no longer the central reference point, nature becomes a mere functional entity to serve the people, and the ecological crisis, including global warming, is the logical consequence. He reminds his readers that as a *khalifa* of God on earth, man does not only have rights but also carries responsibility for his environment.[34]

Another idea he postulates is reviving tradition, which is gaining popularity in a few places. Referring to the appreciation of nature, he mentions traditional societies as positive examples, which, in his opinion, regard nature as their wives, whereas modern Western countries treat nature as property to be bought, sold, and consumed.[35] Being aware that ordinary people often do not know much about environmental problems, including global warming, he further argues that classical Persian literature reflecting an Islamic view towards man, nature and God, might be used to raise their ecological awareness. In this context, he takes the Persian author Saadi as an example, who lived in the 13th century and was acquainted with Sufi practices. Among the different aspects of this topic treated in his *The Rose Garden* (*Gulistan*), written in 1258, Saadi emphasises that all creatures, including humans, should perform their duty to praise God:

> 'Twas but a bird at early dawning wailed
> Yet over strength, sense, power her note prevailed;
> My friend to whom my cry came marveled much,
> I could be moved to tears and prayers by such,

33 "The Islamic Perspective on the Environmental Crisis: Seyyed Hossein Nasr in Conversation with Muzaffar Iqbal" (transcript of a conversation between Professor Seyyed Hossein Nasr and Muzaffar Iqbal, recorded in Sherwood Park, Canada, on February 19, 2007), published by TheFreeLibrary.com, last modified June 22, 2007, accessed November 20, 2012. http://www.thefreelibrary.com/The+Islamic+perspective+on+the+environmental+crisis%3A+Seyyed+Hossein...-a0164596587.

34 Nasr, "Masalah lingkungan di dunia Islam kontemporer," 57.

35 "The Islamic Perspective on the Environmental Crisis."

> Yea, unto madness; but I gave reply:
> "What! Shall a bird praise Allah and not I?"[36]

The eagerness of all creatures to praise God can be seen as a prerequisite for a harmonious life and a balanced cosmos. However, Nasr does not only consider classical texts appropriate for proselytisation (*dakwah*) but also modern poetry because both can encourage ordinary people to take greater responsibility for environmental protection and think critically about their own role within the cosmos. The philosophical-mystical approach taken by Nasr is also shared by other Islamic environmentalists such as Massoumeh Ebtekar, the first female vice president of Iran,[37] and Seyyed Mohsen Miri, former rector and active lecturer at the Islamic College for Advanced Studies (ICAS) in Jakarta. Both, in accordance with Nasr, propose to seek answers to environmental challenges in Islamic sources, and assume that environmental problems cannot be solved within the realms of the secular. Miri also argues that science and modernisation are inevitably leading to the destruction of philosophical, cultural and spiritual elements, so that the essence of humans is reduced to purely materialistic and scientific matters. The consequence is a downgrading of culture:

> Such a perspective will gradually make spiritual life, aims, hopes, happiness and holiness disappear, since the natural substance is being cut off from the soul, from the invisible world, and from the perfect purity and permanent truth of the creators of high philosophy.[38]

Both Miri and Ebtekar take the thoughts of the Persian philosopher Mullā Ṣadrā, a representative of post-Avicennan Islamic philosophy,[39] as a model of

---

36 Arnold, Edwin, trans., *The Gulistan. Being the Rose Garden of Shaikh Sa'di* (London, New York: Harper & Brothers 1899), 117.

37 She was also the head of the Department of the Environment in Iran from 1997 to 2006 and is currently working as an associate Professor for Cellular and Molecular Immunology at Tarbiat Modares University in Tehran.

38 Miri, Seyyed Mohsen, "Prinsip-prinsip Islam dan filsafat Mulla Shadra sebagai basis etis dan kosmologis lingkungan hidup [Islamic Principles and the Philosophy of Mulla Shadra as Ethical and Cosmological Basis for the Environment]," in *Menanam sebelum kiamat* [Planting before Doomsday], ed. Fachruddin M. Mangunjaya, Husain Heriyanto and Reza Gholami (Jakarta: Conservation International 2007), 25.

39 His birthname was Muhammad ibn Ibrahim ibn Yahya al-Qawami al-Shirazi. For further details of his life see: Kalin, Ibrahim, *Knowledge in Later Islamic Philosophy: Mulla Sadra on Existence, Intellect and Intuition* (Oxford, New York: Oxford University Press, 2010), xivff.

Islamic environmentalism. The reason Miri provides for the significance of Ṣadrā's philosophy is that his considerations counter a secularised view of nature. Miri uses Ṣadrā's text as an appeal to his readers to go back to their spiritual roots, to the purity and holiness of God, to their own self and nature.[40] Relating to the four journeys Ṣadrā proposes to undertake in trying to reach four stages of cognition, Ebtekar observes that they are necessary to achieve inner peace and balance, which, according to her, is a prerequisite for good leadership. In her description of these journeys on which the traveller transcends his ego, continues his path to the truth by penetrating the spiritual realm to overcome excess and indifference, and finally returns from God to the people, she highlights the third journey's purpose to install peace in the minds and hearts. For her bringing the spirit and the material world into balance is a prerequisite for solving pressing social and environmental problems:

> It also seeks to regain the confidence that we have lost in their hearts and minds of the youth. It is an indication of the intrinsic ties between the spirit and the material world – an essential message for the leaders and politicians who claim to be searching for solutions to the quagmires of global warming, war and sense of insecurity and lack of confidence.[41]

These examples reveal that Muslim authorities use Islamic philosophical ideas and classical literature as a means of reviving spiritual values and creating a sense of community, with the purpose of addressing environmental challenges across borders. With regards to the Indonesian context, Muslim actors, aware of Indonesia's role as a big carbon producer, have also increased their efforts to counteract global warming. On the one hand Indonesia's Muslim scholars rely on theological arguments commonly related to in other parts of the Muslim world, but on the other hand they make use of special Indonesian institutions[42] to mitigate the effects of climate change. Among others these are pesantren, Islamic boarding schools belonging to one of the largest educational systems in the world, and the deep-rooted Muslim mass organisations Nahdlatul Ulama

---

40 Miri, "Prinsip-prinsip Islam dan filsafat Mulla Shadra," 26.

41 Ebtekar, Massoumeh, "Leadership for Peace and Sustainable Development: The Spiritual and Feminine Factor," *Farzaneh* 15.5 (2008): n.p. http://en.farzanehjournal.com/index.php/articles/no-15/6-no-15-5-leadership-for-peace-and-sustainable-development-the-spiritual-and-feminine-factor.

42 Indonesia plays a significant role in Islamic environmentalism. For instance, Schwencke calls Indonesia a 'hot spot for conservation activities and civil society action'. See: Schwencke, A. M., "*Globalized Eco-Islam: A Survey of Global Islamic Environmentalism*" (Leiden: Leiden Institute for Religious Studies, 2012), 62.

(NU) and Muhammadiyah[43], which are larger than any other Muslim welfare organisations. Both refer to themselves a 'moral movement', characterised by upholding and spreading Islamic values. They officially declared themselves a moral movement in 2006, as a response to the vibrant discourse on pornography and widespread corruption.[44] One difference between the two is that traditional elements such as the veneration of saints play a role for NU, whereas Muhammadiyah strongly rejects this practice. Although in recent years both organisations have lost members, who felt attracted by Islamist groups, especially the Welfare and Justice Party (Partai Keadilan Sejahtera, PKS) and Hizbut Tahrir Indonesia (HTI), the Indonesian branch of Hizbut Tahrir (HT), they still claim a membership of millions.[45] These institutions have the potential to influence the community, due to the great number of pesantren and members in the mass organisations. The following paragraph starts with a sketch of pesantren as important traditional educational institutions. Afterwards, it presents local approaches of pesantren affiliated with the largest Indonesian Muslim mass organisation, NU, towards environmentalism, Islamic education and entrepreneurship.

3.2 *Local Responses: 'Eco-Pesantren' and Muslim Mass Organisations*

Pesantren have been an integral part of Indonesian culture for a long time. There are different theories on the origin of pesantren, for instance that some of the nine saints (*wali sanga*) have established them or that they were learning centres inspired by pre-Islamic institutions.[46] Bruinessen observes that there is

---

43 Muhammadiyah is the second-largest Indonesian Muslim mass organisation. Its members mostly belong to the urban middle class, and it has a large network of universities, and schools.

44 "Warta NU-Muhammadiyah Akan Deklarasikan Gerakan Moral Anti-Pornografi [Reports that NU and Muhammadiyah will Declare an Anti-pornography Morality Movement]," published by NU Online, last modified January 31, 2006, accessed December 7, 2012. http://www.nu.or.id/a,public-m,dinamic-s,detail-ids,1-id,4077-lang,id-c,warta-t,NU+Muhammadiyah+Akan+Deklarasikan+Gerakan+Moral+Anti+Pornografi-.phpx.

45 The problem of losing members to PKS has already given rise to intensive discussions in Muhammadiyah circles since 2008. One example worth mentioning here is a meeting of Nasyiatul Aisyiyah, the young women's body of Muhammadiyah, which was held in the Al Ahzar Suryowijayan mosque in Yogyakarta, Central Java, to tackle the problem of a loss of members. Author's fieldnotes, May 25, 2008.

46 The first theory relies on *The History of the Land of Java* (*Babad Tanah Jawi*), the official chronicle of the kingdom of Mataram; the second mentions *mandala* and *ashrama* as institutions that have influenced pesantren. See: Pigeaud, Theodore G. Th., *Literature of Java: Synopsis of Javanese Literature 900–1900 A.D.*, vol. 1 (The Hague: Nijhoff 1967), 76; De Graaf, H. J., and Theodore G. Th. Pigeaud, *De eerste moslimse voorstendommen op Java:*

little information about the origins of pesantren, although he suggests that colleges at the Al Azhar University in Cairo (*riwaq*) provided models for pesantren in the late 18th and 19th centuries, and the Sawlatiyya[47] in India did so at the turn of the 19th and 20th centuries.[48] According to him, institutionalised education was an exception before the 18th century, and passing *hajis*, pilgrims to Mecca, or Arab traders provided informal basic education:

> The most interested students would visit the *'ulama* at home and even stay there, and the really ambitious would seek more learning in Java or, when possible, Mecca. It seems highly likely to me that this was also the situation in Java and Sumatra during the first centuries of islamisation, and that the first *pesantren* proper were not established before the 18th century.[49]

Since then, however, the number of pesantren has been steadily increasing,[50] and they have become a significant part of Indonesian culture and the educational system. Since Indonesia gained independence, pesantren, Muslim religious schools (*madrasah*),[51] and Western-style governmental schools have been shaping Indonesia's education system. Whereas the state has taken over part of the *madrasah* and schools, pesantren have remained private institutions usually run by a kiai. Although classical texts such as the yellow books (*kitab kuning*)[52] determine the religious contents taught in pesantren, everyday

---

studien over de staatkundige geschiedenis van de 15de en 16de eeuw ('S Gravenhage: Nijhoff, 1974), 246–247. For a summary of assumptions on the origin of pesantren see: Qomar, Mulamil, *Pesantren: Dari transformasi metodologi menuju demokrasi institusi* (Jakarta: Erlangga, 2005), 8.

47 A *madrasah* based on educational reform in Indian Islam.

48 Bruinessen, Martin van, "*Pesantren* and *kitab kuning*: Maintenance and Continuation of a Tradition of Religious Learning," in *Texts from the Islands: Oral and Written Traditions of Indonesia and the Malay World* [*Ethnologica Bernica* 4], ed. Wolfgang Marschall (Bern: University of Bern, 1994), 131, 137.

49 Bruinessen, "*Pesantren* and *kitab kuning*," 133.

50 Azra, Afrianty and Hefner estimate the number of pesantren at 10,000. See: Azra, Azyumardi, Dina Afrianty, and Robert W. Hefner, "*Pesantren* and *madrasa*," in *Schooling Islam: The Culture and Politics of Modern Muslim Education*, ed. Robert Hefner and Muhammad Qasim Zaman (Princeton: Princeton University Press, 2007), 173.

51 *Madrasah* combine the traditional focus on religious education with non-religious subject matters.

52 Due to the colour of the paper on which early Middle Eastern editions found in Indonesia were printed, these classical commentaries compiled by Muslim theologians are known as "yellow books". See: Bruinessen, "*Pesantren* and *kitab kuning*," 121.

practice shows that there is more flexibility in handling teaching material. On the whole, pesantren, traditionally located in rural areas, are relatively open to educational innovations.[53] Their main purpose is to convey profound knowledge about Islam, make children familiar with Islamic values through *kitab kuning*, and foster social cohesion.

With the increasing rise in consciousness of ecological problems in pesantren in the 1980s and the 1990s, and the advent of cultural Islam, some progressive pesantren have started to address topics such as environmentalism,[54] religious tolerance, gender equality and human rights.[55] As Ibu Zaimah, Head of the Subdivision for Social Organisation in the Ministry of Environment, indicates, pesantren Abah Anom in Suryalaya, which focused on irrigation and environmental irrigation,[56] was the first 'eco-pesantren' the Ministry of Environment cooperated with in 1980, although the term was not yet used. As in other pesantren, the *santri*, Islamic boarding school students, and the local population highly esteem and respect the figure of the kiai, and his teaching strongly influences them. This is in line with the fact that, according to Javanese concepts of power, the social environment attributes charismatic power to a kiai. As a role model for the community, he is first and foremost an example for careful interaction with the natural environment when tackling ecological issues.[57]

Factors such as the authority of kiai, their public visibility, charisma, and the possible influence pesantren can exert on environmental issues because of their strong connection with nature have prompted former Environment Minister Muhammad Hatta to transform existing boarding schools into eco-pesantren. When Hatta and Sultan Hamengkubuwono X debated this issue in November

53 Azra, Afrianty, and Hefner assert that this is also true for other Muslim schools in Indonesia. See: Azra, Afrianty and Hefner, "*Pesantren* and *madrasa*," 174.

54 However, since pesantren may be Salafist, modernist, traditionalist and/or Sufi in nature, they are not always open-minded towards concepts of environmentalism. In Salafist pesantren there is no curriculum, and students of pesantren will only hear the words of the kiai and the *kitab kuning*.

55 NGOs such as the Paramadina Foundation and the Muslim mass organisations also promoted such ideas in the 1980s and 1990s. See: Fealy, Greg, and Virginia Hooker, ed., *Voices of Islam in Southeast Asia: A Contemporary Sourcebook* (Singapore, Institute of Southeast Asian Studies, 2006), 48.

56 A pesantren of the Sufi order Qodiriyyah-Naqsyabandiyyah, founded in 1915. The good cooperation in terms of ecological issues might be explained with the fact that pesantren Abah Anom Suryalala's branch of the Qodiriyyah-Naqsyabandiyyah order strongly supported President Suharto's government in the New Order era. See: Howell, Julia Day, "Sufism and the Indonesian Islamic Revival," *Journal of Asian Studies* 60.3 (2001): 714.

57 Gade, Anna M., "Tradition and Sentiment in Indonesian Environmental Islam," *Worldviews* 16 (2012): 269.

2009, they reached an agreement to rename and develop 90 pesantren into eco-pesantren, hoping that these will become centres for environmentally friendly conduct and serve as models for the community.[58] Some pesantren with innovative initiatives for environmental protection have obtained the Kalpataru prize, an award of the Indonesian government for interesting ecological concepts.[59] According to the Indonesian Ministry of Environment, ten pesantren have received this award so far.[60] Ibu Zaimah argues that these award winners are important role models for the community:

> [...] of course, we turn informal leaders, who have already proven their care for their environment, into models, creating pioneers. And it is these leaders we ask to go out the people, saying come on, let's plant trees.[61]

The idea that pesantren are valuable institutions for tackling climate change was also under discussion at the International Muslim Conference on Climate Change in Bogor, West Java, from 9 to 10 April, 2010. The participants passed a resolution to initiate eco-pesantren as pilot schemes for environmentalism,[62] assuming that they encourage values such as a sense of community, responsibility, and honesty. Another reason for this decision was the assumption that pesantren incorporate Islamic teachings about environmental care and resourcefulness in their curricula. The following section shows how three pesantren with ties to the Muslim mass organisation NU in the region of Yogyakarta, Central Java, understand and practice environmentalism.[63] The examples are pesantren located in Bantul regency:[64] Al-Imdad in Kauman hamlet, Wijirejo village, Pandak subdistrict, Islamic Studies Center (ISC)[65] Aswaja

---

58 Susanto, Slamet, "Ministry Launches Eco-'pesantren' Program," *The Jakarta Post*, November 6, 2009, accessed November 16, 2012. http://www.thejakartapost.com/news/2009/11/06/ministry-launches-ecopesantren039-program.html.

59 This prize has been awarded since 1982.

60 Interview with Ibu Zaimah, November 23, 2012. This interview was carried out by Eva Nisa.

61 Ibid.

62 Simamora, Adianto P., "Muslim Countries to Lead on Green Issues," *The Jakarta Post*, April 12, 2010, accessed November 26, 2012. http://www.thejakartapost.com/news/2010/04/12/muslim-countries-lead-green-issues.html.

63 Nahdlatul Ulama was established in Surabaya in 1926 by a group of ulama, Hasjim Asj'ari (d. 1947), Wahab Chasbullah (d. 1971) and Bisri Syamsuri (d. 1980). NU exerts influence on pesantren through kiai, who are often NU officials.

64 Bantul regency is located in the southern part of Yogyakarta Province. Research in these pesantren was carried out in August 2013 and February 2014.

65 Aswaja (*Ahlus Sunnah Wal Jamaah*) is the NU ideology; "Lintang Songo" (nine stars) stands for the nine saints (*wali songo*).

Lintang Songo in Pagergunung hamlet, Piyungan subdistrict, and pesantren Al Furqon in Murtigading village, Sanden subdistrict. These Islamic boarding schools have in common that they have ties with NU and are located in rural areas. Moreover, environmental issues have become part of their everyday life[66] in the last years, when the local Yogyakarta government, in particular the Regional Environment Agency DIY Yogyakarta,[67] put more emphasis on implementing environmental education. One measure taken was inviting kiai, informing them about ways to protect the environment and providing pesantren with necessary tools such as composters.[68] Therefore, it is not surprising that the selected pesantren cooperate with the Regional Environment Agency DIY Yogyakarta, and kiai H. Heri Kuswanto bin KH Muhammad Zaidan[69] even works together with six ministries.[70]

Another avenue the heads of the pesantren explored to widen their scope and increase their opportunities was entering the formal structure of NU or becoming a party member of PKB (Partai Kebangkitan Bangsa National Awakening Party), a party with close ties to NU. Kiai Heri is affiliated with PKB, and he was a member of the provincial parliament in Bantul. Muhammad Irfan Chalimy,[71] head of pesantren Al Furqon, is a NU member, and kiai Habib A Syakur, the head of pesantren Al-Imdad, is also the head of the Association of Islamic Boarding Schools (Rabithah Ma'ahid Islamiyah, RMI) in Yogyakarta. This institution is connected to NU and serves to implement its policies within pesantren.

The religious leaders' affiliation with NU shapes their attitude towards environmentalism in several respects. They believe in the importance of *kitab*

---

66 It should be pointed out here that by no means all pesantren are active in environmentalism. According to Isti Zusrianah, the current chairwoman of Fatayat NU Bantul, most pesantren in Yogyakarta do not yet have a similar programme. Private communication with Isti Zusrianah, August 26, 2013.

67 Badan Lingkungan Hidup Pemerintah DIY (Regional Environment Agency). DIY stands for: *Daerah Istimewa Yogyakarta* (Special Province of Yogyakarta).

68 Interview with Muhammad Irfan Chalimy, August 26, 2013, pesantren Al Furqon.

69 As he is usually called kiai Heri, this term will be henceforth used in the article.

70 The pesantren works together with the Kantor Wilayah Kementerian Agama DIY (Regional Office of the Ministry of Religion), Dinas Kehutanan dan Perkebunan DIY (Agency for Forestry and Horticulture), Dinas Pertanian DIY (Agency for Agriculture), Dinas Kesehatan Pemerintah Kota Yogyakarta (Health Agency), Dinas Pendidikan Kota Yogyakarta (Agency for Education), Dinas Perindustrian, Perdagangan, Koperasi dan UKM Provinsi DIY (Agency for Industry, Trade, cooperatives and Small and Medium Enterprises). It has received several awards, for example for its land usage and environmentalism in August 2010, issued by the State Ministry of Environment.

71 He does not yet carry the title kiai as he is still too young. He now manages pesantren al Furqon because his father died.

kuning for the education of the santri and they emphasise the significance of tradition. Their aim is to 'preserve good things from the past'[72] and accommodate modern elements through a combination of entrepreneurship, environmentalism, and traditional Islamic education. Thus, the pesantren continue providing Islamic training such as *fiqh* (jurisprudence), *tajwid* (science of Qur'an recitation), *hadith* and *tasawuf*[73] but they now also impart practical knowledge about recycling, composting waste, building fish ponds and planting various trees species and vegetable crops. The kiai agree that santri and the local communities should learn to not throw away or burn rubbish but rather collect and compost it. Each of the pesantren acquired several composters, partly with the support of the Regional Environment Agency, thus acting as role models for the community. Habib A Syakur[74], kiai of pesantren Al-Imdad in Kauman hamlet, explains that he wishes to develop santri into entrepreneurs with a strong Islamic background. His pesantren attracts santri from the area of Yogyakarta but also from other provinces, for instance West Java, Lampung (South Sumatra), and West Kalimantan. The reason, as kiai Habib asserts, is that his institution provides an inexpensive but solid Islamic education.[75] His pesantren has become more accepted in the last few years; in 2010 only 30 santri had been enrolled, and at present 215 santri stay in Al-Imdad. For this purpose, the santri are involved in several activities, for instance waste management. They collect non-organic waste, which the school then sells for recycling purposes. Moreover, they separate organic waste from plastic rubbish, pile it up for several days, and add bacteria to it. Afterwards, they sift it and leave it to stand for some more days. It is then ready to be used as fertiliser and soil for crops such as eggplants, chilli, and trees, in particular teak trees and yellow mimosa planted on the land of the pesantren.[76] The santri also assist in activities related to working on the wet rice fields in the afternoon, filling polybags with eggplants, chilli and other vegetables and cleaning the pesantren grounds. Kiai Habib says they grow teak trees for their own needs but they have also planted trees on the slopes of volcano mountain Merapi to prevent erosion. He narrates that he also visited numerous mosques and encouraged the *ustadz* (Islamic clerics) to plant trees on its grounds, five teak trees, five mahogany and ten

72 Several informants have mentioned this characteristic.

73 Mysticism or mystical practice itself, which is guided by the concepts of Islamic mystics and regulated by particular orders or brotherhoods (*tarekat*).

74 As people usually call him kiai Habib, this term will henceforth be used in this paper.

75 Interview with kiai Habib, pesantren Al-Imdad, August 22, 2013. If not mentioned otherwise, the author conducted the interviews.

76 Ibid. Kiai Habib also points out that he has encouraged the local community to give vegetable seeds to the pesantren.

yellow mimosa, but not all of them still take part in these activities. Moreover, he asserts that such environmental efforts are useful for the following purposes: the santri learn life skills, the pesantren itself becomes more efficient, and they have a positive impact on the environment.[77]

In a similar way, kiai Heri highlights the benefits for the santri. He recounts that his pesantren provides them with deep Islamic knowledge, life skills, and it teaches them a high degree of social responsibility. According to him, a mixture of religiosity and practical skills is vital for the future perspectives of the santri.[78] Despite the fact that this Islamic boarding school does not offer any regular school education, several practice programmes aim to ensure that santri will become small or medium scale entrepreneurs or farmers.[79] For instance, the santri can acquire experience in raising animals and producing local products such as snacks. Moreover, on five hectares of land the santri learn how to plant wet rice and grow vegetables. According to kiai Heri, it would be ironic if the rice for the pesantren needed to be bought, although there was enough land for planting.[80]

As can be expected from the textual culture of the pesantren, readings of traditional texts do complement environmental practices. Muhammad Irfan Chalimy, for example, head of the pesantren Al Furqon, located in the village Murtigading in district Sanden, refers to concrete Islamic sources as a basis for his environmentalism. Referring to Surat Ar-Rum (30th surah of the Qur'an), he claims that Islam forbids humans to destroy the earth. He argues that although God has envisioned people to perfect nature, many humans cause destruction. However, by reconsidering their actions they could return to the right path. Another example he provides is a debate between God and an angel, where God explains that humans should be his vice regents on earth. The angel disagrees, adding for consideration that humans are not necessarily reliable because they are passionate and wish to control others. The angel does not understand why God does not choose angels to become his vice regents on earth as they do not know *nafsu* (desire, yearning, passion) and thus, in his opinion, are more trustworthy. God closes the conversation, arguing that having *nafsu* is not an

77 Interview with kiai Habib A Syakur, pesantren Al-Imrad, August 22, 2013.

78 Interview with kiai Heri, pesantren Aswaja Lintang Songo, August 23, 2013.

79 In the morning the 61 santri who currently stay in the pesantren go to a near-by school and return to the pesantren in the afternoon, where they participate in evening prayers, discussions and instructions, for instance on *fiqh*, *aqidah* (creed), *akhlak* and *tarikh* (Muslim history).

80 According to kiai Heri, it would be ironic if the rice for the pesantren needed to be bought, although enough land was available to plant it. Interview with kiai Heri, pesantren Aswaja Lintang Songo, August 23, 2013.

impediment for humans becoming his vice regent. If they learn to control their feelings, thus the argument, they will be superior to angels.[81]

Kiai Habib A Syakur argues in a more general way when he puts forward that Islam requires humans to protect the environment. Kiai Heri argues that he does not use any particular texts on environmentalism in his pesantren but his approach is more pragmatic: which measures can pesantren take to become more environmentally friendly? One method he refers to is collecting water from ablution and re-using it for fish ponds. In line with the *hadith* on wastefulness of water mentioned above,[82] kiai Heri rejects wastefulness and emphasises that as water is precious, it needs to be re-used. He describes pesantren as institutions that are able to cause a change in the attitude of the community and the santri towards the environment.[83] This, in kiai Heri's opinion, is not possible without a solid religious education, which should begin when children are one or two years old and continue until they are adults. An essential aspect of this education, he suggests, is to teach children to respect and correct Islamic behaviour. Muhammad Irfan Chalimy also points to the importance of pesantren as a holistic educational system, when he refers to his school as an umbrella for long-lasting common learning.[84] Thus, he shares the assumption that becoming a santri is a life-long process of learning and teaching, which also carries an element of Indonesian nationality.[85] Thus, staying true to one's nation and its educational system is a declared aim of the pesantren under study. Another goal is to preserve cultural elements. According to Muhammad Irfan Chalimy, one of these is Sufism, which is deeply rooted on Java and exerts a strong influence on his pesantren. Practices such as *zikir* (chanting in praise of Allah) and the development of the pupils' secondary virtues, politeness and respect, are indicative of this.[86] However, he thinks he is still too young and inexperienced and his "heart is not yet steadfast" to take the consequences for his own emotional life. In contrast to what Anna Gade has observed for

81 Interview with kiai Heri, pesantren Aswaja Lintang Songo, August 23, 2013.

82 Sunan Ibn Majah, chapter no. 2, *hadith* no. 425, accessed March 3, 2013. http://ahadith.co.uk/chapter.php?cid=147&page=16.

83 "*Pesantren* can be agents of change; they can change the character of the *santri* and the society," Interview with kiai Heri, pesantren Lintang Songo, August 23, 2013.

84 Interview with Muhammad Irfan Chalimy, August 26, 2013, pesantren Al Furqon.

85 This idea was put forward by Baso. See: Baso, Ahmad, *Pesantren Studies 2a. Buku II: Kosmopolitanisme Peradaban Kaum Santri di Masa Kolonial* [Pesantren Studies 2a. Book II: The Cosmopolitan Culture of Muslim Seminarians in the Colonial Era] (Jakarta: Pustaka Afid, 2012), 87.

86 His personal relation to Sufism is marked by the influence of his own father, who had a Sufi teacher. Interview with Muhammad Irfan Chalimy, August 26, 2013, pesantren Al Furqon.

pesantren Al-Ittifaq,[87] he does not appeal to the community to change their hearts by "realising the consequences of one's positive and negative feelings".[88] However, his pesantren is an example of preserving religious and cultural elements and giving santri the prospect to become independent and develop the necessary skills to be future entrepreneurs or farmers. Results from a questionnaire reveal that almost a third of the santri consider becoming an entrepreneur an attractive career option, whereas almost none of them have career aspirations in farming. They would like to work in a job where they can earn more money than their parents, most of whom are seasonal workers. Another explanation they give is that they want to be useful for society.

Within wider NU circles, traditional practices are also kept alive, as Ibu Susiana, Secretary of the Muslimat NU Foundation for Education (Yayasan Pendidikan Muslimat NU), and secretary of its Research and Development Department narrates. She points out that planting a coconut tree when a baby is born, a practice of NU families, is meant to revive spiritual values to foster environmentalism. As a symbol of the people's strong ties to nature, coconuts or other gardening products are offered when the child has grown up and the wedding ceremony is taking place. Thus, the child gets an understanding of caring for the environment, which, according to Ibu Susiana, is an element of Islamic teaching.[89] She highlights that if people do fell a tree, a new one must be planted, and thus conservation is enabled. Ibu Susiana claims that in this sense women, particularly in rural areas, are important multipliers because they have a close relationship with nature. For her, women combat the negative effect of modernisation by preserving tradition.[90]

These examples showed that although religious leaders do not explicitly address climate change as a topic, they carry out environmental practices or in the case of Ibu Susiana's example, local rituals and encourage the local Muslim communities to do something good (*amr ma'ruuf*). However, NU also established special departments to tackle climate change. An important trigger certainly was the tsunami that devastated the North Sumatran province Aceh in December 2004. In 2005 NU established its Community Based Disaster Risk Management (Penanggulangan Bencana Berbasis Komunitas, CBDRM), launching community projects such as 'Pesantren and community involvement in managing disaster risks in coastal areas through mangrove planting' in Brondong district, Lamongan, East Java, in 2010. In this project NU involved coastal

87 This pesantren is located in Ciburial Village, near Bandung in West Java and also has a Sufi background.

88 Gade, "Tradition and Sentiment in Indonesian Environmental Islam," 284.

89 Interview with Ibu Susiana, November 20, 2012, Muslimat NU, Jakarta. This interview was carried out by Eva Nisa.

90 Ibid.

communities in conservation efforts, mobilised pesantren and taught people how to plant and nurse mangroves.[91] One of CBDRM's prominent characteristics is that it uses local knowledge to tackle environmental problems, as Avianto Muhtadi, the current head of the Nahdlatul Ulama Climate Change and Disaster Management Institution (Lembaga Penanganan Bencana dan Perubahan Iklim Nahdlatul Ulama, LPBI NU), CBDRM's successor organisation, emphasises:

> This [community based][92] concept does not come from outside but elicits one's own local and traditional knowledge, since it is the communities themselves that know the state of their environment, the topography, geography and everyday practices. Religion and norms are expressed, they unite and converge and become a communal lifestyle in these areas.[93]

CBDRM, however, regarded as an 'ad hoc response' to disasters, with the purpose to curb the most harmful effects,[94] was substituted by LPBI NU subsequent to its 32nd Muktamar held in Makassar from 22–27 March 2010. LPBI NU continues the community-based approach but sets different priorities than CBDRM, with a more pronounced focus on disaster management and advocacy. Projects launched so far relate to waste management, advocacy, and policy. The first one, 3R (Reuse, Reduce, Recycle), implemented in West Java, familiarised the local people with waste management machines and informed them how to use compost efficiently, similar to the case studies of pesantren. The second programme, Institutional Advocacy for Disasters (Advokasi Kelembagaan Bencana, AKB), aimed at informing local communities about existing regional laws on disaster management through announcements (*wara-wara*), as well as gatherings at which Islam is discussed (*pengajian*). Moreover, it pushed regional governments in eight East Java regencies[95] to enact new regional laws on this

91 "Pesantren and Community Involvement in Managing Disaster Risks in Coastal Areas through Mangrove Planting," published by Mangroves for the Future, accessed December 4, 2012. http://www.mangrovesforthefuture.org/assets/Repository/Documents/012-SG-ID-01-06-07.pdf.

92 Author's addition.

93 "LPBI NU, garda depan penanganan bencana di NU [LPBI NU, the Avant Garde of Dealing with Disasters in NU]," published by NU Online, last modified June 28, 2010, accessed December 4, 2012. http://www.nu.or.id/a,public-m,dinamic-s,detail-ids,5-id,23553-lang,id-c,halaqoh-t,LPBINU++Garda+Depan+Penanganan+Bencana+di+NU-.phpx.

94 Ibid.

95 The regencies were Trenggalek, Tulungagung, Malang, Lumajang, Pasuruan, Mojokerto, Lamongan, and Bojonegoro.

issue.[96] In cooperation with the National Board on Disaster Management (Badan Nasional Penanggulangan Bencana, PNPB) the third programme combined activities in managing disasters. In addition to raising disaster management and climate change issues, NU also launched a more general programme in 2007, the National Movement for Forestry and the Environment (Gerakan Nasional Kehutanan dan Lingkungan, GNKL), which mainly deals with forest conservation in areas close to pesantren. NU-affiliated kiai consider reforestation and other conservation efforts a part of a Muslim's service to God (*ibadah*) and a responsibility of man's function as *khalifa*.[97]

Many activities of the NU environmental departments[98] are similar in that they assist the government in increasing the communities' awareness of environmental problems by training local authorities, holding workshops, and disseminating brochures for different target groups, including children. NU has often been criticised for 'political opportunism', due to its support of the government. In this context, however, Bruinessen convincingly argues that in the case of NU political accommodation is not based on expedience but might rather be considered a matter of principle. Moreover, he points out that many kiai consider "one hour of political chaos (*fitna*) worse than a century of tyranny, in line with the Sunni tradition of political conservatism".[99] Although the latter statement might sound somewhat exaggerated, I agree with Bruinessen that

96 "Langkah LPBI NU dalam penanggulangan bencana [Measures Taken by LPBI NU to Handle Disasters]," published by NU Online, last modified July 17, 2012, accessed December 6, 2012. http://www.nu.or.id/a,public-m,dinamic-s,detail-ids,5-id,38862-lang,id-c,halaqoh-t, Langkah+LPBI+NU+dalam+Penanggulangan+Bencana-.phpx.

97 For instance, kiai KH Noor Nasroh Hadiningrat, head of the pesantren Wali Sembilan Gomang in Singgahan, East Java, emphasises this aspect. See: "Kiai NU se-Tuban Sepakat Lestarikan Hutan [Kiai of NU in Tuban Agree to Preserve the Forest]," published by Shvoong, last modified April 20, 2012, accessed December 7, 2012. http://id.shvoong.com/society-and-news/opinion/2284646-kiai-nu-se-tuban-sepakat/.

98 It should be pointed out here that the second-largest Indonesian mass organisation, Muhammadiyah, has also set up their own departments for tackling environmental problems, including climate change. It has founded its first environmental institution in 2002, the Institute for the Study and Empowerment of the Environment (*Lembaga Studi dan Pemberdayaan Lingkungan*, LSPLH). Furthermore, it established the Environmental Institute (*Lembaga Lingkungan Hidup*, LLH) in 2005, and the Muhammadiyah Disaster Management Center (MDMC) in 2007. Subsequent to its 46th Muktamar from July 3 to 8, 2010, Muhammadiyah changed LLH to Environment Council (*Majelis Lingkungan Hidup*, MLH), with the purpose of 'accelerating the transformation of the environmental movement within the Muhammadiyah body'. One of MDMC's roles is to support the victims of disasters.

99 Bruinessen, "*Pesantren* and *kitab kuning*," 125.

stability, as well as nationalism, are guiding principles of NU scholars. KH Said Aqiel Siradj, the current chairman of the Central Board of Nahdlatul Ulama (Pengurus Besar Nahdlatul Ulama, PBNU), has put forward in this context that the Indonesian nation is a gift from God people should be thankful for.[100]

The Indonesian Ministry of Environment also co-published a major publication of NU's eco-theology,[101] *Jurisprudence of the Environment: Environmental Ethics*,[102] another indication of NU's readiness to assist the government in implementing its programmes. This book resulted from a meeting of religious scholars in Lido, Sukabumi, West Java, in May 2004. The purpose of this meeting was to bring a group of experts together with the purpose of creating a theology of the environment and subsequently using the results as a reference work in the community. This text contains source collections of classical texts, in addition to commentaries and summaries. In their debates, the ulama discussed how *kitab kuning* deal with environmental issues and how to use natural resources in a more sustainable way. One important work they referred to is *Islamic Jurisprudence and its Proofs*,[103] a text published by Wahbah al-Zuhaily, an expert of contemporary *fiqh* and a well-known Islamic scholar born 1932 in Damascus. His works have become compulsory reading for students of *tafsir* (exegesis of parts of the Qur'an), and Islamic law in Indonesia. He received his degree from the *Syariah* department[104] of the Al Azhar University in Cairo in

100 Quoted in "Meneguhkan Visi Kebangsaan Nahdlatul Ulama [Strengthening the Vision of Nationalism in Nahdlatul Ulama]," published by NU Online, last modified February 6, 2013, accessed September 18, 2013. http://www.nu.or.id/a,public-m,dinamic-s,detail-ids,4-id,42323-lang,id-c,kolom-t,Meneguhkan+Visi+Kebangsaan+Nahdlatul+Ulama-.phpx.

101 Amri points out that, in 2007, Muhammadiyah also formulated its eco-theology. See: Amri, Ulil, "Divine Power: Faith-based Mass Groups Take on Climate Change," *Strategic Review* 1.1 (2011): 99. Muhammadiyah's Environment Council MLH has also published books based on their eco-theology, for instance: Majelis Lingkungan Hidup, Kementerian Lingkungan Hidup, ed., *Akhlaq Lingkungan: Panduan Berperilaku Ramah Lingkungan* [Environmental Ethics: A Guide to Environmentally Friendly Conduct] (Jakarta: Majelis Lingkungan Hidup PP Muhammadiyah, Kementerian Lingkungkan Hidup Republik Indonesia, 2011), together with the Indonesian Ministry of Environment. This book argues that the present environmental destruction is a result of human misperception and wrongdoings and provides examples from Islamic sources on how Muslims should behave in an environmentally friendly manner.

102 Muhammad, Ahsin Sakho et al., ed., *Fiqh lingkungan; fiqh al-bi'ah* [Jurisprudence of the Environment: Environmental Ethics], 2nd edition (Jakarta: Conversation International, 2006).

103 Zuhaili, Wahbah, *Al-fiqh al-Islami wa adillathu* [Islamic Jurisprudence and Its Proofs] (Beirut: Dar al-Fikr, 1984).

104 Department for Islamic Law.

1956. The book *Islamic Jurisprudence and its Proofs* is based on religious scholars who represent the four *madhab*.[105] In addition to matters of legal issues and worship, it also discusses contemporary socio-political and economic issues. Relating to the fourth volume of this work,[106] the ulama concluded that humans may only use their rights as long as they are in accordance with Islamic rules and do not result in damage or danger for individuals or groups. Furthermore, based on volume three[107] they stated that an individual may construct any building on his property as long as it does not obstruct the neighbours' sun and air. A target of criticism, however, is the government's land use policy. Kiai KH Asyhari Abta of pesantren al-Munawwir, Krapyak, Yogyakarta, criticises the government for taking away land use rights from the people, especially for traditional land (*tanah adat*) and cemeteries, as it opposes *fiqh*. He asserts that in this way enterprises can easily exploit land resources.[108] The kiai suggests that the government only issues user licenses for businesses (Hak Guna Usaha, HGU) on a case-by-case basis.

4 Concluding Remarks

One idea Muslim scholars share on environmentalism is that man's role is to keep the balance and harmony of the universe by performing good deeds. Thus, in relation to the theological framework of environmentalism, many of them highlight the prominent role of humans as *khalifa*, which not only contains rights but also the duty of basing his thoughts and actions on *akhlak* to preserve nature. They argue that only this Islamic moral approach mitigates climate change effects and ensures environmental protection. Across the borders between Islamic religious currents such as Sufism and Islamism, the wish to preserve one's identity and revive traditional cultural values guides the ideas on environmentalism in the Muslim world. In this context, pieces of classical Persian literature such as the *Gulistan* by Saadi, or Islamic philosophy like Mullā Ṣadrā's, serve to create a vision of what an ideal relationship between God, man, and nature should be. As selected examples indicate, several ulama propose that these sources are important to reject a secular worldview, rediscover one's spirituality, and find inner equilibrium and peace. They see secularism,

105 The Hambali, Hanafi, Maliki and Syafi'i are the four schools of thought concerning the interpretation of Islamic law.

106 Volume IV, 30.

107 Volume III, 13.

108 Muhammad, Ahsin Sakho et al., *Fiqh lingkungan*, 82.

connoted with Western capitalism and claims to power, as a primary cause of environmental degradation. In a world of increasing insecurity Muslim scholars construct Islam as a moral-religious counter-model, which may restore morality, prevent further destruction and raise ecological awareness. However, in Indonesia, this does not only relate to the increasing influence of Islamist groups such as HTI or the Muslim Brotherhood. Rather, also adherents of traditional Islam and Sufis tend to use this argument, thus distinguishing themselves from the West and creating their own identity through elements of traditional culture.

Both the government and Muslim actors are becoming increasingly aware that environmental protection is necessary. Several Muslim CSOs hosted the International Muslim Conference on Climate Change in Bogor in 2010 to counteract environmental degradation. Moreover, the government turned a considerable number of pesantren into eco-pesantren with the aim to mitigate the damages of climate change and ensure sustainable development. Governmental authorities realised that they needed to address local communities to increase awareness of environmental problems. Thus, they contacted local religious leaders and tried to persuade them of the benefits of implementing waste management programmes, for instance. There are several reasons why Muslim scholars in NU and affiliated pesantren have welcomed such efforts. First, political accommodation is strong within NU circles. The article showed that the Ministry of Environment co-authored part of NU's eco-theology, and NU also carries out its activities in close cooperation with the local government. At this point, I would like to re-iterate Amri's statement that the government lacks programmes at the grassroots level and so has to rely on mass organisations such as NU.[109] Acknowledging their support of governmental initiatives, I further argue, however, that the Muslim mass organisations also serve as moral guides for the government by advising authorities and proposing their own ideas. One example is when NU's National Movement for Forestry and the Environment (GNKL) at a national gathering on sustainable forest management and civil society empowerment in July 2012 issued advice (*tausiyah*) urging the government to take forceful measures to prevent enterprises from destroying the forests and the environment. It further reinforced that destroying the environment is prohibited (*haram*) and must be regarded as a criminal act (*jinayat*).[110]

---

109 Amri, Ulil, "Divine Power," 106.

110 Mangunjaya, Fachruddin Majeri, and Jeanne Elizabeth McKay, "Reviving an Islamic Approach for Environmental Conservation in Indonesia," *Worldviews* 16 (2012): 303.

Another reason why religious scholars support the environmentalist cause is that it adds to the common good of the nation, the people, local communities and the santri. Based on verses of the Qur'an and *hadith*, they consider that in this way santri and local people are encouraged to live in harmony with nature. The practice programmes, tree and crop planting, animal husbandry and waste management serve to solve environmental problems, teach santri practical skills and make the pesantren more profitable. The kiai believe that they provide attractive perspectives for their santri as they teach them how to live in accordance with Islamic regulations, provide them with entrepreneurial skills and guide them towards independence. The fact that the number of pupils has been rapidly increasing in these pesantren in the last years is indicative of the fact that this combination is well received among the pupils' parents.

An additional element worth mentioning here is that although religious authorities play an important role in conveying ideas about climate change, they are not the only ones who encourage an environmentally sensitive way of life. Rather, traditional local practices such as planting coconut trees when a child is born into an NU family also serve to foster sustainable development. While it is safe to assert that the organisations and actors presented in this article have increasingly become aware of their responsibility towards nature, further research is required to analyse the impact of local practices, eco-theology and proselytisation.

PART 2

Past

CHAPTER 4

Irrigation in South Thailand as a Coping Strategy against Climate Change: Past and Present

Janice Stargardt

Abstract

This chapter first summarises data on the climatic history of South Thailand in the past 1,500 years, covering events that have influenced hemispherical climate, including South and Southeast Asia, with particular attention to localised impacts. With the context of past climate change thus established, the structure and operation of the ancient irrigation works of the Satingpra Peninsula, South Thailand are discussed in the specific environment of the Kra Isthmus and ecotone. The long history of expansion and contraction of the ancient irrigation system of the Satingpra area, 6th to 14th century, provides our only record of the ancient farmers' efforts to cope with climate change in the past. In conclusion I present the results of the current programme to restore parts of this system and the impact it again has on modern farmers in the area.

1 Introduction

This chapter deals with an ancient irrigation system in South Thailand, discovered and studied by the author and her research group in the decades of the 1970s and 1980s. In order to understand its role as a set of coping mechanisms, I place its stages of expansion and decline in the context of what is known about

* The author offers sincere thanks to the following institutions and individuals for professional and financial support: The Royal Society (Programme on Science-based Archaeology), the British Academy, the University of Cambridge Foreign Travel Fund, the Department of Biology and the Faculties of Engineering and Natural Resources, Prince of Songkla University (PSU) Hat Yai Campus, the National Research Council of Thailand and its Remote Sensing Division, and the Lee Foundation of Singapore; Professor Clive Oppenheimer, Professor Tim Bayliss-Smith and Mr Philip Stickler, Department of Geography, University of Cambridge, Professor Puangpen Siriruksa, Assistant Professor Choathip Puintavaragul (PSU), their colleagues and graduate students.

 | DOI 10.1163/9789004273221_006

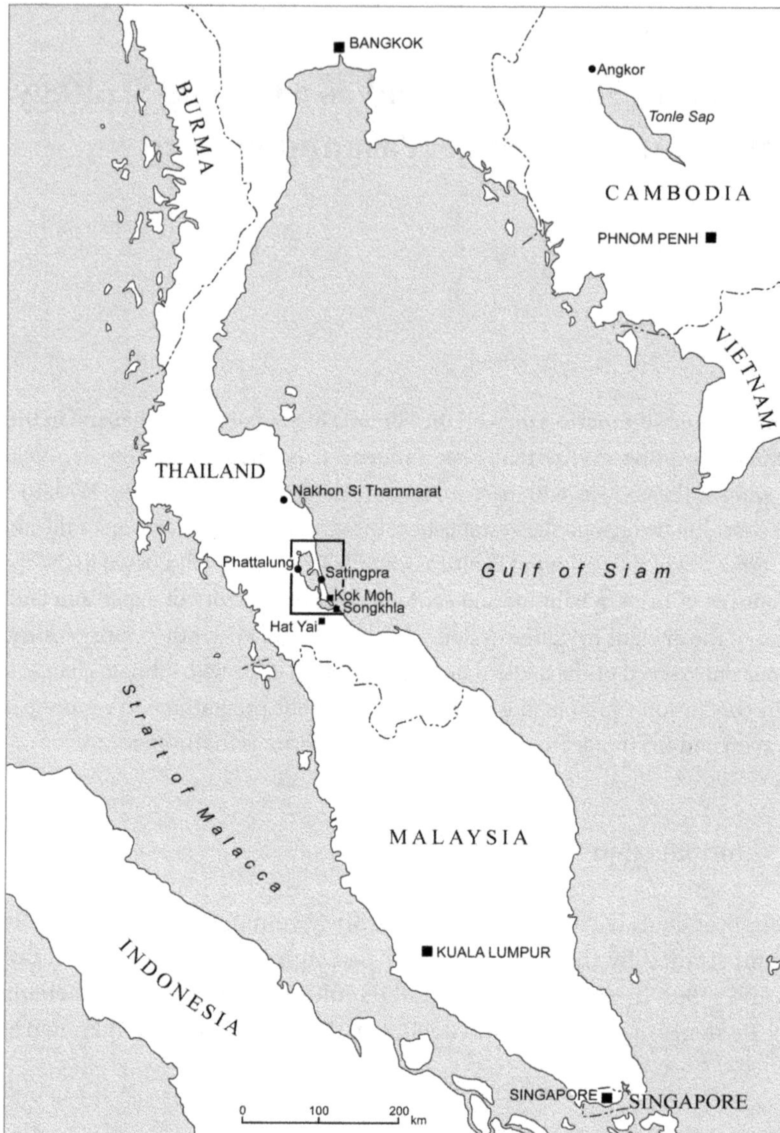

FIGURE 4.1 *Map of Thailand showing the location of the Great Lakes and the Satingpra Peninsula*[1]

climate change in Southeast Asia in the first and early second millennium CE. Recent research has revealed very complex conditions of weak monsoons affecting much of Asia over long periods, but punctuated by extreme and sudden

1 Redrawn by Philip Stickler, 2012 from data © Janice Stargardt.

reversals. To succeed, the ancient Satingpra irrigation technology had to cope with both these extremes of climate. The archaeological traces of success and failure discussed here are among the most eloquent records of the vicissitudes of human economic history available for this part of ancient Thailand. All the evidence, archaeological and ethnographic, points to the central importance of local communities for the specialist knowledge, decision-making and organising of reciprocal labour to create and maintain irrigation works.[2] After some 700 years of neglect since the 14th century, parts of this irrigation system have been rehabilitated since the 1990s and the process continues. The renewal marks a major change of government policy, away from a misguided attempt to impose the field, crop and canal 'modernisation' of Central Thailand upon a quite different environment in the south. The permitting of local farmers to revive the ancient irrigation system, ponded rice fields and traditional rice types (as traced by our research) also takes place within the context of the latest phase of climate change: a gradual warming and drying of the atmosphere since 1970.[3]

2 Reconstructing the Past Climate

The factors influencing the climate of Southeast Asia, and specifically South Thailand, range from those operating on a hemispherical, processual scale, such as the slow millennial decrease in summer insolation in the northern hemisphere as a result of the precession of the equinoxes,[4] to Asia-wide failures or weakening of the monsoonal system since 800 CE.[5] There are three schools of thought among contemporary climatologists to explain these climatic phenomena: in the first, they are considered to be directly linked to the spontaneous effects of the El Niño/Southern Oscillation (ENSO) events taking place in the Pacific with some impact on world climate, but particularly strong repercussions for India, Thailand, Indonesia and the rest of Southeast Asia.[6] In the

2 Stargardt, Janice, *Satingpra I: The Environmental and Economic Archaeology of South Thailand* (Oxford, Singapore: BAR International Series 158 in Association with ISEAS, 1983), 58–64.

3 Cook, Edward R. et al., "Asian Monsoon Failure and Megadrought during the Last Millennium," *Science* 328 (2010): 486–489.

4 Miller, Gifford H. et al., "Abrupt Onset of the Little Ice Age Triggered by Volcanism and Sustained by Sea-Ice/Ocean Feedbacks," *Geophysical Research Letters* 39 (2012): 1–5.

5 Cook et al., "Asian Monsoon Failure."

6 "El Niño-Southern Oscillation," published by Wikipedia, last modified July 23, 2012, accessed May 3, 2012, http://en.wikipedia.org/wiki/El_Nino-Southern_Oscillation; Grove, Richard H., and J. Chappell, *El Niño, History and Crisis: Studies from the Asia-Pacific Region* (Cambridge: White Horse, 2000).

second, they are thought to be linked to the effects of major volcanic eruptions in heightening the probability of the occurrence of ENSO events and their impacts on South and Southeast Asia in complex ways, bringing floods or droughts

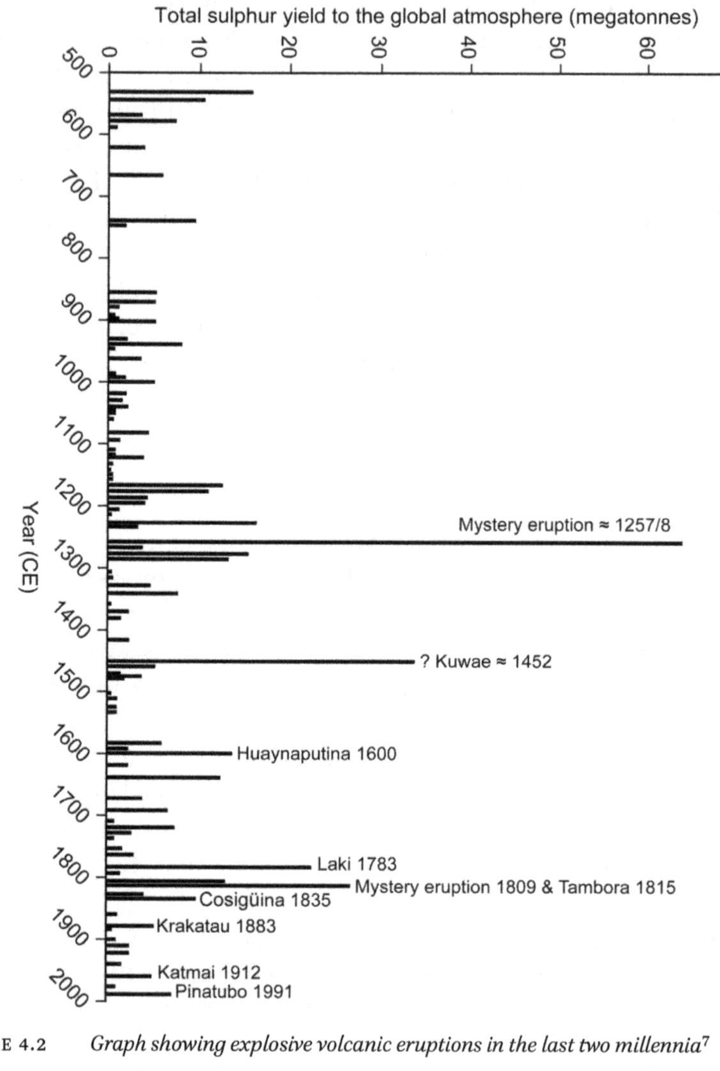

FIGURE 4.2 *Graph showing explosive volcanic eruptions in the last two millennia*[7]

7 Courtesy of Oppenheimer, Clive, "Ice Core and Palaeoclimatic Evidence for the Timing and Nature of the Great Mid-13th Century Volcanic Eruption," *International Journal of Climatology* 23 (2003): 418.

over brief or decadal lengths of time.[8] In the third, they are thought to be connected with the immediate and mid-term effects of a large volume of aerosol particles in the tropical atmosphere on the Asian monsoon following a major volcanic eruption.[9] As the eruptions studied took place close to ENSO events, it is not always easy to separate the volcanic from ENSO signals in this last school of thought.

Tree ring data, ice and sediment cores, glacial retreats and advances, corals and speleothems among others are studied as proxies for climate data in the era before written records. For many scholars, the impetus to study the past climate lies in the desire to predict more accurately climate changes taking place in the present and near future. The proxies can provide insights into periods of drought and exceptionally heavy rainfall or frost in Eurasia and Asia but are not without their difficulties. Studies of tree ring data in Siberia, Mongolia, India, Vietnam, Cambodia and Thailand provide an impressive record of the constant refinement of research techniques while areas of uncertainty or disagreement still remain.[10] A key area of disagreement concerns the causes of ENSO events. Some authors consider that the ENSO warm/dry and cool/wet phases alternate irregularly but spontaneously (i.e. the first school of thought). They consider that ENSO events can, in more recent times, be predicted up to

8 Emile-Geay, Julien et al., "The Volcanic Eruption of 1258 A.D. and the Subsequent ENSO Event," *Geophysical Research Letters* 321 (2006): 1–5; Emile-Geay, Julien et al., "Volcanoes and ENSO Over the Past Millennium," *Journal of Climate* 21 (2008): 3134–3148.

9 Personal comment by Clive Oppenheimer, 2012; Trenberth, Kevin and Aiguo Dai, "Effects of Mt Pinatubo Volcanic Eruption on the Hydrological Cycle as an Analog of Geoengineering," *Geophysical Research Letters* 34 (2007), L15702; Robock, Alan, "20 Reasons why Geoengineering May Be a Bad Idea," *Bulletin of the Atomic Scientists*, May–June (2008): 14–18, 59.

10 D'Arrigo, Roseanne D. et al., "Spatial Response to Major Volcanic Events in or about AD 536, 934 and 1258: Frost Rings and Other Dendrochronological Evidence from Mongolia and Northern Siberia: Comment on Stothers, R. B., 'Volcanic Dry Fogs, Climate Cooling, and Plague Pandemics in Europe and the Middle East' (*Climate Change* 42, 1999)," *Climate Change* 49 (2001): 239–246; Emile-Geay et al., "The Volcanic Eruption of 1258;" Emile-Geay et al., "Volcanoes and ENSO;" Timmreck, Claudia et al., "Limited Temperature Response to the very Large AD 1258 Volcanic Eruption," *Geophysical Research Letters* 36 (2009): 3–4; Cook et al., "Asian Monsoon Failure;" Buckley, Brendan M. et al., "Climate as a Contributing Factor in the Demise of Angkor, Cambodia," *Proceedings of the Natural Academy of Sciences* 107.15 (2010): 6748–6752; Day, Mary B. et al., "Paleoenvironmental History of the West Baray, Angkor (Cambodia)," *Proceedings of the Natural Academy of Sciences* 109.4 (2012): 1046–1051; Mann, Michael E. et al., "Global Signatures and Dynamical Origins of the Little Ice Age and Medieval Climate Anomaly," *Science* 326 (2009): 1256–1260 (among many others).

two years in advance based mainly on changes in sea surface temperatures in the Pacific.[11] In this view, extreme variations in the climate of Southeast Asia, in the period c. 800–1465 CE, with weak monsoons and drought punctuated by heavy rains, were caused by ENSO phases taking place in the much larger context of the transition from the Mediaeval Climate Anomaly to the Little Ice Age, combined with the effects of a movement of the Intertropical Convergence Zone to a position some 5° southwards before resuming its present position during the 19th century.[12] Other scholars (i.e., the second and third schools of thought) emphasise the possible role of volcanic forcing on climate, with or without associated ENSO events. Numerous sulphur-rich (SO_2) eruptions over the past two millennia are known or suspected to have influenced climate at hemispheric to global scale.[13]

3 Phases of the Ancient Irrigation Works on the Satingpra Peninsula

This study of the irrigation works of the Satingpra Peninsula is mainly concerned with four phases: an initial small-scale activity – 4th–6th century, followed by three ancient phases – 6th–9th, 9th–13th, 13th–14th century and one contemporary revival, 20th–21st century. They will be placed, where possible, in the context of what is known about their climatic conditions. Despite areas of disagreement concerning their cause, there is general agreement that ENSO events can be extreme, that they impact on the western Pacific and Southeast Asia with particular severity, and that they cause periods of significant climatic

11 Chen, Dake et al., "Predictability of El Nino over the past 148 years," *Nature* 428 (2004): 733–736; Cane, Mark A., "The Evolution of El Nino, Past and Future," *Earth and Planetary Science Letters* 230 (2005): 227–240.

12 Grove, Jean M., *Little Ice Ages: Ancient and Modern*, 2nd edition (London: Routledge, 2004); Buckley et al., "Climate as a Contributing Factor," 6750.

13 Stothers, Richard B., "Climatic and Demographic Consequences of the Massive Volcanic Eruption of 1258," *Climate Change* 45 (2000): 361–374; D'Arrigo et al., "Spatial Response to Major Volcanic Events," 239–246; Oppenheimer, "Ice Core and Palaeoclimatic Evidence;" Mann, Michael E. et al., "Volcanic and Solar Forcing of the Tropical Pacific over the Past 1000 Years," *Journal of Climate* 18 (2005): 447–456; Emile-Geay et al., "The Volcanic Eruption of 1258;" Emile-Geay et al., "Volcanoes and ENSO;" Timmreck et al., "Limited Temperature Response," 3–4 and figures 2, 3; cf. Mann, Michael E. et al., "Underestimation of Volcanic Cooling in Tree-ring-based Reconstructions of Hemispheric Temperatures," *Nature Geoscience*, Advance Online Publication (2012): 1–4, accessed July 5, 2012. http://www.nature.com/ngeo/index.html; Miller et al., "Abrupt Onset of the Little Ice Age."

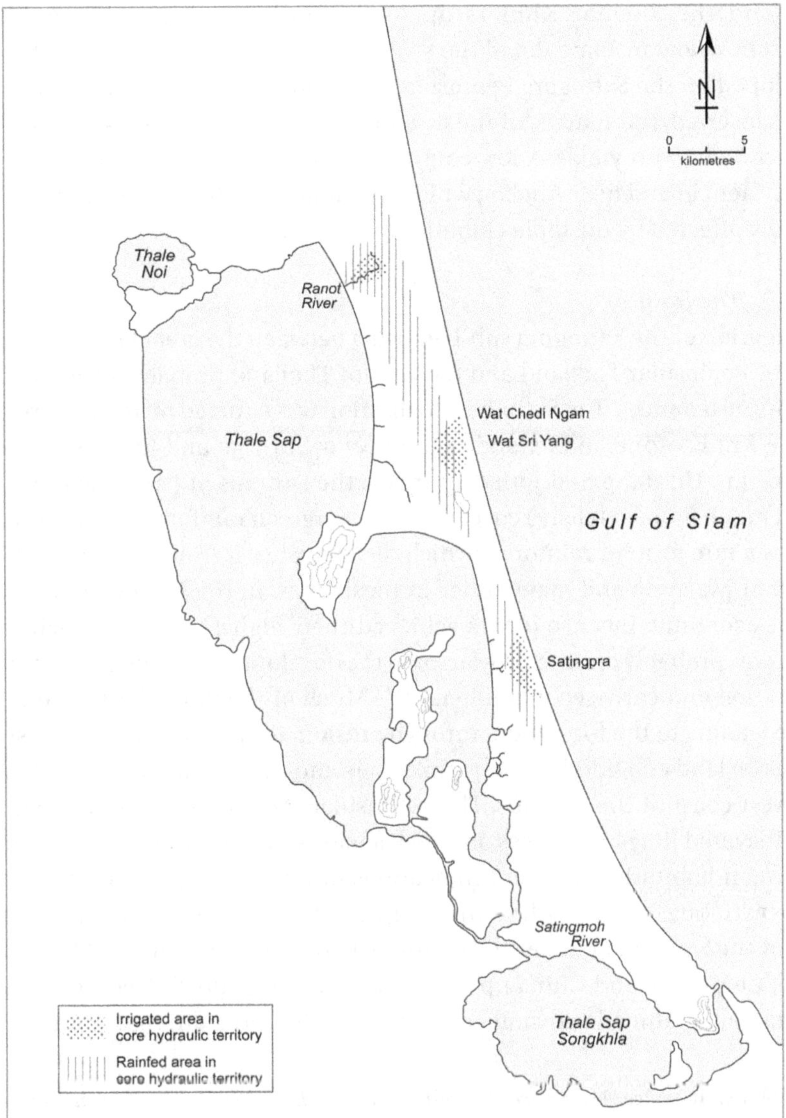

FIGURE 4.3 *Map of the Satingpra Peninsula, Phase 1. Irrigated and rainfed areas 4th–6th century*[14]

instability. Irrigation works are viewed in this chapter as major coping strategies employed in the face of that instability by past and present communities on the Satingpra Peninsula, by the Pyu in central Burma throughout the first millennium

14 All figures below redrawn by Philip Stickler, 2012 from data © Janice Stargardt.

CE[15] and the ancient Khmers up to the mid-15th century at Angkor.[16] I present below in some detail the structure of the irrigation system that was developed on the Satingpra Peninsula, the geomorphic conditions into which it was inserted, the sources of the water it stored and distributed, and its likely impact on harvest yields. A few comparisons are made with the Khmer irrigation system in and near Angkor, which was almost contemporary and also profoundly affected by unstable climatic conditions.

3.1 *The Habitat*

Occupation of the Satingpra sub-Peninsula between the great lakes of the Isthmus of Peninsular Thailand and the Gulf of Thailand probably owed much to its natural defences. The Satingpra civilisation was situated on the eastern plain of the Kra Ecotone and extended from N7°07′ to 7°50 and E100°18′ to 100°36′ (figure 4.1). Until the mid-19th century CE, the Isthmus of Peninsular Thailand had a much more extensive cover of semi-evergreen rainforest than at present. This is a rare form of rainforest which distinguishes it from the wet evergreen forest of Malaysia and many other tropical areas. In the Kra Ecotone, several major economic incense plants achieved their highest quality. Proximity to them was probably another key factor in the development of the Satingpra sites in the first and early second millennia.[17] Much of the upland area is on limestone leading to the formation, through erosion, of spectacular turmkarst pinnacles on land and hundreds of turmkarst islands in the Andaman Sea flanking the west coast of the isthmus. The peninsular ranges are honeycombed with multi-layered limestone caves and rock shelters, which preserve a long record of human habitation. Ancient indications of involvement in maritime trade begin with fragments of 4th century imperial Roman glass found in the palace area of the Satingpra citadel and continue with artefacts from the Middle East, India, Cambodia and China up to the early 14th century.[18] A combination of natural and manmade advantages meant that from the 9th–13th centuries CE,

15 Stargardt, Janice, *The Ancient Pyu of Burma Vol. I: Early Pyu Cities in a Man-Made Landscape* (Cambridge, Singapore: PACSEA in Association with ISEAS, 1990).

16 Groslier, Bernard-Philippe, "La cité hydraulique angkorienne: exploitation ou surexploitation du sol," *Bulletin de l'école française d'extrème-orient* 66 (1979): 161–202; Fletcher, Roland et al., "The Water Management Network of Angkor, Cambodia," *Antiquity* 82 (2008): 658–670.

17 Stargardt, Janice, *Satingpra I*; Stargardt, Janice, "Behind the Shadows: Archaeological Data on Two-way Sea-trade between Quanzhou and Satingpra, South Thailand, 10th–14th Century," in *Emporium to the World: Quanzhou and Maritime Trade, 10th–14th Century*, ed. Angela Schottenhammer (Leiden: Brill, 2001): 358–363.

18 Stargardt, "Behind the Shadows," table 1; 345.

the ancient settlement pattern attained greater intensity on the Satingpra Peninsula than elsewhere around the lakes, as shown by the development of ceramic, iron and tin smelting industries and the construction of numerous monuments (figure 4.5).

The Satingpra population was obliged from the beginning of permanent settlement to create supplementary sources of water because of perennial rainfall deficits. Neighbouring settlements at these latitudes commonly experience two rainy seasons per annum. This is not the case, however, on the Satingpra Peninsula which, together with the modern provincial capital, Songkhla, experiences a rain shadow effect from the peninsular ranges lying in the pathway of the south-west monsoon. For instance, Phattalung, on the western edge of the great lakes and Nakorn Sri Thammarat just 80 km north of Satingpra are unaffected by this phenomenon (figures 4.1, 4.4). Phattalung and Nakorn Sri Thammarat receive 2000–2500 mm rainfall p.a. Both receive well-distributed rain from the southwest and the northeast monsoons, the latter being heavier and the usual cause of seasonal flooding. The Satingpra Peninsula, by contrast, receives only 1200–1700 mm per annum, which is largely confined to one season: the north-east monsoon, with the heaviest downpours concentrated in January. The minimum annual precipitation for rain-fed rice cultivation is usually set around 1800 mm of fairly well distributed rainfall per annum. Many ancient and modern settlements have flourished under lower local rainfall regimes if they benefitted from river or canal flooding as a result of rains falling elsewhere.[19] The poor drainage system of the Satingpra Peninsula initially denied this possibility to the local population. They did, however, have access to two significant water resources: the generally high water table and the water in the great lakes. The history of the Satingpra irrigation system depends on the local people's increasing skill in exploiting these two resources. Figures 4.3, 4.4, 4.5 and 4.6 indicate, in four chronological stages, how the inhabitants of the Satingpra Peninsula added canal digging to their repertoire of water storage in tanks and ponds. The significant stages of the expansion and intensification of the ancient irrigation system have been aggregated as: 4th–6th century, 6th–9th century, 9th–13th century, and 13th–14th century. The chronology is based on datable artefacts found in stratified locations in canals and ponds during the intensive coring of the ancient hydraulic works.

This programme was carried out between 1976 and 1982, while satellite, aerial and surface surveys (including lake coring) covered a much longer period from 1971–1990s. Ancient tank and canal digging showed a thorough knowledge

19 Stargardt, *The Ancient Pyu of Burma*, 18–19.

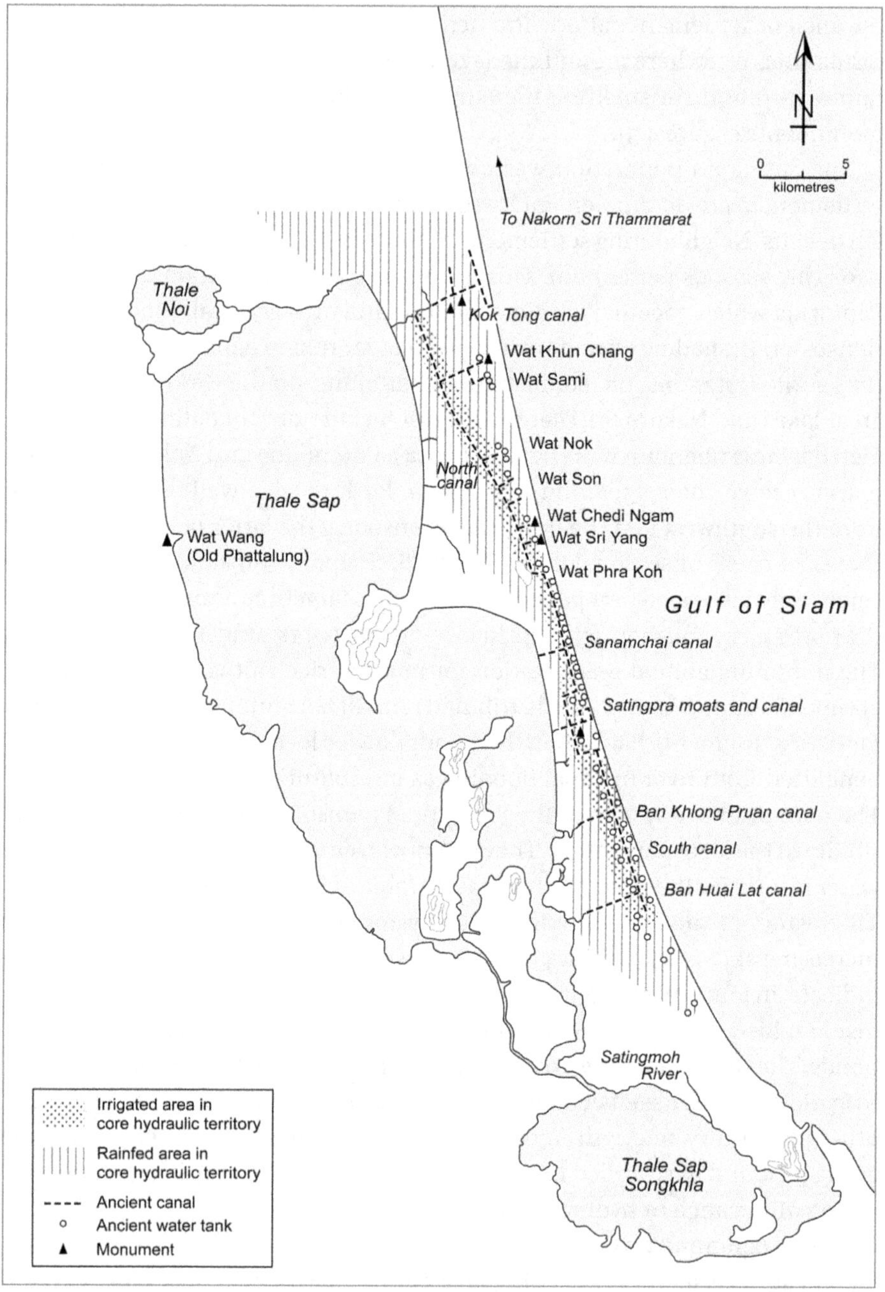

FIGURE 4.4 *Map of the Satingpra Peninsula, Phase 2. Irrigated and rainfed areas 6th–9th century*

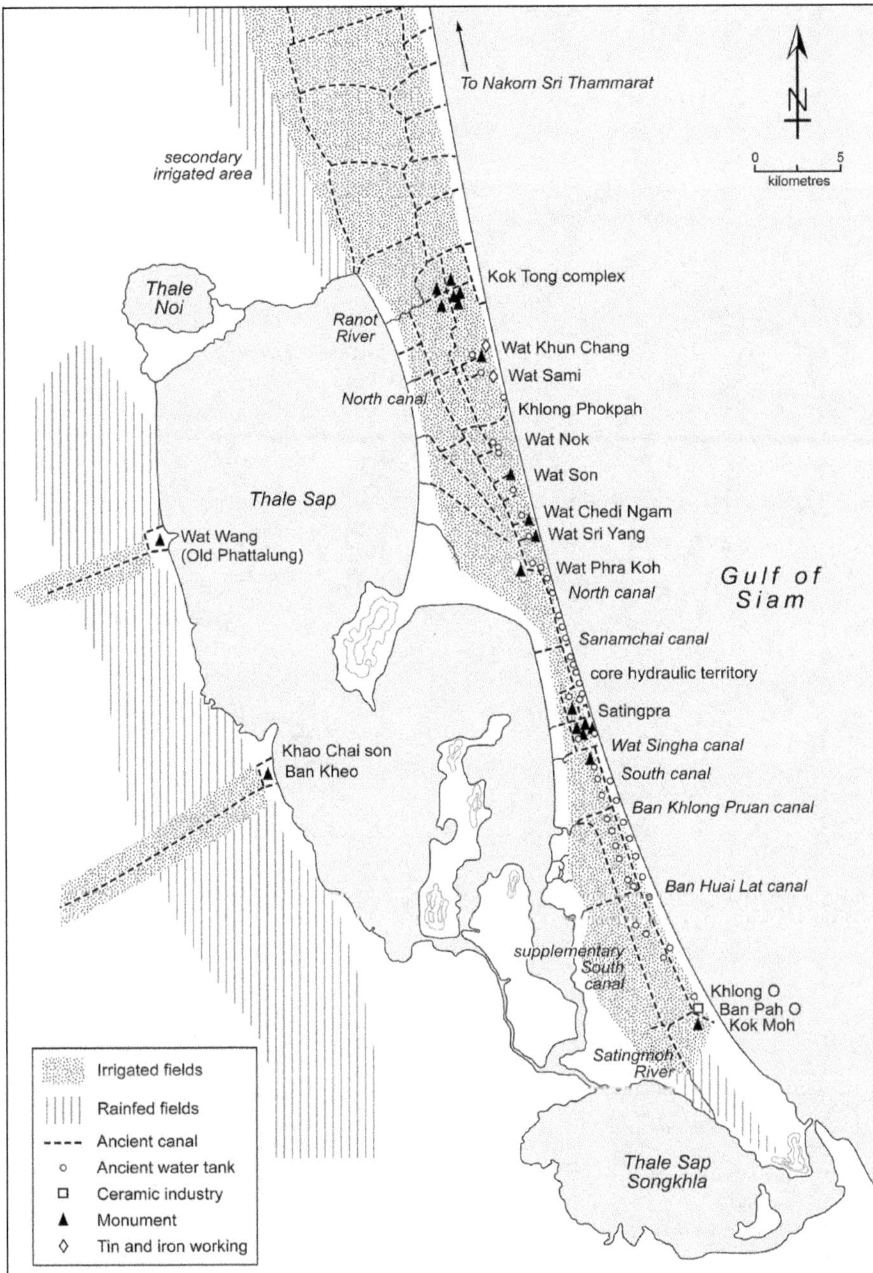

FIGURE 4.5 *Map of the Satingpra Peninsula, Phase 3. Irrigated and rainfed areas 9th–13th century*

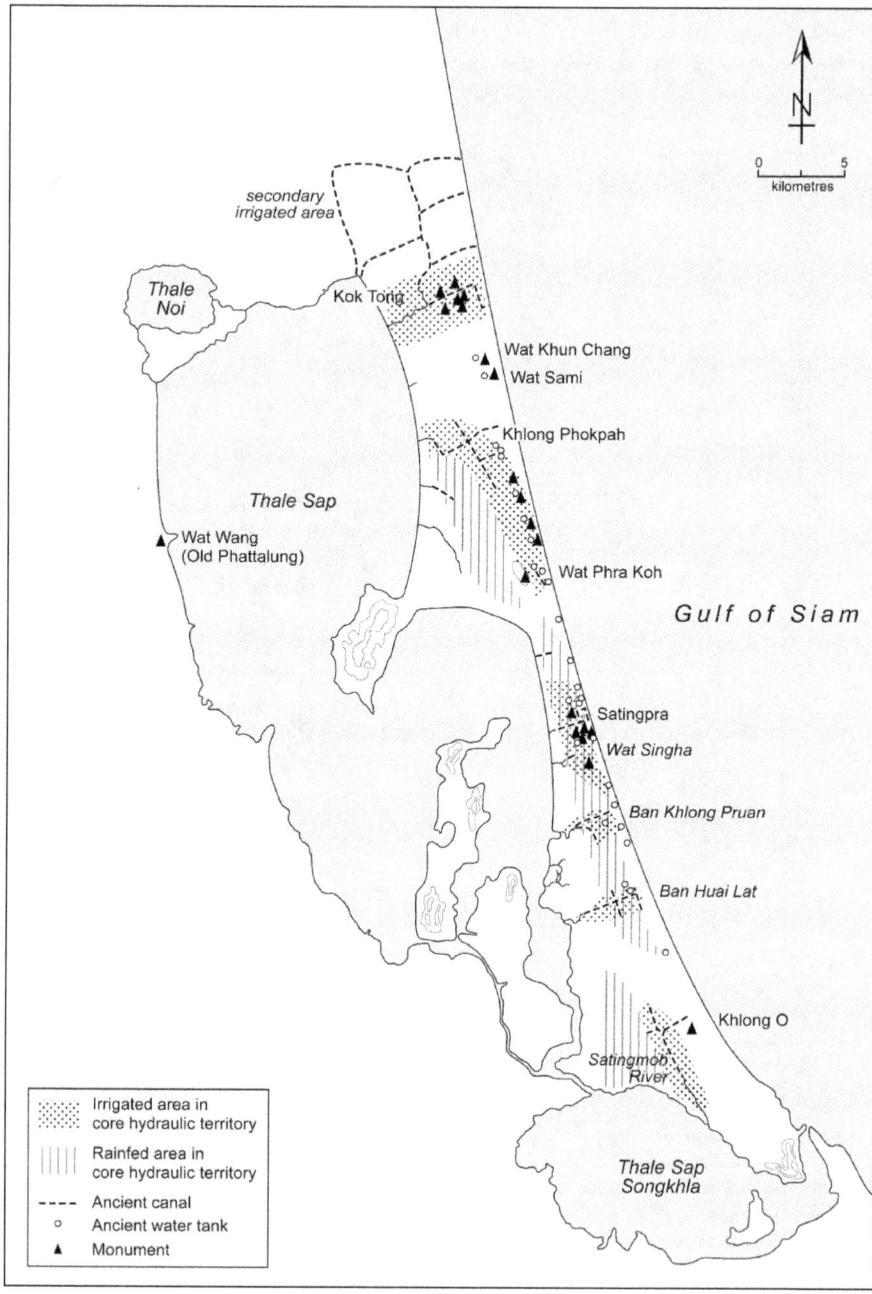

FIGURE 4.6 *Map of the Satingpra Peninsula, Phase 4. Irrigated and rainfed areas 13th–14th century*

of the local soils by favouring the blocky solid clays of marine origin for the locations of tanks and the alignments of the main canals. Three categories of such soils were identified in our environmental surveys: the Bangkok series, the Samut Prakhan series and the Thon Sai series. These three soil types have an excellent capacity for water retention and a very high potential rice rating.[20] They cover the greater part of the land area of the Satingpra Peninsula.

3.2 *Irrigation Tanks*

The Satingpra water tanks are the oldest form of irrigation works on the Satingpra Peninsula, and the first to be described here. All the water tanks were simple depressions dug into the solid clays just named. They were not lined as this would have inhibited the processes of sub-surface percolation they largely relied on. The soil coming from the original excavations and from seasonal cleaning of sediments thereafter was heaped up around the tank in the form of a bund, especially on the upslope sides. Tanks were thus a simple construction and vulnerable, in a landscape of minimal slope towards the lakes, to seasonal inwash of humus and debris. The coring programme indicates that they were subject to seasonal clearing for a long period. Some tendency towards standardisation in tank dimensions has been detected. Small tanks were typically c. 20×20×3 m deep; medium-sized tanks ran from c. 20×40×3 m to 60×70×3 m. The mean dimensions of this category have been calculated at c. 40×50×3 m. Large tanks ran from 100×100×3 m to 170×220×3 m. The mean dimensions of this category have been calculated as 120×140×3 m. Some 90 to 95 per cent of the ancient tanks traced on the Satingpra Peninsula belonged to the medium category, while large tanks were found only with major monumental sites like the urban area of Satingpra or Wat Phra Koh.[21] The water retention capacities of the Satingpra tanks of medium size were as follows: in the wet season from October to February they filled to their maximum, each with an average water holding of c. 6000 m^3; in the sensitive mid-season period of March to April, this dropped by c. one-third to c. 4000 m^3, while in the dry season from May to October the water volume dropped again to c. 3000 m^3. With an estimated total number of c. 150 medium-sized tanks in existence in the phase 9th–13th century, it is estimated that they could hold 630,000 m^3 of water in the wet season, 420,000 m^3 in the critically important intermediate season and 315,000 m^3 in the dry.

At first sight these dimensions and capacities appear insignificant when compared with the *barays* of Angkor, whose construction probably began

20 Stargardt, *Satingpra I*, 46–48.

21 Figure 4.5 here and Stargardt, *Satingpra I*, 134–137.

before the 10th–11th century and which appear to have declined in the 14th–15th century.[22] The largest of these, the western baray, measured 8×2 km and if filled to capacity could have held an estimated 53,000,000 m^3.[23] All other barays were smaller but still with a much larger surface area than any tank on the Satingpra Peninsula. Appearances are to some extent deceptive, however, owing to the strikingly different technology involved in each of these types of water storage tank and the sources of their water. The barays constructed at Angkor and a very few other ancient Cambodian urban centres were all built above ground by means of surrounding a flat or almost flat area with earthen bunds flanked by ditches along the inner edges. Originally, barays were provided at the highest point with a single intake channel from a natural river or canal. Water flowed into the bunded space, initially along the interior ditches and then rose to cover the floor of the baray to a varying depth according to its angle of slope and the volume of water flowing through the intake channel. This water supply only functioned once a year, in the summer monsoon (in this case July-October), bringing a finite stock of water into the baray, which might fill it to overflowing (during a heavy rainfall) or might not hold enough to meet irrigation needs up to the following wet season (during a severe drought). Seasonal fluctuations of the monsoons provided the motivation to enclose a large space within a baray but it also meant that a large surface area of shallow water was vulnerable to evaporation during the dry season. The interior ditches of the barays may, I suggest, have preserved enough moisture in the soil on the occasions that these tanks were almost empty to prevent the base of the baray from cracking open – as did happen in the ancient tanks of Sri Lanka.[24]

All the water tanks of the Satingpra Peninsula were initially dug to a depth of c. 3 m, i.e. well below the lowest level of the water table, which sank during the dry season to Datum minus 1.5 m (Datum being the ancient surface level). Tanks were maintained close to that depth until the second half of the 13th century.[25] This meant that during both the intermediate and dry seasons, the water drawn out of each tank replaced itself according to the level of the water table within 24 hours of being used. Environmental surveys showed that this was the length of time that sub-surface water took to percolate through the

---

22 Groslier, "La cité hydraulique angkorienne;" Buckley et al., "Climate as a Contributing Factor," 6749; Day et al., "Paleoenvironmental History," 1050.

23 Fletcher et al., "The Water Management Network;" Day et al. "Paleoenvironmental History."

24 Results from my field studies in 1981.

25 Table 4.1 here and Stargardt, *Satingpra I*, 139–141.

blocky clay and occupy the water tanks.[26] The small surface area of these tanks also meant that very little of the water stock was exposed to evaporation during the dry season. The mechanism governing the level of the water table on the Satingpra Peninsula was the volume of water in the chain of great lakes just to the west: Thale Sap Songkhla, Thale Sap and Thale Noi. Although, as already mentioned, the Satingpra Peninsula itself received a relatively low, single-season rainfall compared with surrounding areas, this was not true of the lakes. They benefited from the 2000–2500 mm p.a. of local rain falling on their western edge around Phattalung; they also received the benefit of non-local rain from the south-west monsoon falling on the ranges to the west and draining down to the lakes through short rivers and canals. This combination of local and non-local natural water sources meant that the great lakes of the Isthmus of Peninsular Thailand remain the largest permanent lakes in mainland Southeast Asia, whereas Tonle Sap downstream from Angkor is a highly seasonal lake.

Although the barays and canals of Angkor were constructed on a larger scale than the Satingpra system, the former may have been somewhat more vulnerable than the latter to any weakening of the summer monsoon because they were dependent upon a single season of rainfall in their catchment area to the northeast of the Angkor plain. In spite of their small individual size, the collective contributions of the water tanks on the Satingpra Peninsula were larger than they at first appear and diverse in kind: each could provide only limited amounts of irrigation water to the rice fields and garden crops in any 24 hours but provided a renewable source of water in the following days, if managed within those constraints. Some tanks also served the needs of humans and domestic animals. Their distribution as represented on figures 4.4 to 4.6 is schematic and meant to indicate trends during the aggregated phases of irrigation development and decline rather than the absolute number and location of tanks, which are impossible to depict at this scale. The tables provide some concrete data to supplement this outline. Table 4.1 supplements data on the estimated average capacity of eight medium-sized water tanks given in *Satingpra I*, calculated for the period from the 6th to the mid-14th century in the wet, intermediate and dry seasons.[27] The data are based on the stratigraphies of maintenance and sedimentation obtained by coring during the environmental surveys, with layers numbered from bottom to top. Standardised soil colours (e.g. 7.5 yr 3/6) are taken from the standard international Munsell Soil Color Code.

26 Stargardt, *Satingpra I*, 135–136.

27 Ibid., 142–145.

TABLE 4.1 *Average stratigraphy of ancient water tanks of middle size on the Satingpra Peninsula*

Layer	Depth	Data
I	D-3.30 m	Dark grey clay of marine origin, belonging to the Bangkok series, the original base of the tank, 7.5YR 3/6
II	D-3.29-2.15 m	Coarse brown sand with small fragments of freshwater molluscs; at D-3.09 m a sherd of green-glazed Chinese earthenware pottery belonging to the Tang dynasty; at D-2.30 m a second sherd of green-glazed Chinese stoneware of Longquan type from the Song dynasty.
III	D-2.14-2.02 m	A mixed layer of coarse brown sand and fine red sand, 5YR 4/4
IV	2.01-1.59 m	Fine red sand, 5YR 3/4
V	1.59-1.46 m	Grey clay and red sand together 5YR 3/6-3/4; broken bricks; small sherds of cream-glazed Chinese stoneware of the Yuan dynasty.
VI	1.45-1.24 m	Heavy, medium-grey clay, 7.5YR 3/2
VII	1.23-1.16 m	Heavy, pale-grey clay with traces of iron 7.5YR 3/2
VIII	1.15-1.06 m	Reddish-brown clay, 5YR 4/4
IX	1.06-0.00 m	Dark grey clay with root traces, 7.5YR 3/6

The stratigraphy of table 4.1 reveals an active programme of clearing of the inwashed sediments that was sustained for some six centuries. This made for a very slow accretion of sediments during Layer I and the first part of Layer II, from c. the 6th to the late 12th century: c. 65 cm of sedimentation in c. 600 years. The second part of Layer II (coming from the same geological source) reveals a more rapid rate of sedimentation. Thus the average depth of these tanks rose to little below the water table by the end of the 13th century: c. 45 cm sedimentation in c. 100 years. Such a rate of sedimentation indicates stress among the population who were unable to maintain the previous levels of tank cleaning and management. The following layers, IV to IX, were thin but geologically diverse, suggesting turbulent environmental conditions, man-made or natural. Their accumulation carried the beds of the tanks above the critical threshold of the water table. From the late 13th to the mid-14th century the increasingly shallow depressions of the water tanks could only retain rainwater when seasonally available, and no longer drew upon the perennial resource of subsurface water. In this reduced state, however, the ancient water tanks still serve the population of the Satingpra Peninsula up to the present. They capture and retain a small body of water when the first showers of the north-east monsoon arrive and are still often used as rice seedling beds. After the rains, the fertile sediments filling the tank depressions are sought-after locations for short-season horticulture.

The ancient irrigation works of the Satingpra Peninsula were unevenly distributed (figures 4.4 and 4.5). There was a clear trend for the ancient water tanks to be concentrated in clusters just to the west of the highest ridge of the peninsula. This is the most recent beach ridge and reaches an average altitude of only c. 4 m, running approximately north-south parallel to the present coast. The land to the west of it slopes down very gradually towards the lakes. While the ridge itself consists of compacted regosols (sands) covered by humus of varying depth, the soils immediately to the west belong to the three types of heavy marine clays already identified, with their excellent capacity to hold water and serve as rice fields. Water taken out of the water tanks and the canals dug in these clays could be passed along small channels into the rice fields as a gravity-fed operation. The fields were minimally terraced in relation to each other, presenting a flat surface within each ponded field but together forming a slope of at most 1:1000 towards the west. These apparently narrow parameters of slope nonetheless sufficed to enable the irrigation water to flow from field to field towards the lakes, as convincingly demonstrated by the revival of these fields today (figures 4.7 and 4.8).

3.3 *Canals*

The canals of the Satingpra Peninsula were of several types and affected by more complex hydraulic conditions than the water tanks. As figures 4.4 to 4.6 show, the larger canals of the system consisted of both lateral and longitudinal channels. Most of the former had lake entrances which were of great importance in the water supply of the whole canal system. Very few connected with the Gulf of Thailand, where there was a risk of saline invasion. The lateral and longitudinal canals together formed a man-made drainage and irrigation system embracing, at its peak, a primary irrigation area of c. 840 km^2 (cf. 1,000 km^2 at Angkor). In addition, a secondary canal area extended to the north as far as Nakorn Sri Thammarat (figure 4.5). Around Nakorn Sri Thammarat, the main function of the canals was to drain and regulate the seasonal flooding caused by the abundant two-season rains that area enjoyed. It is impossible to include on the maps the small-scale distributary channels and runnels that brought water from the main canals and tanks into the ponded rice fields so characteristic of this area in the past and being revived again in the present. In figure 4.7, showing a mature, rehabilitated field of the ancient type, a small cleft through the massive bund can be seen, permitting water to pass from field to field. In

FIGURE 4.7 *Photograph of a renewed remnant of the traditional ponded rice fields of the Satingpra Peninsula with heavy bunds planted with sugar palms*[28]

FIGURE 4.8 *Seedling beds, coconut and sugar palms under the impact of reconstructed traditional irrigation in a ponded rice field of the Satingpra Peninsula*[29]

28 Photo: © Janice Stargardt, 2002.

29 Photo: © Janice Stargardt, 2002.

figure 4.8, a rehabilitated channel is bringing water through a culvert into two ponded rice fields serving as seedling beds.

The stratigraphic record of the central canal of the whole system is recorded in table 4.2. In some ways its long phase of maintenance (Layers I and II, 6th–13th century) is typical of the system as a whole. But from the late 13th into the 14th century (Layer III), it was the focus of exceptional efforts to maintain some of its functions at a time when most of the irrigation system was being allowed to collapse (cf. figures 4.5 and 4.6 and tables 4.1 and 4.2). In Layers IV and V, it too was overwhelmed by processes of rapid, unreversed sedimentation.

TABLE 4.2 *Stratigraphy of the Central Canal of the Satingpra system, 6th to early 14th century*

Layer	Depth	Data
I	D-4.60-4.40 m	Grey clay of marine origin; 7.5YR 3/6
II	D-4.39-3.40 m	Grey sediments mixed with clay and sand; 7.5YR 3/6 + 5YR 3/4
III	D-3.39-3.15 m	Heavy grey clay; a sherd of Chinese celadon belonging to the late Song period at 3.25m; 7.5YR 3/6
IV	D-3.14-1.20 m	Dark grey clay mixed with sediment and sand; fragments of brick at 1.40m; 7.5YR 3/6
V	D-1.20-0.4 m	Brown sediment with sand, 7.5YR 3/2

As table 4.2 shows, the canals were originally constructed to levels far below the lowest depth of the water table (Datum minus c. 1.50 m). They were dug to over 4 m depth so as to be able to draw upon this resource and also be navigable to small ships. In addition, however, during the long dry period on the Satingpra Peninsula itself, the lakes – as already noted – were swollen by rains from the southwest monsoon. This raised them above the western edge of the land of the Satingpra Peninsula causing localised flooding and a rise in the level of the water table over most of the peninsula. The lake entrances of the lateral canals provided man-made channels allowing the swollen lake waters to enter the canals creating a slow reverse flow. Lake water was thus able to replenish water levels in a considerable part of the ancient canal system, both lateral and longitudinal during the intermediate season, when it was critically important for the rice crop. It provided a seasonal supplement to the water resources of the peninsula, which could be widely distributed by the canal system for six or seven centuries.

The central section of the Satingpra Peninsula was the area with the greatest concentration of irrigation works. They included water tanks, moats around the citadel or palace area of the capital and the central canal which had multiple functions: it bisected the capital city, provided a transport link between the Gulf of Thailand and the biggest lake, Thale Sap, and provided irrigation water. There was a danger of saline penetration into this canal from the gulf and particular attention was paid to examining this possibility in our coring programme. It was found that salinity in the cores dropped sharply within 200 m of the ancient entrance of this canal into the gulf. No explanation for this has as yet been found but the following features may have been significant: the mouth of the central canal forms an angle of 18° to the south as it passes through the highest beach ridge. This may have protected the canal to some extent from the wind-driven incursion of saline water during the north-east monsoon. By manipulating the levels of the bed of this and other lateral canals, it was possible to bring water from the lake towards the east, forming a slow current, up to the western edge of the ancient beach ridge. Exactly how the canal was protected beyond that point against saltwater incursion is not clear, but possibly a movable barricade was used. The initial sections of the main longitudinal canals branched towards the north and south out of the central canal in the central section. The central area was served by seventy-three water tanks, four moats and 27.9 linear km of large canals suitable for transport as well as irrigation. The stratigraphies obtained by coring these works indicate that, at the peak period of maintenance, they introduced a maximum volume of c. 1,402,200 m^3 of freshwater into an area of c. 3,196 ha.[30] This is the equivalent of 438 m^3/ha. The disposable water available in the intermediate season, when it was needed most for irrigation, was only approximately two-thirds of this amount but still abundant. Of fundamental importance is the fact that these volumes of water renewed themselves within a short time for centuries. The stratigraphic data presented in tables 4.1 and 4.2 indicate that the Satingpra irrigation system was able, under optimal conditions, to permit some three or four applications of supplementary water during the growing season of the traditional rice plants. When the irrigation system was functioning well, this would have sufficed to stimulate very significant increases in rice harvests, as our experiments have demonstrated (below). Table 4.1 shows that the water tanks were typically maintained to depths below the water table from the 6th to the late 12th century and were residually useful thereafter. Table 4.2 shows that the central canal was maintained for still longer than the water tanks, while many of the other canals of the system were declining at about the same time as the water tanks (figure 4.6). The details of

30 Stargardt, *Satingpra I*, 145–147.

climatic oscillations between dry and wet conditions outlined below imply that irrigation conditions were often not optimal. Its access to two seasons of water resources, even in times of weakened monsoons, conferred some degree of resilience on the Satingpra irrigation system, which was matched by the capacities for drought resistance of its traditional rice varieties.

I now turn to the last two elements in this ancient irrigation system: the rice fields and rice plants that the water tanks and canals were designed to serve. On the Satingpra Peninsula, there are botanical indicators of the existence of a pronounced seasonal aridity (see the contemporary local rainfall deficits already mentioned). Poor rainfalls may have been equally or even more marked in the period from the 6th to the 14th century. The indicators are the sugar palm (*Borassus flabellifer L.*), the cashew nut (*Anarcardium occidentalis*), and kapok (*Ceiba pentandra*), all of which flourish locally but not in the neighbouring areas receiving a heavier precipitation. Some of these are present in the oldest pollen cores as well, dating to 1600 BP. Cropmarks in aerial photographs of the Satingpra Peninsula reveal an ancient field pattern quite unlike that of central Thailand which dominated the ancient landscape of the Satingpra Peninsula (figure 4.9).

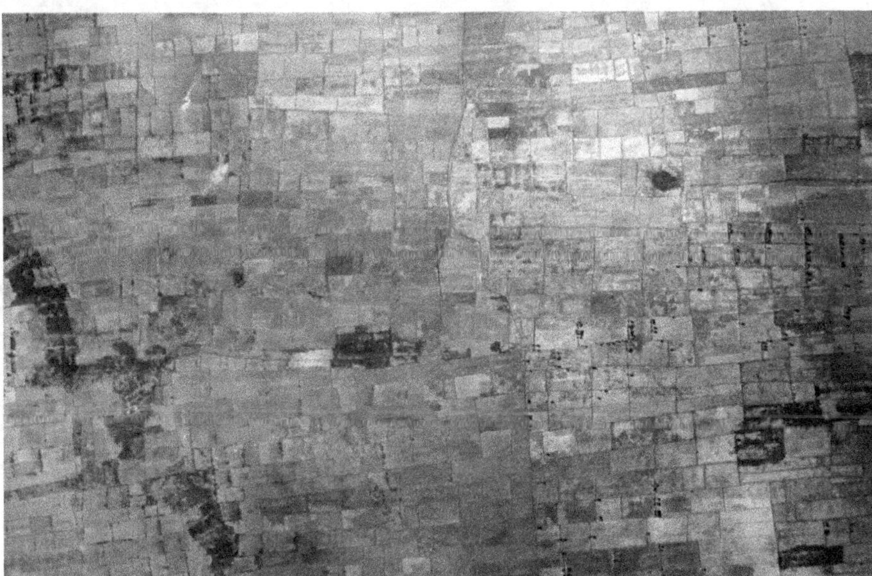

FIGURE 4.9 *Aerial photograph showing soil- and cropmarks of ancient ponded rice fields in the extended rice fields of the Satingpra Peninsula*[31]

31 Reprinted courtesy of Royal Thai Airforce, scale 1:600.

Its characteristics indicate that some aridity existed in the past. The ancient fields were small, almost square and surrounded by massive bunds of rammed earth, designed to be higher than flood level. Their bulk was supported and probably strengthened by the cultivation of sugar palms (*Borassus flabellifer* L.). Pollen cores from the Satingpra Peninsula confirm the association of rice with the sugar palm c. 1600 years BP, i.e. in the 4th century CE.[32] Strong official pressures in the 1960s and 1970s were exerted on the population of the Satingpra Peninsula to convert their agricultural landscape to the extended rice fields and gridded canals of central Thailand, which are the basis of the modern rice export industry (figure 4.10), and to adopt the cultivation of short season hybrid rice types. Unfortunately, the botanical indicators of aridity were ignored and the transformation of the Satingpra field system resulted in serious crop failures and much local suffering (figure 4.11).

FIGURE 4.10 *Google Earth image of irrigated extended rice fields of Central Thailand, E100°45′ N14°09′*[33]

32 Stargardt, *Satingpra I*, 53, 57–58, figures 26b, 28–32.

33 Accessed May 17, 2012 by Philip Stickler; reproduced in accordance with the terms and conditions of Google Earth.

FIGURE 4.11 *Photograph of extended rice fields on the Satingpra Peninsula with dying rice plants and sugar palms*[34]

Fortunately, a small remnant of the ancient ponded fields had been tenaciously preserved, which continued to yield good harvests of traditional rice types in this period (figure 4.7). To some extent they mitigated the negative consequences of the conversion to the field type of central Thailand while demonstrating how accurately the ponded fields and traditional rices were adapted to the climatic conditions of the Satingpra Peninsula. They were studied in detail by our environmental survey group during the 1970s–90s. The ponded rice fields enable maximum use to be made of limited amounts of water, allowing it to pass around each field in small perimeter channels as well as percolate across them with minimal loss to evaporation. Narrow cuts are made seasonally through the high bunds to allow water to pass directly from field to field (figure 4.7). The first outliers of the northeast monsoon serve to soften the soil. Water buffaloes and cattle are driven into the fields, and their hooves break up the crust of dry surface soil, allowing oxygen and moisture to penetrate below the surface. Ploughing (by men) follows so that when the first substantial showers arrive, farmers are ready to broadcast seeds into the highest fields, which, together with ancient tank depressions, serve as the seedling beds for that season (figure 4.8). Small amounts of irrigation water are raised from tanks and canals by means of simple hand-powered scoops and, in combination with the first rains, can maintain

34 Photo: © Janice Stargardt, 1982.

very good growth conditions in the seedling beds. When the real rains arrive, (female) collaborative working groups of each village are ready to transplant, having gained three to four critically important weeks of plant growth in advance of the real wet season.

This is important because the preferred rice varieties of this area are the long-season, photosensitive rices, which produce heavy yields but take between five to seven months to do so. Harvesting is done by women only, using traditional rice knives with wooden backs and small iron blades, held with the palm of the hand cupped around them. Two fingers reach out to grasp a small bundle of rice stalks just below the heads of grain, which are simultaneously cut by the blade. Stone blades of similar shape and size have been found near the turm-karst pinnacles on the western side of the lakes. It was noted that this harvest technique is adapted to the fact that the long-season rices of the Satingpra Peninsula ripen unevenly; thus each field will be visited several times before the harvest is completed.

4 Yields of Traditional Rice with and without Irrigation

Analyses carried out in the International Rice Research Institute at Los Banos campus near Manila on samples of the thirty-four varieties of traditional long-season rices collected on the Satingpra Peninsula, indicate that they probably are ancient types.[35] Government pressures to prioritise the cultivation of the modern hybrid rice varieties which mature in three months are resisted by farmers on the Satingpra Peninsula because they lack the fine flavours and drought resistance of the traditional long-season varieties. Action-based research was undertaken by my group in the Satingpra district to observe the performance of traditional rice varieties per hectare in response to no irrigation water and variable amounts of irrigation water. The data gained are directly relevant to the motives involved in creating and maintaining irrigation works and, by extension, to the past and present roles of irrigation works as coping mechanisms in the face of climatic variations. To our surprise, it was found that the traditional rice varieties required almost twice the amount of seed grain as the modern hybrids. Thus on average, 40 kg of traditional rices were needed as seed grain per ha in contrast with as little as 25 kg per ha of the modern hybrids. As rainfed rice, the traditional varieties of the Satingpra Peninsula produced c. 700 kg per ha, or a ratio of seed sown to seed harvested of over 1:16. This figure throws a valuable light on the capacities of the traditional rice varieties under

35 Stargardt, *Satingpra I*, 89–96.

poor conditions in the past. With even limited supplies of supplementary water during the intermediate season, the plants responded very strongly with increases of c. 100 per cent in grain yields, giving a sown to harvest ratio of 1:34. Under improved conditions of irrigation and drainage further very strong responses produced sown to harvest ratios of 1:40, which were probably not the maximum. Additions of fertiliser made little impact; the cardinal factor governing the scale of harvests was clearly water. Other striking insights already foreshadowed were that the traditional rice varieties had a very high resistance to drought. Under conditions of prolonged aridity, the plants survived and produced enough grain to replace themselves but only a reduced harvest for consumption of 1:16 or less.

Strikingly similar results were gained from experimental studies of the traditional rice varieties cultivated in the 1960s from rural areas around Tonle Sap in the Angkor region.[36] There the traditional rice fields and small water tanks are quite similar to those on the Satingpra Peninsula: small squarish fields with sugar palms growing on the bunds and small square or rectangular water tanks, unlike the barays which seem to have been associated closely with urban areas.[37] Tonle Sap itself is larger at its peak than the great lakes of the isthmus but extremely seasonal: it sinks to one-third of its maximum in the dry season. It is only implicated in the ancient or modern irrigation of the Angkorian plain to a limited degree – as drainage basin for Angkor's rivers and canals, as the locus of flood retreat cultivation and for the cultivation of floating rice.

5 Satingpra Irrigation and Regional Climate Change

Let us now try to locate the major phases of the Satingpra irrigation system in the larger context of what climatic data are available for the region. The initial phase was of very modest dimensions in the pre-urban period, 4th–6th century, mainly confined to the digging of water tanks and the creation of some ponded fields, which probably supported a small population (figure 4.3). A significant expansion of irrigation works, and presumably of the population who needed, created, and maintained them, took place in the 6th–9th century (figure 4.4). The motives for excavating the first large canals from the 6th century onwards and intensifying the network of water tanks were probably complex, including the emergence of an urban centre, Satingpra, on a site bisected by the

36 Legris, Pierre, Francois Blasco, and Tichit, *Notice de la Carte, Cambodge, Carte Internationale du Tapis Végétale Series,* 156 (Pondichéry: Institut français de Pondichéry, 1972).

37 Aerial photograph in Stargardt, *Satingpra I*, figure 52.

central canal (figures 4.4, 4.5). But the approximate coincidence of this 6th century development with a major volcanic eruption in, or shortly before, 536 is noteworthy. D'Arrigo et al. discuss an extensive range of tree ring and ice core data for the early to mid-sixth century from sites as far apart as Greenland, Europe, China, North and South America before focusing on tree ring data from Mongolia. They conclude that the evidence points to a major volcanic eruption with consequences for the climate and humanity extending into Asia. Notably it coincides with an unusually cold decade, 536–545 CE with a brief recovery in 538. They consider this a delayed climatic response linked to one volcanic eruption or possibly two separate events.[38]

6 Satingpra Irrigation and Climatic Instability

It is commonly assumed that the impetus to irrigation works is created by drought, but the data from all the climatic studies mentioned so far show that the Asian climate in the first two millennia CE has been marked by extremes of aridity and pluviality. Research that might show whether the eruption of 536 impacted on the Asian monsoon and, if so, how it manifested itself in South Thailand – is lacking at present. But it is important to recognise that the Satingpra irrigation works must have been created to meet multiple needs: to provide a drainage system for excessive rainfalls, to allow lakewaters to recharge the canals and water table seasonally, and also to relieve decadal conditions of aridity. It is as essential to get water out of rice fields at a certain point in the growth cycle as it is to get it into them for germination and initial growth to occur. Figure 4.2 shows major volcanic eruptions – identified and unidentified – that have taken place in the past two millennia. The atmospheric models created by Emile-Geay and his colleagues suggest that tropical, sulphur-rich explosive eruptions, such as that of Mount Pinatubo in 1991 or greater, raise the probability of an ENSO event by 75 per cent, while they acknowledge that ENSO events also occur without volcanic forcing.[39] If this is so, then some of the other eruptions identified in figure 4.2 may have caused ENSO events with major consequences for the climate of southern Thailand. Or they may have exerted climate forcing directly on the Asian monsoon.[40] D'Arrigo and colleagues have found tree ring evidence in Mongolia showing frosts in 938 CE and cite signifi-

38 D'Arrigo et al., "Spatial Response to Major Volcanic Events," 239–242.

39 Emile-Geay et al., "Volcanoes and ENSO," 3141–3142.

40 Personal comment by Clive Oppenheimer, June 2012; Trenberth, and Dai, "Effects of Mt Pinatubo Volcanic Eruption."

cantly reduced rates of growth in Siberia in 933–935 and 940. As they comment, these appear to coincide with the great Eldgja eruption in Iceland in 934, but could also result from an unrelated local episode of extreme cold. The same group has recovered tree ring evidence from Mongolia and Siberia showing frosts and reduced growth in the years around the time of the 1257–1258 eruption.[41] Oppenheimer considers that among the eruptions listed in figure 4.2, those in 1028–1032, 1100, 1171, 1229, 1257, 1341 and Kuwae c. 1452 were all potentially climate-forcing events. In none of these cases do we know for sure that they affected an ENSO event, nor in what ways they may have impacted directly on the Asian monsoon and, in particular, on the climate of the Satingpra Peninsula. The possibility has to be considered, however, that the second and third phases of irrigation development on the Satingpra Peninsula, 6th to 9th and 9th to 13th century, took place in conditions of some considerable climatic instability. During the latter phase, the irrigation system reached its maximum extent of c. 840 km^2.

Turning to research on tree rings found closer to the Satingpra Peninsula, the findings of Buckley and colleagues for the Angkor region have a direct relevance for the Satingpra Peninsula, although the main emphasis of their study falls on a slightly later period. They argue that there was a period of prolonged drought over mainland Southeast Asia in the transition from the Mediaeval Climate Anomaly into the Little Ice Age.[42] This view is also held by Mann and colleagues, modelling climate on a world rather than regional scale,[43] and is accepted by Day and colleagues.[44] Buckley also notes, however, that within this prolonged pattern of weak monsoons, there were several abrupt reversals from drought to very intense monsoons. For instance, he notes that the two wettest single years in his entire record were 1258 and 1453 with a further six very wet years occurring at the end of the 14th and beginning of the 15th century.[45] Buckley and colleagues note that the earlier of these dates coincides with the major eruption in 1257 or 1258.[46] The Buckley group consider, however, that it is unnecessary to

41 D'Arrigo et al., "Spatial Response to Major Volcanic Events," 242–243; Oppenheimer, "Ice Core and Palaeoclimatic Evidence," 422–424 and figure 4.

42 Buckley et al., "Climate as a Contributing Factor," 6749.

43 Mann et al., "Global Signatures," 1258–9 and figure 3.

44 Day et al., "Paleoenvironmental History," 1046.

45 Buckley et al., "Climate as a Contributing Factor," 6749, 6751.

46 Figure 4.2 here; cf. Stothers, "Climatic and Demographic Consequences;" D'Arrigo et al., "Spatial Response to Major Volcanic Events;" Oppenheimer, "Ice Core and Palaeoclimatic Evidence;" Emile-Geay et al., "The Volcanic Eruption of 1258;" Emile-Geay et al., "Volcanoes and ENSO," among others. The later date, 1452 or 1453, is often claimed to be Kuwae: figure 4.2 here; Gao, Chaochao et al., "The 1452 or 1453 A.D. Kuwae Eruption Signal Derived

pursue the possible role of volcanic forcing when this kind of variability in climate can occur 'as part of coupled ocean-atmosphere dynamics'.[47]

7 Climate and the Decline of the Satingpra Irrigation System

Without pursuing the controversy over volcanic climatic forcing on the ENSO or Asian monsoon, it is evident that the Buckley data provide insights into the climatic conditions prevailing on the Satingpra Peninsula in the 9th to 13th century that are otherwise lacking. The expanding construction of canals and water tanks on the Satingpra Peninsula in this period must have taken place in prevailing conditions of drought and weak monsoons but they must also have had to cope with sudden onsets of very intense rainy seasons, perhaps clustered around the end of the first millennium[48] and certainly occurring in the course of the 13th and 14th century. These would have been major climatic events, whatever their ultimate cause. A picture of past climate in the region begins to crystallise showing that a critical event or series of events occurred in mainland Southeast Asia in the second half of the 13th and early 14th century, which overwhelmed the successful operation of the entire Satingpra irrigation system. Tables 4.1 and 4.2 show the signs of serious rapid sedimentation occurring in water tanks and canals. Special efforts were devoted to keeping the central canal operational, even to a reduced degree, at the turn of the century but were abandoned in the early decades of the 14th century. Since earlier centuries of maintenance showed that both manpower and technology were present, the abrupt decline of the irrigation system at this time suggests that the manpower supply was diminished and thus unable to cope with the extreme rates of sedimentation.

The dramatic decline in the number of irrigation works still operating on the Satingpra Peninsula in the 13th–14th century (figure 4.6) may well have been connected with the extreme event of 1257/1258. There was a significant phase of sedimentation in the central canal at that time which was reduced, but not completely removed, when maintenance work resumed (table 4.2). A *coup de*

---

from Multiple Ice Core Records: Greatest Volcanic Sulphate Event of the Past 700 Years," *Journal of Geophysical Research* 111 (2006): 11.

47 Buckley et al., "Climate as a Contributing Factor," 6751, citing Meehl, Gerald et al., "Factors Affecting Climate Sensitivity in Global Coupled Models," *Journal of Climate* (2006): 1584–1596.

48 Cf. figure 4.2 here; D'Arrigo et al., "Spatial Response to Major Volcanic Events;" Oppenheimer, "Ice Core and Palaeoclimatic Evidence."

grâce was delivered by a later heavy rainfall event c. 1350.[49] By the time of the eruption in 1452/1453, the Satingpra canals had ceased to operate and the irrigation system had contracted to a few small ponds and tanks. From that period onwards, the only mitigating factor in the Satingpra landscape against the effects of aridity and extreme pluvial events were the small ponded rice fields surrounded by their massive bunds. These bunds exceeded flood levels and, being held in place by the strong roots of the sugar palms, were able to withstand the destructive force of exceptionally strong rainfall events, while providing the means of survival in droughts. The total extent of such rice fields is likely, however, to have contracted post-14th century with the final destruction of the urban centre, Satingpra, before 1340, while the decline in the population seems to have begun from the late 13th century onwards. Cook and colleagues, basing their research on multiple sources – tree rings, corals, ice cores, speleothems and ocean sediments – to provide proxy records of the Asian climate,[50] conclude that there were persistently weak monsoons in tropical South and Southeast Asia in the mid-14th and early 15th centuries, especially the seventeen years from 1351–1368. Buckley and colleagues have recovered evidence of serious and irreversible sedimentation of a part of the Angkor canal system during the 14th century, possibly as the result of a single flood event.[51] This echoes the kind of extreme events that seem to have destabilised the Satingpra irrigation system in the mid- and late 13th and early 14th century. Cook and colleagues identify and analyse four later Asian droughts, which took place on a regional scale in 1560–1587, 1682–1699, 1756–1768 and again since the mid-1970s.[52] Their data extend our knowledge of Asian climate instability from the late first and early second millennium to cover most of the second millennium and bring it up to the present period. It is clear that, with the collapse of the complex irrigation mechanisms on the Satingpra Peninsula for coping with climate instability, the population that could be supported by rainfed riziculture alone must have been much reduced.

8 The Rehabilitation of the Satingpra Irrigation System

The harvest ratios given above provided crucial insights into the responses of traditional rice varieties of the Satingpra Peninsula to varying volumes of

49 Studied in: Buckley et al., "Climate as a Contributing Factor," 6749; perhaps of short duration. Cf. Cook's 1351 data below.

50 Cook et al., "Asian Monsoon Failure," 486, figure 1/B.

51 Buckley et al., "Climate as a Contributing Factor," 6749.

52 Cook et al., "Asian Monsoon Failure," 487–488.

irrigation water and drainage at the right times in the growth season. They thus reveal the motivations for creating and intensifying irrigation works on the Satingpra Peninsula in the first and early second millennium CE. So too do they provide the key to a reversal of government policy following the publication of our monograph on the irrigated agriculture of the Satingpra Peninsula in 1983.[53] Pressures to impose the extended field types of central Thailand and the modern hybrid rice varieties were slowly relaxed. This policy reversal arose from the intense interest of the Crown of Thailand in our 1983 report, with its obvious relevance to agricultural development, new mapping and the details of traditional rural technology – longstanding concerns of the Crown in Thailand. The policy change began to have practical results from the late 1980s through the work of the Ministry of Agriculture and the Irrigation Department, who sought our technical input on canal alignments and original depths and adopted our maps. By 2010, a substantial part of the ancient longitudinal canal system had been re-excavated to the original depths and alignments but not as yet the ancient water tanks. Figure 4.12 (E100°22′30″, N7°38′42″) shows a sample section of a major canal situated on the northern half of the Satingpra Peninsula that has been re-excavated and is now bearing a considerable volume of water (the cropmark of a second, completely sedimented canal can be seen to the west). Because the former was re-excavated to the depth revealed by our coring programme, this canal is water-bearing all year round. It is transforming agricultural livelihoods for the local farming communities by once again making possible greatly enhanced harvest yields and, above all, making it possible for farmers to make rational plans about labour investments and anticipated harvest yields for years ahead.

The area in figure 4.12 was chosen for illustration because it reveals a dual process taking place: in response to the rehabilitation of the canal, the agricultural landscape is also undergoing a recovery. By subdividing the unsuccessful field type of central Thailand, farmers are recreating the ancient system of ponded rice fields whose thick bunds are planted with young sugar palms. This process can be seen taking place along both sides of the canal. In two decades, the bunds of these small fields will become more massive as the long-lived sugar palms come to maturity. Their roots will add to and also stabilise the clay in the bunds, so that the mature field-bund-tree system will resemble that depicted in figure 4.7. In fact, rehabilitation of the main ancient north-south canals began in the south; one canal has now been re-excavated almost all the way from the Satingmoh River in the south (N7°19′ 7.68″, E100°28′ 46.56″) to the Ranot River in the north of the Satingpra Peninsula. The process of

53 Stargardt, *Satingpra I*.

FIGURE 4.12 *Google Earth image, E100°22′30″ N7°38′42″, showing a reconstructed ancient canal on the Satingpra Peninsula and the continuing process of conversion from the extended rice field type of Central Thailand back to the traditional ponded rice fields with sugar palms*[54]

redeveloping the ancient ponded fields is even more advanced in the southern and central sections of the peninsula than in the area depicted. Conditions of seasonal aridity still prevail on the Satingpra Peninsula and it has now been demonstrated persuasively that the traditional irrigation techniques, ponded rice fields, associated sugar palms and traditional rice varieties provided better coping strategies than the innovations which suited the conditions of central Thailand.

54 Accessed May 17, 2012 by Philip Stickler; reproduced in accordance with the terms and conditions of Google Earth.

9 Conclusion

In conclusion, let us situate the microdata on the irrigation system of the Satingpra Peninsula from the 6th to the 14th century in the perspectives provided by the macrodata on climate and climate change in Southeast Asia. The climate from the 8th to the 14th century seems to have been a period of predominantly La Niña-type drought and monsoon failure,[55] within which the Satingpra irrigation system apparently operated with some degree of success from 6th–13th/14th century. The motivations for creating the system are clear, but were adequate water resources available? The term 'monsoon failure' employed by Cook and colleagues has to be understood as a relative not an absolute failure, in the sense that rainfalls over long periods were delayed or reduced but not absent. Our experimental research revealed two important facts that conferred a degree of robustness on the Satingpra system in the face of such conditions: firstly the drought resistance of the traditional rice varieties of the area, and secondly the fact that the lakes of the isthmus received rainfalls from two monsoons, not one, even when these were functioning at a reduced level. There is a strong probability that these centuries were punctuated by episodes of heavy rainfalls of shorter or longer duration. Consideration has to be given to their causes even if current research does not permit firm conclusions to be reached: the millennial record of major volcanic eruptions (figure 4.2) probably had significant effects on the Asian monsoon and possibly on ENSO events throughout this period of c. 800 years. ENSO events also occurred spontaneously, at least some of which had significant effects on the climate of Southeast Asia. Buckley and his colleagues recovered tree ring evidence showing abrupt and heavy phases of drought reversal, supported by in situ signs of extreme sedimentation that put parts of the Angkor irrigation system out of operation. Similar but earlier evidence of heavy rainfalls was revealed by the Satingpra stratigraphy. Comparing the Satingpra system and its water sources with the Angkor irrigation system, although the latter embraced vast individual works such as the great barays, it was a storage and distribution system fed by the rains of one monsoon only, falling locally and on its catchment area in two bursts from July to October.[56] Inasmuch as the lakes of South Thailand potentially captured rains and the runoff from rains of two monsoon seasons, they may have been slightly more resistant to drought than the single-season system at Angkor. It is clear, however, that both at Angkor and on the Satingpra Peninsula, the ancient

55 Cook et al. "Asian Monsoon Failure."

56 Fletcher, R. et al., "The Water Management Network of Angkor;" Buckley et al. "Climate as a Contributing Factor;" Day et al. "Paleoenvironmental History."

irrigation systems had to be able to cope with sudden onset, heavy rains as well as general conditions of drought. The contemporary rehabilitation of a substantial part of the Satingpra canal system is taking place in a climatic context of another phase of atmospheric warming since 1970. This may be punctuated by rainfall events of general or quite limited extent: e.g., the flooding of central Thailand in 2011, which did not affect the long-standing drought of the northeast provinces or the slow, sustained warming of the southern provinces. This study of the vicissitudes of the ancient irrigation system of the Satingpra Peninsula provides insights available from no other sources into the lives and works of rural society in one part of Southeast Asia over a period of some 700 years, in particular their varying degrees of success in coping with conditions of climatic instability.

CHAPTER 5

The Emergence of the Cryoscape: Contested Narratives of Himalayan Glacier Dynamics and Climate Change

Marcus Nüsser and Ravi Baghel

Abstract

Himalayan glaciers and changes in their size and mass balance have recently been at the centre of the global climate change debate. These dynamic and fragile ice bodies have not only become prominent topics of scientific research but have also received international media attention as both indicators and icons of climate change. Despite their crucial importance to freshwater storage and supply, detailed monitoring of contemporary changes only exists for a limited number of these glaciers. Hence, the uncertainties of spatial and temporal extrapolations of local studies on a Himalayan scale remain a research challenge. Beyond their status as physical landscape features composed of ice, snow and debris, Himalayan glaciers have increasingly become contested and controversial objects of knowledge, susceptible to cultural framing as both dangerous and endangered landscapes. Against this background, the term *cryoscape* is proposed as a conceptual framework to analyse the emergence of Himalayan glaciers in the context of a dynamic, globally imagined mediascape. This chapter retraces the Himalayan glacier debate in the aftermath of the 2007 report of the Intergovernmental Panel on Climate Change (IPCC) and attempts to put the "vanishing glacier discourse" in a larger picture.

1 Introduction

The fundamental hydrological importance of mountain glaciers as sources of the world's freshwater supply has led to them being labelled "water towers"[1] and

* We gratefully acknowledge the support of the Cluster of Excellence "Asia and Europe in a Global Context" at Heidelberg University, Germany for funding our project "Himalayan Glaciers: Endangered and Dangerous Cryoscapes of Knowledge".

1 Viviroli, Daniel et al., "Mountains of the World, Water Towers for Humanity: Typology, Mapping, and Global Significance," *Water Resources Research* 43, W07447 (2007): 1–13, accessed December 23, 2012. doi:10.1029/2006WR005653; Viviroli, Daniel, and Rolf Weingartner, "'Water

 | DOI 10.1163/9789004273221_007

has made the possibility of their loss or massive retreat especially threatening. This holds especially true for Himalayan glaciers due to the large number of downstream populations directly dependent upon a perennial flow of water from rivers originating in these glaciers. Against this background and due to their retreat, Himalayan glaciers have recently become powerful icons in the debate on global climate change.

However, the investigation of glaciers has a much longer history. Some of the earliest scientific interest in glaciers coalesced around Swiss alpine glaciers, especially through the work of Louis Agassiz[2] and Johann von Charpentier[3] that inaugurated a "glacier fever" among geoscientists in Europe. Interest in glaciers began to spread worldwide and the era of the "scientist explorer" was extended to Himalayan glaciers when Alexander von Humboldt sent the three Schlagintweit brothers to investigate them in 1854–1857. Their detailed records in the form of paintings, maps and measurements of glacier velocities continue to be useful today.[4] The subsequent period of colonial exploration of the Himalayas was led by people like the Workmans, Younghusband and Godwin-Austen, to name just a few. Their motivation to visit inaccessible glaciated areas was as much for scientific investigation, as a form of conquest of nature. Through their narratives glacier landscapes became an arena of confrontation of civilisation and a savage wilderness. In later decades the glaciers of specific Himalayan regions received more systematic attention, which included photogrammetrical measuring of ice movement by researchers.[5]

In the context of the current climate change debate, the region's glaciology has received intense scrutiny and aroused public concern. This is because subtropical and tropical glaciers are expected to clearly establish the impact of climate change as they respond to climate variation on the decadal scale or less;

Towers' – A Global View of the Hydrological Importance of Mountains," in *Mountains: Sources of Water, Sources of Knowledge, Volume 31: Advances in Global Change Research*, ed. Ellen Wiegandt (Dordrecht: Springer, 2008), 15–20.

2 Agassiz, Louis, *Etudes sur les glaciers* (Neuchatel: Jent et Gassmann, 1840).

3 Charpentier, Johann von, *Essai sur les glaciers et sur le terrain erratique du bassin du Rhone* (Lausanne: Marc Ducloux, 1841).

4 Kick, Wilhelm, *Adolph Schlagintweits Karakorum-Forschungsreise 1856, Forschungsberichte, Band 6* (München: Deutscher Alpenverein, 1993); Kick, Wilhelm, "Forschung am Nanga Parbat: Geschichte und Ergebnisse," in *Forschung am Nanga Parbat: Geschichte und Ergebnisse, Band 8: Beiträge und Materialien zur Regionalen Geographie,* ed. Wilhelm Kick (Berlin: Institut für Geographie der Technischen Universität Berlin, 1996), 1–133.

5 See e.g. Finsterwalder, Richard, *Die geodätischen, gletscherkundlichen und geographischen Ergebnisse der Deutschen Himalaja-Expedition 1934 zum Nanga Parbat* (Berlin, Karl Siegismund Verlag, 1938).

thereby serving as an easily observable indicator of human transformation of earth systems. However, ongoing academic and policy debates have tended to neglect the diversity within the South Asian mountain belts, as witnessed in the discussions of the impacts of global climate change on Himalayan glaciers and the extent of land degradation. A better understanding of these multi-scalar interactions in the Himalayan arena is possible through a greater attention to the historical dimension and external interventions.

2 The Himalayan Mountain System

The name "Himalaya", originating in the Sanskrit term *hima-ālaya*, literally means "abode of snow" and symbolises the importance of snow and ice since ancient times when pilgrims travelled in these regions (figure 5.1).

FIGURE 5.1 *The south face of Kangchenjunga (8,586 m) with clean ice glaciers in the accumulation zone. View from 4,940 m, near Gocha La, April 6, 2011.*[6]

The highest mountain system on earth forms the transition zone between South and Central Asia over approximately 2,500 km and spans more than 20° of

6 Photo: © M. Nüsser. Literally meaning "The five treasures of great snow", the third highest mountain in the world, located along the India-Nepal border is considered sacred and may symbolise the multifaceted meanings of Himalayan glaciers.

longitude and 13° of latitude. Due to this subcontinental dimension, the orogenetic belt between the massifs of Nanga Parbat (8,126 m) in the Northwest and Namcha Barwa (7,756 m) in the southeast separates the alluvial Indo-Gangetic plain from the Tibetan plateau. Adjacent to the northwestern part of the Himalaya are the mountain ranges of the Karakoram and Hindu Kush, sometimes together referred to as the Hindu Kush-Karakoram-Himalaya (HKH) mountain system. Whereas the southern declivity of the Himalayan arc is characterised by a monsoonal regime with dominant summer precipitation, the northwestern parts are dominated by precipitation from westerly disturbances throughout the year, with single monsoonal events during the summer months. The whole mountain system depicts considerable climatic and ecological diversity and holds the planet's largest body of ice outside the polar caps. The numbers of individual glaciers in the region is not known with great accuracy and the total perennial snow and ice cover is thought to exceed 31,500 km^2 in the Himalaya and 15,140 km^2 in the Karakoram.[7] Based on remote sensing analyses, new estimates of glacier occupied areas are 22,829 km^2 for the Himalaya and 17,946 km^2 for the Karakoram.[8]

3 The IPCC Glacier Controversy

> Glaciers in the Himalaya are receding faster than in any other part of the world [...] and, if the present rate continues, the likelihood of them disappearing by the year 2035 and perhaps sooner is very high if the Earth keeps warming at the current rate. Its total area will likely shrink from the present 500,000 to 100,000 km^2 by the year 2035.[9]

Apart from the unsubstantiated figures for the extent of glacier areas in the mountain system, the unfounded prediction of disappearance of Himalayan glaciers by 2035, made in the chapter on Asia in the 938 page contribution of Working Group II to the Fourth Assessment Report of the Intergovernmental

7 Wissmann, Hermann von, *Die heutige Vergletscherung und Schneegrenze in Hochasien mit Hinweisen auf die Vergletscherung der letzten Eiszeit: Abhandlungen der Mathematisch-Naturwissenschaftlichen Klasse 14* (Mainz: Akademie der Wissenschaften und der Literatur, 1959).

8 Bolch, Tobias et al., "The State and Fate of Himalayan Glaciers," *Science* 336 (2012): 310–314.

9 Cruz, Rex Victor et al., "Asia," in *Climate Change 2007: Impacts, Adaptation and Vulnerability: Working Group II Contribution to the Fourth Assessment Report of the Intergovernmental Panel on Climate Change*, ed. Martin Parry et al. (Cambridge: Cambridge University Press, 2007), 493.

Panel on Climate Change (IPCC) (Cruz et al. 2007) caused mounting furore in the media and in the general public. This single prediction became a bone of contention, was questioned and shown to be erroneous.[10] Apparently an outcome of a typographical error and uncritical integration/citation of non-peer reviewed documents, the claim was sufficient to bring the *entire* IPCC assessment of climate change into popular disrepute. This gaffe may have contributed to a strengthening of climate scepticism and became part of a politicised debate. On the other hand, the controversy instigated the current boom in Himalayan glacier studies.[11]

This incident underlines not only the importance of examining environmental knowledge, but also points to a distrust of "expert" environmental knowledge in the popular domain. Most importantly, this incident highlights the iconic role that glaciers play in conveying diffuse threats like global warming, and how they take on the role of rhetorical devices within discussions of climate change.

Glaciers stand at the intersection of multiple strands of environmental knowledge: scientific, cultural, temporal, spatial and political. They have become imbued with meanings which extend far beyond their status as topographical features composed of snow, ice, and debris. Prominent examples of this are the African glaciers on Kilimanjaro and Mt. Kenya, which were seen as

10 Cogley, J. Graham et al., "Tracking the Source of Glacier Misinformation," *Science* 327 (2010): 522; Cogley, J. Graham, "Himalayan Glaciers in 2010 and 2035," in *Encyclopedia of Snow, Ice and Glaciers*, ed. Vijay P. Singh, Pratap Singh and Umesh K. Haritashya (Dordrecht: Springer, 2011), 520; Bagla, Pallava, "No Sign yet of Himalayan Meltdown, Indian Report Finds," *Science* 326 (2009): 924–925.

11 E.g. Bolch et al., "The State and Fate;" Diodato, Nazzareno, Gianni Bellocchi, and Gianni Tartari, "How do Himalayan Areas Respond to Global Warming?" *International Journal of Climatology* 32.7 (2012): 975–982; Kääb, Andreas et al., "Contrasting Patterns of Early Twenty-first-Century Glacier Mass Change in the Himalayas," *Nature* 488 (2012): 495–498; Kargel, Jeffrey S. et al., "Himalayan Glaciers: The Big Picture is a Montage," *Proceedings of the National Academy of Sciences of the United States of America (PNAS)* 108.36 (2011): 14709–14710; Miller, James D., Walter W. Immerzeel, and Gwyn Rees, "Climate Change Impacts on Glacier Hydrology and River Discharge in the Hindu Kush-Himalayas," *Mountain Research and Development* 32.4 (2012): 461–467; Scherler, Dirk, Bodo Bookhagen, and Manfred R. Strecker, "Spatially Variable Response of Himalayan Glaciers to Climate Change Affected by Debris Cover," *Nature Geoscience* 4 (2011): 156–159; Schmidt, Susanne, "Der Himalaya bald ohne „Hima"? Was wir über die Gletscherentwicklung im Himalaya wissen,"*Geographische Rundschau* 64.4 (2012): 10–16; Schmidt, Susanne, and Marcus Nüsser, "Fluctuations of Raikot Glacier during the Last 70 Years: A Case Study from the Nanga Parbat Massif, Northern Pakistan," *Journal of Glaciology* 55.194 (2009): 949–959; Schmidt, Susanne, and Marcus Nüsser, "Changes of High Altitude Glaciers from 1969 to 2010 in the Trans-Himalayan Kang Yatze Massif, Ladakh, Northwest India," *Arctic, Antarctic and Alpine Research* 44.1 (2012): 107–121.

anatopisms, and whose symbolic meaning has changed over time from objects of colonial conquest to resonant icons of global warming.[12]

Why did the topic of Himalayan glacier dynamics resonate so widely in the case of the IPCC glacier controversy, and not just with scientists ('experts') but with a global public as well? The answer to this question has consequences for understanding the process of communication of a complex (and controversial) phenomenon like global warming to the lay public. In this chapter we address this question firstly by examining the ways in which Himalayan glaciers are known and imagined. In the following section we identify some broad differences in the ways in which different sets of actors make sense of these glaciers. Using the idea that different epistemic practices lead to different, and at times conflicting, understandings of glaciers we emphasise the broader cultural and scientific dimensions that frame glaciers as objects of knowledge. Building upon this expanded understanding of glaciers as more than mere physical landforms, we then propose the concept of cryoscapes as a way of incorporating epistemic, cultural and scientific practices that in association with physical phenomena produce the distinct realities of Himalayan glaciers as understood by different actors. We conclude by pointing to the advantages of using the idea of cryoscapes to identify the different ways in which glaciers are framed as a rhetorical and epistemic device within the broader debate on climate change.

4 Knowing and Imagining Glaciers

Some of the most important reasons for the resonance of glaciers in the understandings of the environment are their visibility, their susceptibility to cultural framing[13] and their role as repositories of scientific data on the environment.[14] The relative visual accessibility of glaciers, as compared to, say, the proportion of greenhouse gases such as carbon dioxide or methane in the atmosphere, makes them a ready referent for understanding and communicating the effects of climate change. The observable nature of glaciers also makes them

12 Nüsser, Marcus, "Kilimanjaro and Mount Kenya: Colonized Mountains and their Rediscovery as Symbols of Global Climate Change," *Geographische Rundschau – International Edition* 5.4 (2009): 26–32 and supplement (map).

13 Orlove, Ben, Ellen Wiegandt, and Brian H. Luckman, "The Place of Glaciers in Natural and Cultural Landscapes," in *Darkening Peaks: Glacier Retreat, Science, and Society*, ed. Ben Orlove, Ellen Wiegandt, and Brian H. Luckman, 3–19 (Berkeley: University of California Press, 2008).

14 Carey, Mark, "The History of Ice: How Glaciers Became an Endangered Species," *Environmental History* 12.3 (2007): 497–527.

relatively easily measurable through techniques like remote sensing, ground penetrating radar and repeat photography. Apart from scientific observation, there is also a long tradition of visits to Himalayan glaciers for purposes ranging from pilgrimage to tourism, which create a visual and often also an oral record of changes to glaciers. This suggests that knowledge about glaciers is a negotiated outcome of various forms of circulation like scientific claims, media images, personal memory, cultural memory, witnessing and exploration.

The IPCC glacier controversy is not only a miscarriage of science but also casts an interesting light on the asymmetry between three different spheres of distinct but partially overlapping domains of environmental knowledge, namely scientific ('expert'), popular ('common') and local knowledge. Generally, expert knowledge is the most authoritative knowledge form, as well as the most narrowly distributed one. Its power is indeed derived from a mixture of its narrow distribution and its combination of general, specific, and experiential ways of understanding. The expert has a differentiated view of various glacier nourishing systems; he or she needs to measure the mass balance while considering distinct accumulation and ablation zones with an annual net gain and net loss of ice; (s)he differentiates between clean ice and debris-covered ice (figure 5.2) while analysing annual ablation rates and (s)he understands the limitations of remote sensing approaches while assessing glacier retreat over certain time periods based on multitemporal satellite imagery.

Common knowledge is merely an understanding of a particular phenomenon, like glacier retreat, that is widespread, and derives its legitimacy from this ubiquity. Building upon the intuitive understanding of ice melting with heat, the melting of glaciers became a shorthand for global warming. The two actors who were awarded the Nobel Peace Prize in 2007 for communicating the threat of climate change to a global public, the IPCC, and Al Gore had both made use of the metaphor of "vanishing glaciers". This simplistic idea of rapid disappearance with rising temperatures was contradicted by empirical and scientific evidence, which showed varying responses of glaciers depending on their location and scale of measurement. However, due to the entrenched perception of a direct correlation of climate change with melting glaciers, even in the aftermath of the IPCC controversy, its Chief, Rajendra Pachauri continued to insist that Indian glaciers were "melting away": "The lay public [...] can see with their eyes what is happening to our glaciers."[15] This statement can be interpreted as

15 Bagla, Pallava, "Climate Science Leader Rajendra Pachauri Confronts the Critics," *Science* 327 (2010): 510.

FIGURE 5.2 *The Rupal face of Nanga Parbat (8,126 m) in Northern Pakistan is characterised by a steep wall over a vertical distance of more than 4,500 m. View from 5,570 m, ascent to Rupal Peak, August 29, 2010.*[16]

a direct reaction to the report by Raina,[17] released by India's Ministry of Environment and Forests, where the idea of a rapid retreat of Himalayan glaciers is rejected. Bagla[18] opened the scientific and public debate, when he confronted the results of Raina with those from the IPCC report[19] and exposed the obvious contradictions between them.

Local knowledge may often be at odds with the other two categories. The most important attribute of this kind of knowledge is its situatedness. One example: In the Indian Trans-Himalaya of Ladakh, Chewang Norphel, a local engineer, has led a movement to construct "artificial glaciers" to address water scarcity.[20] These artificial glaciers build upon the traditionally used local irrigation structures called *Zing*, and add to it the experiential knowledge of glaciers

16 Photo: © M. Nüsser. The South Face of the mountain ridge depicts a mosaic of rock walls, clean ice glaciers and avalanche tracks in the accumulation zone and debris-covered glaciers in the ablation zone.

17 Raina, Vijay Kumar, *Himalayan Glaciers: A State-of-Art Review of Glacial Studies, Glacial Retreat and Climate Change* (New Delhi: Ministry of Environment and Forests, Government of India, 2009).

18 Bagla, Pallava, "No Sign."

19 Cruz et al., "Asia."

20 Bagla, Pallava, "Artificial Glaciers to Help Farmers," *Science* 282 (1998): 619.

and icing. The artificial glaciers are sited in shaded mountain slopes above the cultivated areas and below natural glaciers at altitudes between approximately 3,900 and 4,600 m. Here, melting in spring starts earlier and increases water availability for the irrigation of fields. These constructions make use of stone embankments across the watercourse to store water in the form of ice during winter. The principle of these constructions is that water freezing can be induced by reducing the runoff velocity. Some of these "artificial glaciers" for water management are constructed in a cascade designed to melt as per harvest requirements.[21] These are all examples of knowledge that is only available in this particular locale. In contrast to previous examples, the melting of glaciers here is not a threat, but a necessity (figure 5.3).

FIGURE 5.3 *The cultivated fields of the village of Hankar (3,950 m) with the Kang Yatze massif (6,401 m) in the Upper Markha Valley, Ladakh, India. View from 3,950 m, September 5, 2009.*[22]

21 Nüsser, Marcus, Susanne Schmidt, and Juliane Dame, "Irrigation and Development in the Upper Indus Basin: Characteristics and Recent Changes of a Socio-hydrological System in Central Ladakh, India," *Mountain Research and Development* 32.1 (2012): 58–59.

22 Photo: © M. Nüsser. Cultivation of barley in the semiarid Trans-Himalayan environments solely depends on irrigation systems fed by glaciers and seasonal snow cover.

An improved understanding of glacier dynamics continues to be necessary here, but for very different reasons. The difficulty of reconciling these practical local requirements with epistemic practices of expert knowledge is perhaps the reason for the lack of academic attention to these phenomena, in spite of requests of the engineer now known as "Glacier Man", for "scientific help from specialists".[23]

Different kinds of environmental knowledge, each with a different epistemic community, are distinguished by their distinct knowledge practices and strategies. Examining the process of contestation in the always unstable domains of environmental knowledge is not only desirable, but imperative, in the light of the consequences it has for our ability to address the looming threat of global climate change. This points not just to the global connectedness of environmental issues, but also points towards an emergent globality in how the environment is imagined.[24] In the context of threatened Himalayan glaciers, it has become more important than ever to focus on the ways in which human perceptions about global environmental change, as well as possible strategies to deal with it, are shaped.

Human interactions with the environment are mediated through knowledge, mental constructions of that environment, and the material reality that exists beyond these. The process through which this knowledge of the environment is produced is clearly influenced by asymmetrical power relations within human societies. This is reflected in, for instance, the greater relative credence given to expert knowledge. The epistemological stances towards environmental knowledge range from a positivist belief in ever more accurate, objective scientific knowledge to a radical disbelief in the possibility of any incontrovertible knowledge and an epistemological relativism. Neither of these stances is useful in addressing global climate change; the positivist stance fails through its disregard for non-expert knowledge, whereas the relativist stance fails in its disregard for mounting evidence of environmental change. This becomes apparent, for instance, in the schism between the so-called climate alarmists and climate sceptics.

Against this background it has become imperative to find new positions through a renewed examination of the 'black box' of environmental knowledge, especially across various cultural boundaries and in different geographical locations. One of the most important entry points for this endeavour is an examination of the dynamic processes of circulations and encounters of knowledge.

---

23 Vince, Gaia, "Glacier Man," *Science* 326 (2009): 661.

24 Appadurai, Arjun, *Modernity at Large: Cultural Dimensions of Globalization* (Minneapolis: University of Minnesota Press, 1998).

This calls for an attention to the role of space, flows, agency, scale, and asymmetry in the production of knowledge.

Contemporary interest in the epistemology of Himalayan glaciers can be seen to emerge from a few major attributes. Although the reasons for the wide resonance of the phenomenon of glacier retreat are many, Orlove et al.[25] identify two as the most important. The first attribute of glaciers that makes them easily *imaginable* is their visibility. They can be painted, photographed, and seen with the naked eye, as well as with many instruments, by common people, locals and experts alike. By thus observing them over a period of time, the retreat of glaciers may become readily apparent. In addition, "the direct accessibility of glaciers to human vision has helped to make them a topic of personal and public concern".[26] This accessibility of glaciers to human vision, *regardless of the knowledge practice*, has additionally made it possible for them to act as a shared referent between the three domains of environmental knowledge, the common, the local and the expert. The second important attribute identified by Orlove et al.,[27] is the susceptibility of glaciers to cultural framing. This again applies across all domains of environmental knowledge, with different framings apparent in different domains, and at different scales. Many glaciers are considered sacred, as the origin of holy rivers (the most prominent example is the Gangotri Glacier in Uttarakhand, India), or due to their association with other religious symbols. This means that when such symbols are affected, due to a mix of human and non-human influences, the impact is not just physical. The cultural framing of glaciers has also led to local practices like classification of glaciers into male and female, and of mating pieces of male and female glaciers to *birth* new glaciers. This practice from northern Pakistan illustrates how glaciers are given different meanings depending on the particular cultural context in which they are interpreted.

A third obvious reason for the importance given to glaciers within environmental knowledge is the gravity of the consequences of glacier melting for lives and livelihoods in downstream communities. This can create various types of problems ranging from increased runoff during the initial phase of glacier retreat to decrease and resulting water stress in the long run. Additionally, potential hazards from glacial lake outburst floods (GLOFs) are always present (figure 5.4).

25 Orlove, Wiegandt, and Luckman, "The Place of Glaciers."

26 Ibid., 5.

27 Ibid., 7.

FIGURE 5.4 *The proglacial Imja Lake in the Khumbu Himalaya of Nepal is considered prone to outburst floods. View from the summit of Imja Tse, 6,189 m, October 27, 2012.*[28]

A further reason for the easily imagined, and ostensibly easily communicated, knowledge of glaciers is the intuitive understanding that most people have of ice and snow. As temperature increases, ice melts and becomes water. This analogy applied to the case of global warming and glaciers gives rise to the commonsensical but erroneous assumption that global warming will melt all glaciers, even though the effects of global climate change are mediated by regional climatic conditions, leading to varying outcomes, including growth in glacier size (as in parts of the Karakoram[29]). This means that the role of glaciers as icons in communicating and gaining public legitimacy for the scientific understanding of climate change is a double edged sword, and areas where glaciers are advancing, or not retreating drastically, are used as evidence to counter the reality of man-made Global Warming. This simplistic notion is far from

28 Photo: © M. Nüsser. The area of the lake in front of the main valley glacier has increased significantly during the last decades. The photograph, taken in the early morning light, shows massive glacier retreat which correlates with the growth of the lake and may symbolise the duality of endangered and (potentially) dangerous glacier landscapes.

29 See e.g. Hewitt, Ken, "The Karakoram Anomaly? Glacier Expansion and the 'Elevation Effect', Karakoram Himalaya," *Mountain Research and Development* 25.4 (2005): 32–40; Hewitt, Ken, "Glacier Change, Concentration and Elevation Effects in the Karakoram Himalaya, Upper Indus Basin," *Mountain Research and Development* 31.3 (2011): 1–13.

being just a common assumption, and is even used by Nobel Prize winning scientists:

> Nature's best thermometer, perhaps its most sensitive and unambiguous indicator of climate change, is ice. When ice gets sufficiently warm, it melts. Ice asks no questions, presents no arguments, reads no newspapers, listens to no debates. It is not burdened by ideology and carries no political baggage as it crosses the threshold from solid to liquid. It just melts.[30]

Our argument to the contrary proceeds with the understanding that glaciers do not *just* melt; they are imbued with cultural, scientific, political and aesthetic meanings. Any understanding of the environmental knowledge of Himalayan glaciers, therefore must take the complex negotiations between these different meanings into account – meanings that emerge from the constant contestation between various actors and overlapping domains of knowledge. We therefore focus on opening the black box of the dynamic, and at times contradictory, body of knowledge about Himalayan glaciers.

5 From Cryospheres to Cryoscapes

In order to analyse Himalayan glaciers as products and producers of environmental knowledge, this paper works with the concept of *cryoscapes*. This neologism is derived from the *cryosphere* as an object of scientific study, glaciers as a *living landscape* (as opposed to wilderness), and finally from the emergence of Himalayan glaciers as a dynamic, globally imagined *mediascape* in the sense of Appadurai[31]. The spread of imagined cryoscapes points to the iconic role of glaciers in communicating diffuse threats like global warming. Glaciers exemplify the multiple dimensions of environmental knowledge: scientific, cultural, temporal, spatial and political. In this sense the cryosphere is something that is affected by human actions, but exists independently from it. Cryospheres may also exist where humans do not, as for instance, in case of the planet Venus. However *cryoscapes* are produced through the coming together of human epistemic practices with the physical phenomena that constitute the cryosphere.

The most important framings of cryoscapes and the resulting questions that arise through the study of different dimensions of cryoscapes can be broadly identified as:

30 Pollack, Henry, *A World Without Ice* (New York: Avery, 2009), 114.

31 Appadurai, *Modernity at Large*, 35.

Spatial: The location of the glacier is important not only in understanding the geophysical properties of the space that influence the glacier (longitude, latitude, altitude, slope, shape, etc), but also in determining the cultural response and the symbolic meanings ascribed to glaciers. How can the results of local and regional studies be extrapolated on a Himalayan scale? How do the properties of the site of knowledge production affect the form and content of the environmental knowledge produced?

Temporal: This dimension puts emphasis on change/flux in glaciers. Changes in glaciers over time vary depending on the temporal scale at which they are measured (daily, seasonal, annual, decadal, centennial). Though the availability of data varies widely depending on which scale is chosen, the accuracy of measurement is high, through the use of techniques like remote sensing and repeated field visits (including repeat photography, figures 5.5 and 5.6) to areas under study. How can the different methods of change detection be integrated and standardised to get comparable time series? How can the changing perception of glaciers as dangerous and endangered landscapes be integrated in the research design?

Symbolic/Cultural: This dimension lays emphasis on how glaciers are given meaning, or the kinds of meaning they are given. How do certain environmental or topographical features like glaciers, rivers, mountains, air, etc. become imbued with symbolic meanings like sacredness, purity, and vitality, or become icons of environmental change, etc., which extends beyond their physical properties?

Political: The best example of this is the controversy over the predicted retreat of Himalayan glaciers by 2035 and the resulting politicisation of this scenario. How does relative power influence the negotiations through which consensual environmental knowledge is arrived at, as in the case of the IPCC's consolidation of Climate Change knowledge?

Scientific: The nature of the scientific enterprise, which makes it such a powerful form of knowledge, at the same time turns an admission of ignorance (or worse, error) into a sign of weakness. This power of the scientific framing of the environment makes it essential to examine not just the processes but also the philosophy and practice of scientific production of environmental knowledge, including the economic and competitive peer pressures in research. In addition, due to the previously described resonance of the phenomenon of glaciers, this knowledge must be communicated in easily understandable and generalisable form. Further, the scientific framing of glaciers results from diverse strains of scientific research, with divergent foci, and the resulting understanding of glaciers may often appear fragmented. How can a substantial database for more reliable extrapolations and prognostic scenarios be generated? The scientific framing of glaciers becomes especially important (and difficult), as glaciers are

FIGURE 5.5 *The debris-covered Tap glacier in the Upper Rupal Valley, Nanga Parbat, Northern Pakistan. View from 4,100 m, July 18, 1993.*[32]

seen as an "endangered species"[33] or icons of global environmental change. What are the underlying reasons for paradigm shifts in the interpretation of glacier dynamics?

As is obvious, Himalayan glacier dynamics are prominent features of the current climate change discourse. In this context, the cultural framing does not only include the indicative function of a thermometer but also of a symbol for rapid changes with far-reaching consequences. "The ice-free Himalaya" is a very powerful icon of global warming. Against this background one can observe the extended use of expressions like "dramatic" change or the "fate" of the glaciers, even in scientific papers.[34] Hence, glaciers become personalised and may be perceived to have individual or collective destinies as threatened species. The use of metaphors like "rivers of ice" suggests a similar attempt to bridge epistemic divides and make glacier changes tangible to a global public (figure 5.7).

32 Photo: © M. Nüsser. The proglacial lake in front of this pedestal glacier is located high above the valley bottom and dammed by a terminal moraine.

33 Carey, "The History of Ice," 520.

34 E.g. Bhambri, Rakesh, and Tobias Bolch, "Glacier Mapping: A Review with Special Reference to the Indian Himalayas," *Progress in Physical Geography* 33.5 (2009): 672–704; Bolch et al., "The State and Fate."

FIGURE 5.6 *The replicate of the photo in figure 5.5 shows no significant glacier change over a time period of 17 years. View from 4,100 m, August 18, 2010.*[35]

An excellent example of such an attempt is David Breashears, an American mountaineer and photographer, who has organised a web project called "Rivers of ice: vanishing glaciers of the Greater Himalaya",[36] which uses repeat photography from some prominent Himalayan glaciers.

His idea is to "both translate climate change to the layperson and to provide valuable new data sets for scientists". As an explanation for his efforts he states: "everywhere we go, we see the same message in the glaciers; they were once healthy and 'relative stable' but now they're under a lot of stress. People and scientists are going to have to look at the imagery and decide for themselves". This is yet another example where glacier response to climate change is expected to provide self-evident and easily understandable imagery. Such unequivocal knowledge is unlikely, but importantly there are epistemic permutations like the coming together of powerful economic interests and experts which actively exploit every perceived contradiction to discredit and distort climate change knowledge. The best example of such practices are think

35 Photo: © M. Nüsser.
36 GlacierWorks, accessed December 31, 2012. http://www.glacierworks.org/.

tanks funded by the petroleum industry, which have also suggestively been labelled "sceptic tanks".[37]

FIGURE 5.7 *Impressive or 'fascinating' glacier scenery in the Khumbu Himalaya, Nepal. View from 5,650 m, October 21, 2012.*[38]

6 Conclusion

Glaciers have always been an object of fascination not just for scientists but for many others. In the context of the climate change debate, Himalayan glaciers were rediscovered and captured the popular imagination as threatened landscape features whose disappearance would have devastating consequences. In the context of these "darkening peaks",[39] one can perceive a transformation of glaciers into "scientific laboratories" as part of the "endangered glacier narrative", according to Carey.[40] In spite of numerous research activities over the last years, uncertainties on a Himalayan scale are still prominent, making ex-

37 Dorsey, Michael K., "Climate Knowledge and Power: Tales of Skeptic Tanks, Weather Gods, and Sagas for Climate (in) Justice," *Capitalism Nature Socialism* 18.2 (2007): 7–21.

38 Photo: © M. Nüsser. The photograph shows the West Faces of Everest (8,848 m) in the centre and Nuptse (7,864 m), both feeding the Khumbu ice stream.

39 Orlove, Wiegandt, and Luckman, "The Place of Glaciers."

40 Carey, "The History of Ice," 520.

trapolations and modelling of future scenarios problematic. As climate change and glacier response are both "heterogeneous, oscillatory, and trending", Kargel et al.[41] point out that the larger picture of Himalayan glacier dynamics "is not yet fully assembled". Apart from that, the IPCC glacier controversy has shed light on the political and epistemic dimensions of the contested narrative. The assertion of an "ice-free Himalayan mountain system" resonated with even those not directly affected due to simplistic and geo-deterministic predictions that the resulting environmental catastrophe would have consequences in the form of environmental refugees and water wars. In this context the discussion resembles selected aspects of the Himalayan Dilemma debate from the 1980s and 1990s.[42]

Glaciers can be seen as phenomena that are neither purely natural nor purely cultural. Our approach breaks down the nature–culture dichotomy by pointing to glaciers as phenomena that are produced through encounters of human agency (knowledge production, greenhouse gas emissions, symbolic meaning making) with non-human agency in the form of dynamic geophysical processes (precipitation, ablation, avalanches, etc.). While recognising human agency, this usage additionally gives attention to the non-human agency that influences this practice of meaning making, without privileging either of them. In accordance with the call of Latour[43] for a politics of the environment that respects the entangled dimensions of environmental knowledge, by recognising it as a coming together of people, things, and concepts, our use of cryoscapes aims for a productive middle. It avoids the extreme of social constructivist studies of science, in recognising that the meaning of glaciers is produced through an interaction of diverse social practices with very real physical phenomena: Mass balances of glaciers are not imagined or constructed, but the *meaning* given to them is a result of identifiable social practices. The other extreme that we attempt to avoid is a realist view that all that is necessary is for the scientific findings to be made publicly available and the truth is then self-evident. As we have argued, the differing epistemic practices that go into producing different kinds of environmental knowledge mean that the same evidence will lead to very different interpretations. Against this background, a better understanding of human-environmental interactions in Himalayan studies strongly depends on a productive interdisciplinary context. Finally, developing an

41 Kargel et al. "Himalayan Glaciers," 14709.

42 Ives, Jack, and Bruno Messerli, *The Himalayan Dilemma: Reconciling Conservation and Development* (New York: Routledge, 1989).

43 Latour, Bruno, *We Have Never Been Modern* (Cambridge, MA: Harvard University Press, 1993).

understanding of cryoscapes is no less important than investigations of the cryosphere, because it is the meaning given to glaciers that affects human responses to these changes.[44]

44 Nüsser, Marcus, "Umwelt und Entwicklung im Himalaya: Forschungsgeschichte und aktuelle Themenfelder," *Geographische Rundschau* 64.4 (2012): 4–9 and supplement (map).

PART 3

Present

CHAPTER 6

A Shift in India's Domestic and International Climate Policy?

Joachim Betz

Abstract

India has long been regarded as a deal-breaker in international climate negotiations, hiding in spite of rapid economic growth and diversification behind poorer developing countries. At the 2009 summit in Copenhagen, India first abandoned this strategic line and made a commitment to reduce carbon emissions voluntarily. This shift was preceded and accompanied by an even more pronounced proliferation of domestic initiatives to save energy, to develop regenerative energies, to develop public transport, to increase afforestation and to safeguard biodiversity. The aim of this article is to explain the causes of this perceptible and uneven policy shift. On the international level there were several causes: India intended to be perceived as a responsible actor, deserving a greater say in global governance matters; India also lost takers of its traditional position among developing and emerging countries and ran the danger of becoming isolated. On the domestic level, shrinking national energy reserves and mounting import dependence made co-benefits of energy saving in reducing greenhouse gas emissions evident. The shift was made easier as important business associations aligned with a more eco-friendly development perspective, as environmental NGOs being in favour of proactive climate policies on the domestic front became more vocal – backed by the higher courts – and – last not least – as the international reduction commitments of the Indian government did not demand stringent emission reductions.

1 Introduction

India was and is still perceived as one of the most obdurate opponents of an effective global climate regime that would also impose responsibilities for early mitigation efforts on emerging economies. The government untiringly repeated the mantra of "common, but differentiated responsibilities" (for reductions of greenhouse gas (GHG) emissions) and the necessary transfer of new and additional financial resources and technology at concessionary terms if cooperation on emission abatement was to be expected from developing

 | DOI 10.1163/9789004273221_008

countries. This hard-line posture was coming under increasing attack, especially at and after the Copenhagen summit in late 2009. Opposition was articulated not only from developed countries, most prominently from European Union members and the United States, but also from poorer developing and island states as well as interest groups and social movements inside the country. As a result, the staunch opposition against compromises on the international level has slightly eroded; India was offering voluntary mitigation commitments for the first time (at Copenhagen), much to the chagrin of leftist and ultra-nationalist forces within India. An even sharper turn is visible with regard to national, domestic initiatives to reduce India's carbon footprint through formulation of new proactive climate change strategies inclusive of efforts to lower energy intensity in industrial production, of household appliances and buildings, of efforts to increase the share of renewable energies in total energy consumption, and of efforts to scale up afforestation rates (see below). This is accompanied by the proliferation of new environmental acts, the launching of official missions and institutions at every political level and juridical activism in environmental matters, most prominently from the side of the Supreme and Higher Courts. If we would simply make a head count of new legislative actions and institutional innovations, India would come out in the last few years (since about 2006) as one of the countries most concerned about the state of the environment (see below).

This change of mind with regard to international climate policy and the big shift with regard to internal eco-friendly policies could be explained by (a) learning effects, namely a serious reappraisal of official strategies through the stage-wise acknowledgement that India will be harder hit by the impact of global warming than most other countries and will have to save energy anyway because of mounting shortages and growing import dependence. These learning effects may have been strengthened by discussions with partners from emerging economies and other developing countries in the last three climate summits; (b) the desire to be perceived as a responsible global power by the international community, therefore reinforced by deserving better representation and a greater say in international organisations, reinforced by the feeling of progressive isolation in the negotiation process; and (c) the strengthening of environmental-friendly groups within India.

This shift may alternatively be characterised at least partly as a kind of symbolic politics, as international commitments and domestic actions were not too ambitious – less than could be achieved by using existing technologies and in line with business as usual mitigation efforts. The government may also have been aware that in India, achievements are traditionally trailing behind announcements in changing the energy mix and other climate-friendly actions

– so far that actual downpayments for climate-saving programmes become somewhat reduced. This argument would, however, suppose a dose of political cynicism on the part of the Indian government.

Before entering into the causal analysis, a brief recapitulation of the official position with regard to climate mitigation responsibilities and its empirical foundation is necessary. This will help to understand why its previous hard-line position lost adherents and also assists us in gauging the extent to which the Indian government moved into the direction of a more climate-friendly position. Next we shall review the slow erosion of this position with regard to international commitments and analyse the initiatives and actions the government has taken on the domestic front, thereby also confronting the legislative and institutional activism with remaining shortcomings and deficiencies in implementation. Finally, we will enumerate and assess the weight of causes for the relative shift in India's position on global greenhouse gas emissions from the Copenhagen summit onwards and for the more pronounced domestic activism. This will include speculations about the remaining roadblocks for still more proactive climate policies.

2 India's Official Position in International Climate Negotiations

India's long held position in international climate negotiations was based on a mixture of principled, equity-based and pragmatic considerations:

Responsible for climate change were – according to official Indian statements – accumulated GHG in the atmosphere, emitted primarily (nearly 75 per cent) by developed countries. These countries had therefore the duty to move first – in line with the stipulations of the Kyoto Protocol – before commitments from other, less developed countries may be expected. Industrialised countries also had the duty to support poorer countries in their mitigation efforts by transfer of relevant technologies and financial assistance. Global equity would demand that every citizen has an equal right to atmospheric emissions. India's foremost aim was to eradicate poverty within the next two decades by maintaining economic growth rates of 8 or 9 per cent p.a. (these are the official targets repeated in numerous documents) and – with special emphasis on power – providing every citizen with a lifeline of safe, clean and convenient forms of energy.[1] With sustained growth and the connection of the remaining nearly 40 per cent of the population without access to the grid, power consumption and GHG emissions will necessarily have to increase markedly, but not indefinitely:

1 Government of India, *National Environment Policy 2006* (New Delhi: Government of India, 2006).

> Despite this, India has already declared that even as it pursues its social and economic development objectives, it will not allow its per capita greenhouse gas emissions to exceed the average per capita emissions of the developed countries [...].[2]

The long-term convergence of per capita emissions are seen as the only equitable basis for a global compact on climate change.[3]

Legally, the Framework of the United Nations for Climate Change and the related Kyoto Protocol does not require commitments for GHG reduction by developing countries. As there is no new binding agreement, the international community must stick to the UNFCC and the Kyoto Protocol and should not allow certain countries (i.e. what Australia and the United States were trying during the Copenhagen summit) to insert a new, ungratified formula for sharing responsibilities.[4] This argument is supported by the fact that Annex I (developed countries) failed to honour their reduction commitments under the Kyoto Protocol.

As far as GHG impact *per capita* is concerned, India must still be regarded as a minor polluter on the world scale. In terms of aggregate emissions, however, it ranks already 5th in the emission of GHG behind the United States, China, the European Union (taken in this argument as one block) and Russia;[5] it ranks 4th if only individual European countries are counted. Currently, India contributes around 5 per cent to worldwide GHG emissions, but has added only 2.3 per cent to the stock of GHG in the atmosphere.[6]

India is also still a poor country. Efforts to promote sustainable development have to be subordinated to improving the lot of the masses[7] or at least "[...] efforts to address climate change [...] should not take away resources from the

2 Government of India, Ministry of External Affairs, *The Road to Copenhagen: India's Position on Climate Change Issues* (New Delhi: Government of India, 2009), 3.

3 Ibid.

4 Ministry of External Affairs, "Remarks by PM at the Informal Plenary of HOS/Gs at the 15th COP at Copenhagen," accessed December 21, 2009. http://www.meaindia.nic.in/speech/2009/12/18ss01.htm.

5 International Energy Agency, *World Energy Outlook 2010* (Paris: International Energy Agency, 2010).

6 Dubash, Navroz K., *Toward a Progressive Indian and Global Climate Politics*. Working Paper No. 1 (New Delhi: Centre for Policy Research, 2009).

7 Mukherjee, Pranab, "India and Global Challenges: Climate Change and Energy Security" (address at the Asia Society, New York, September 30, 2009), accessed January 22, 2010. http://www. meaindia.nic.in/speech/2008/09/30ss03.htm.

core development needs and growth objectives of the developing countries".[8] As a consequence of poverty, a significant share of the population is not connected to the electricity grid at all, whereas the rest is consuming very little per head and is in addition plagued by unscheduled outages and fluctuating voltage. Emission-intensive consumption habits and individual car ownership are rare.[9] Therefore, the sometimes-expressed argument of Western media and governments that India was "hiding behind the poor" is unfounded; the so-called middle class in India is still very small.[10]

In spite of not being obliged to any mitigation commitment, efforts to reduce greenhouse gas emissions have already been taken in India: The emission intensity of India's Gross Domestic Product (GDP) declined by more than 30 per cent during the period from 1994–2007, due to private and public domestic efforts.[11] The energy intensity of India's industry was lower than of comparable countries (China) and only slightly above the average level of the European Union.

The arguments enumerated above are only partly convincing for the following reasons:

First, there is no collective legal obligation on the part of developed countries (or of any other sovereign actor) to compensate past wrongdoings in general, and they therefore are not long-term obliged to make extra efforts for saving the world climate by contributing alone or more than others to the global common good. This compensation argument is strictly of a purely moral order, yet is not without persuasive power and influence. It sounds however less convincing if we do consider the actual emissions of India (and emerging economies in general) as well as the growth in emissions of developing countries in the last few years, more so in the foreseeable future (if economic growth stays buoyant). The emissions of India will increase at least threefold until 2035 (without major abatement efforts), as declining energy intensity of production will be overcompensated by rapid growth.[12] India will by then contribute 9.6 per cent to global GHG emissions (up from around 4.8 per cent today) and be the third largest emitter after China and only slightly behind the United States). The remaining atmospheric carbon space if global warming shall not exceed 2 degrees centigrade on average is very limited (around a quarter of already accumulated

8 Prasad, H. A. C., and J. S. Kochher, *Climate Change and India – Some Major Issues and Policy Implications* (New Delhi: Government of India, Ministry of Finance, 2009), 17.

9 Government of India, *National Environment Policy 2006*.

10 Cf. Dubash, *Toward a Progressive Indian and Global Climate Politics*.

11 Government of India, *Greenhouse Gas Emissions 2007* (New Delhi: Government of India, 2010).

12 Cf. International Energy Agency, *World Energy Outlook 2010*.

emissions). Global warming can therefore not be limited to that extent if emerging economies do not contribute to emission reduction efforts very soon, regardless of relative achievements of developed countries in that respect. In addition, mitigation efforts are cheaper in poorer countries for the same amount of emissions avoided, as mitigation efforts become successively more expensive with approaching the technological frontier for mitigation efforts.[13] Finally, India could lower its carbon footprint by around a third by using only existing technologies and without undue hardship in terms of financial expenditures and foregone growth (see below). Therefore, the free riding of India (on the efforts of other countries) is not a good option from the perspective of its own long-term interests.

Second, the Indian government is indeed somewhat 'hiding behind the poverty' of other developing countries with lower per capita growth, lesser export dynamism, far lower foreign exchange reserves and nonexistent industrial sectors for regenerative energies. This has not gone unnoticed by some groups of poorer (e.g. the island) countries, demanding a more compromising stand by India and other emerging powers during the last climate summits. The government is indeed hiding behind the poor in India: The carbon footprint of the richest one per cent of the population (and, to a lesser extent, also of the next four income classes) is not different from the rich in developed countries and certainly unsustainable from a global point of view.[14] The middle class, whose size is often exaggerated in Indian PR publications for foreign investors (comprising allegedly 300 million people), is suddenly shrinking when it comes to GHG emissions. But even 150 million middle-class people would leave a large carbon footprint. In addition, every forecast of India's future growth trajectory predicts a massive growth in this class and a convergence of their consumption habits to those prevalent in developed countries.[15] It is also not true that any investment for a greener future always takes resources away for development and poverty eradication as some experts in India try to make out.[16] First of all, India is already a big supplier of wind and solar energy instalments, so much so that a shift in India and elsewhere towards regenerative energies will bring the country significant economic benefits. Next, there are strong synergy effects

13 World Bank, *World Development Report 2010: Development and Climate Change* (Washington, D.C.: World Bank, 2010).

14 Cf. Greenpeace India, *Hiding behind the Poor* (Bangalore: Greenpeace, 2007).

15 See, for example, Asian Development Bank, *Asia 2050: Realizing the Asian Century* (Manila: Asian Development Bank, 2011).

16 Cf. Gautam, P. K., "Climate Change and India's Position," IDSA Strategic Comments (published December 5, 2007, New Delhi).

between green technology change and general technological progress.[17] Finally, there are massive co-benefits in India of saving and better handling energy supply by reducing transmission losses, theft and waste (encouraged by extremely low tariffs for poor consumers and farmers). Last but not least, India already is and will remain among the countries to profit most from new financing mechanisms (like the Clean Development Mechanism, of which India is the second largest beneficiary after China), as its economy is presenting a host of low-cost mitigation opportunities.

It is true that the energy intensity of India's growth has come down compared to the 1980s and 1990s, coming closer to the average OECD level, but (a) the emission intensity has fallen a bit less, as energy production still relies predominantly on fossil fuels, especially coal with a high ash content (responsible for more than 50 per cent of the total energy demand);[18] (b) the fall in energy intensity has to be regarded against the backdrop of the still rather minuscule industrial sector in India (counting for about 18 per cent of its GDP), which will certainly grow in the next decades. Furthermore (c), against the backdrop of the massive contribution of the low energy-consuming service sector to overall growth since the 1990s, and the still moderate but rapidly increasing energy intensity of consumption in India if prosperity (and therefore car ownership) spreads to larger groups of the middle class as in China; (d) last but not least, the actual energy intensity of growth in official publications is always calculated by using GDP adjusted by purchasing power parity (instead of using the nominal exchange rate), thereby deflating intensity considerably. If nominal parity is used instead, energy intensity of production in India moves up to about four times the level of the United States.[19]

In sum, the arguments advanced by environmentally conservative circles in India (representative of traditional industrial sectors, the Ministry of External Affairs etc.) are as self-serving as those from other actors. This is increasingly criticised by social movements and environmental groups within India.[20] No wonder that these arguments have found fewer adherents internationally.

17 Ellis, Karen, Bryn Baker and Alberto Lemma, *Policies for Low Carbon Growth*, ODI Research Report (London, Overseas Development Institute, 2009); and International Energy Agency, *World Energy Outlook 2010*.

18 Government of India (Planning Commission), *Eleventh Five Year Plan 2007–2012* (New Delhi: Government of India, Planning Commission, 2008); and International Energy Agency, *World Energy Outlook 2010*.

19 Subramanian, Arvind, Nancy Birdsall, and Aaditya Mattoo, "India and Climate Change: Some International Dimensions," *Economic and Political Weekly* 64 (2009): 31; 43–50.

20 Dubash, *Toward a Progressive Indian and Global Climate Politics*; Rajamani, Layanya, "India's Negotiating Position in Climate Change: Legitimate but Not Sagacious," Issue Brief No. 2 (New Delhi: Centre for Policy Research, 2007).

It was, however, not the intellectual flaws in the official statements which led to the gradual and unfortunately still vacillating shift in India's position with regard to climate change.

3 The Policy Shift Before and After Copenhagen

India left its traditional international hard-line position to some extent during and before the Copenhagen conference (in 2009) where a possible successor to the Kyoto Protocol was discussed and negotiated. As the United States were still unwilling to enter the boat and to make binding reduction commitments, Australia proposed a weaker version of a follow-up treaty, removing the differences between Annex I countries and the rest by calling for individual commitments of every emitter and promising in exchange generous mitigation and adaptation assistance. The proposal obviously aimed at engaging the United States, themselves keen to break up the common front of the G77 (the developing countries) who were insisting on commitments by developed countries alone. In spite of having signed a Memorandum of Understanding with China in 2009, stressing the UNFCC and the Kyoto Protocol as the adequate framework for the solution of the global climate problem and agreeing on closer cooperation between the two countries (also in formulating a common negotiation position),[21] and in spite of the promise not to leave the camp of developing countries during the summit, the Indian Prime Minister announced some days before the summit that India was prepared to voluntarily "reduce the emissions intensity of our growth by 20–25 per cent in 2020 as compared to 2005".[22] As this unilateral commitment was accompanied or preceded by similar promises of other emerging economies (Brazil, China, Indonesia and South Africa) who also wanted to avoid the stigma of a deal breaker, it was not too difficult for the US government to arrange a final arrangement (behind closed doors) with the delegations of these countries and India coming very close to its original intentions. Commentators within India deplored that India had thereby left the camp of developing countries and made concessions without asking for reciprocity.[23]

21 Cf. "India, China Sign MoA on Climate Change."

22 Ministry of External Affairs, "Prime Minister's Statement Prior to his Departure for Copenhagen," published December 17, 2009, accessed December 21, 2009. http://www.meaindia. nic.in/speech/2009/12/17ss01.htm.

23 Jayaraman, T., "Will It Be a US Endgame at Copenhagen?" *Economic and Political Weekly* 44.50 (2009): 13–15.

The representative of India at the Cancún summit, the Minister of the Environment, Jairam Ramesh, went even further in accommodating demands for a more proactive Indian climate position. He argued at the conference that every country should accept "binding 'emission reduction' commitments in an appropriate legal form", and that there should be international consultation and analysis of developing country actions in a non-intrusive and non-punitive manner.[24] This came very close to the demand of developed countries for a legally binding commitment by emerging economies, subject to international verification. Under attack at home from the opposition parties and some 'Third Worldist' NGOs (like the Centre for Science and the Environment in New Delhi, the leading environmental Non-Governmental Organisation in India, which is proactive in domestic environmental issues but clinging to the strict adherence to the Kyoto Protocol), the Minister had to clarify that he was only "nuancing" the known position of the Indian government at the conference and that India would not accept internationally binding cuts "at this stage".[25] Jairam Ramesh, while hailed by other representatives at the conference for his bridge-building role, was at a later cabinet reshuffle replaced by Jayanthi Natarajan, an environmentally more conservative lady.[26] She acted more in the usual Indian way as a deal-breaker (supported by the Chinese delegation), insisting before and at the Durban summit of late 2011 on the enduring importance of the Kyoto Protocol,[27] therefore rejecting legally binding agreements for developing countries. From India's perspective, the argument that the world has changed since the agreement of the Protocol in 1997, as representatives of developed countries argued, and that emerging economies therefore would have to shoulder greater responsibilities, did not apply.[28]

4 Domestic Environmental and Climate-related Policies

The forward-and-backward-changing international posture of the Indian government was, however, internally preceded by a more proactive domestic

24 Pande, Varad, "India at Cancun: The Emergence of a Confident Dealmaker," *Economic and Political Weekly* 56.6 (2011): 14–15.

25 Cf. "On Commitments, Jairam Drops a Line, Eats Words," *Express India*, December 10, 2010.

26 Cf. Gopalakrishnan, Manasi, "Indian PM Criticized for Unsatisfactory Reshuffle," *Deutsche Welle Asia*, July 12, 2011.

27 Cf. "West's Prescription No Panacea," *The Hindu*, December 6, 2011.

28 Mahnke, Eva, "Muskelspiel der Schwellenländer," published by Klimaretter.Info, December 7, 2011.

energy and climate policy. This does not mean that environmental initiatives were completely absent before. India is one of the very few states where environmental protection is a 'duty' of the state and its citizens;[29] this constitutional provision is supplemented by specific laws, ordinances and regulations, prescribing environmental clearance for every major project and accompanied by a multi-tiered institutional framework of central and state ministries for the environment, pollution boards etc. By any benchmark, India had an extensive environmental management system to implement and enforce the respective policy objectives even before the onset of a new wave of activism, from around 2006 onwards. The armada of specific laws and institutions newly introduced, combined with former ones, was absolutely equal to those adopted in developed economies:[30]

– The National Environmental Policy, 2006, provides the basis for the integration of environmental considerations in all public development activities and serves as a guide to action in regulatory reform, programmes for conservation and enactment of legislation. The new policy promises among other actions the streamlining and decentralisation of environmental clearances, a revisiting of coastal regulation zones, the creation of Environmentally Sensitive Zones, a better monitoring of environmental compliance, the promotion of standardised environmental accounting practices (including the introduction of fees for access to specified natural resources), the increase of forest and tree cover to one third of the country's land area until 2012, the expansion of areas for wildlife conservation, the encouragement of efficient use of groundwater (supported by revised electricity tariffs), special protection of the mountain ecosystems, the abatement of air, soil and water pollution and the promotion of energy efficiency, the introduction of an environmental management system in large enterprises, eco-labelling and certification of industrial products and financial support to promote shifts to clean technologies.[31]

Stakeholders from quite a few sections of society did participate in the framing of the National Environmental Policy, which will be examined for revision every three years with a group of researchers, industry associations, and community-based and voluntary organisations. Since 2006, there is also a National Committee to Assess the Impacts of Climate Change, which includes climate

29 Sims, Holly, "Population, Urbanization, and the Environment," in *Understanding Contemporary India*, ed. Sumit Ganguly und Neil DeVotta (New Delhi: Viva Books, 2003), 157–189.

30 Cf. World Bank (South Asia Environment and Social Development Unit), *India: Strengthening Institutions for Sustainable Growth: Country Environmental Analysis* (Washington, D.C.: World Bank, 2008).

31 Cf. Government of India, *National Environment Policy*.

experts, hydrologists and energy economists, as well as representatives of key ministries. This committee is also assessing options to mitigate climate risks. At the national level, the integration of climate change in national development is guided by the Prime Minister's Council on Climate Change, comprising representatives of ministries as well as experts, and representatives of industry associations and the media.[32]

– The Integrated Energy Policy was adopted in 2006. Some of its key provisions are the promotion of energy efficiency in all sectors of the economy, an emphasis on mass transport, on producing biofuels on a far larger scale, the accelerated development of nuclear and hydropower, and the increase of public funding for R&D on several clean energy related technologies. In the aftermath of the submission of the New Integrated Energy Policy, a new codex for energy conservation in official commercial buildings was introduced (2007). Large firms are required to prepare energy audits.[33]

– The National Urban Transport Policy stresses a shift to broad-based public transport facilities and away from motorised modes of individual transport. The government has the intention to allocate road space on an equitable basis, with people at its focus. The shift to public transport by municipalities will be supported by contribution from the central government. The expansion of the metro rail and bus systems in Delhi and Bangalore are steps in its implementation.[34]

– An energy-labelling programme for appliances was launched in 2006; labelling has been introduced for fluorescent tube-lights, air conditioners, and distribution transformers.

– The percentage of electricity that electricity distribution companies must procure from renewable sources at preferential prices was raised in 2006. The management of new power plants was encouraged to adopt more efficient, super-critical technologies.

– An energy conservation building code for new large commercial buildings was launched in May 2007, which aims to optimise the buildings' energy demand.

32 Government of India, *National Environment Policy*. Cf. also Government of India (Ministry of Power, Bureau of Energy Efficiency), *India: Addressing Energy Security and Climate Change* (New Delhi: Government of India, 2007).

33 Government of India, *Integrated Energy Policy: Report of the Expert Committee* (New Delhi: Government of India, 2006).

34 Government of India, *National Urban Transport Policy* (New Delhi: Government of India, 2006).

– The conduct of energy audits was made mandatory in large energy-consuming units in nine industrial sectors (2007). The companies covered must employ certified energy managers, report energy conservation data annually and have to adhere to energy consumption norms specified by the government. Less efficient energy companies have to make a greater effort. A Perform, Achieve and Trade mechanism has been introduced, allowing firms that move beyond targets to sell savings certificates to the others. Fiscal incentives for energy-saving investments were implemented to reduce risks, covered by the Credit Guarantee Trust of India who will also function as a venture capital provider for those investments.[35]

– The National Action Plan on Climate Change (2008) is based upon the National Environmental Policy, but is more explicit by asking for a national focus in combating climate change, and is more implementation-oriented. Eight national missions form the core of the Action Plan; two of them (the National Solar Mission and the National Mission for Enhanced Energy Efficiency) have become active and have already launched specific programmes. Representatives of academia, experts from industry and the civil society have been co-opted into these missions.[36]

– An expert group of the Planning Commission of India prepared a relatively detailed (provisional) plan of low-carbon strategies for specific industrial sectors (steel and cement), appliances (bulbs, air conditioners, refrigerators etc.) and a shift to non-motorised and public transport in cities, sufficient to reduce India's CO_2 emissions by as much as 33 to 35 per cent till 2020 compared to business as usual.[37]

5 Deficiencies in Policy Implementation

This is a really impressive list, which could even be extended. But there is no reason to become overly enthusiastic if we take the fate of past efforts as a yardstick for the future:

---

35 Government of India (Ministry of Power), *National Mission for Enhanced Energy Efficiency: Draft Mission Document: Implementation Framework* (New Delhi: Government of India, 2008); and Government of India, *India: Addressing Energy Security*.

36 Cf. Government of India, *National Action Plan on Climate Change* (New Delhi: Government of India, 2007).

37 Government of India (Planning Commission), *Low Carbon Strategies for Inclusive Growth: An Interim Report* (New Delhi: Government of India, 2011).

(a) The central Ministry of Environment and Forests is still rather weak, although until recently headed by an experienced politician. He was, however, replaced by a politically weaker, more traditionalist successor. Several other ministries are implied in the protection of the environment without provision of any crosscutting coordination mechanism. Institutional fragmentation also characterises India's energy and energy saving policies: At the moment, four different ministries are directly involved in the energy sector. Here, at least, a coordination mechanism was created in 2005.[38]

(b) The aim of starting new eco-friendly initiatives in India is often more the saving of energy than the protection of the environment or saving of the global climate. This is not inherently bad if dual win-win solutions are still possible. In cases where there is severe contradiction of these two goals, however, energy security takes precedence.[39] Reliance on coal (which in India has a high ash content) as the principal energy source, for example, is projected far into the future, and the expansion of nuclear energy sources will be vastly expanded with only a very short interruption after the Fukushima disaster etc.

(c) Protection of the domestic ecosphere was effected until 2010 according to the command and control method alone, with rather blunt instruments at their disposal as, for example, the threat to close down overly polluting industrial units by court order, rather than by incentives like the introduction of tradable certificates and tax rebates/subsidies for mitigation efforts, or by encouraging the participation of civil society (to act as watchdogs or awareness raisers). This has changed, albeit only to some degree; the certification process just started, and participants of national missions from the civil society are still handpicked.[40]

(d) Environmental regulations were not enforced strictly in India; of all the companies under the control of pollution boards, 50 per cent did not comply with regulations. In addition, only the larger factories were monitored, although medium and smaller units are responsible for 70 per cent of the industrial emissions. Sanctioning for flouting rules could only take two forms, namely placing the units under trial (pollution boards cannot prosecute violations on their own) or closing them. Closure is a rather blunt instrument in a country with

38 Betz, Joachim, and Melanie Hanif, *The Formation of Preferences in Two-level Games: An Analysis of India's Domestic and Foreign Energy Policy*, GIGA Working Papers No. 142 (Hamburg: GIGA, 2010).

39 Chikkatur, Ananth P., and Ambuj D. Sagar, *Positioning the Indian Coal-Power Sector for Carbon Mitigation: Key Policy Options*, White Paper Series (Arlington: PEW Center Global Climate Change, 2009).

40 Organization for Economic Co-operation and Development (OECD), *India: OECD Economic Surveys* (Paris: OECD, 2007).

few attractive jobs and poor networks for social protection, and charge-sheeting these units is also ineffective, as Indian courts are notoriously overburdened and therefore extremely slow to settle cases. Therefore regulating bodies often prefer to do nothing.

(e) Monitoring bodies for environmental regulations are inadequately staffed and financed, their employees insufficiently trained and legally not well versed.[41]

(f) Program targets for the replacement of fossils by renewable energy sources for energy saving, reforestation etc., are not very ambitious if we compare them with their Chinese counterparts. To give only one example: The Planning Commission (2008) projected the share of regenerative energies in the energy mix to rise only to 5 or 6 per cent by 2032,[42] which is less than half of China's target. Additionally, achievements have often been trailing far behind schedule, especially with regard to the development of hydro and nuclear power, the reduction of transmission losses in electricity or the passing on of higher international oil and gas prices to the Indian consumers.[43] Some of the new initiatives simply remained stuck; from the eight missions envisaged in the National Action Plan on Climate Change (of 2008), only two – the National Solar Mission and the Mission for Enhanced Energy – became really operational until 2011. Delays were explained by the lack of a strong agency in charge of renewable energies, multiple and overlapping incentives, uncoordinated state policies, protests by environmental organisations and affected groups in the case of hydropower and overly cumbersome clearances for new projects.[44]

Therefore, the rhetorical ambition of the Indian government with regard to its environmental and climate strategy has to be qualified by the lower actual achievements and mentioned institutional deficiencies, which cannot be cured so easily. Therefore, we should not expect that the new programmes have the quick and drastic effect of lowering emissions compared to business as usual.

41 World Bank, *India: Strengthening Institutions.*

42 Government of India (Planning Commission), *Eleventh Five Year Plan.*

43 Group Centennial, *India 2039: Unleashing an Energy Revolution*, Policy Paper 4 (Manila: Asian Development Bank, 2009); Ramana, M. V., *The Future of Nuclear Power* (Philadelphia: Center for the Advanced Study of India, India in Transition, 2010); and Government of India (Planning Commission), *Low Carbon Strategies.*

44 Alagh, Yoginder K., "Energy and High Growth in India: Sustainability Issues," in *India's Economic Future: Education, Technology, Energy and Environment*, ed. Manmohan Agarwal (New Delhi: Social Science Press, 2009), 219–246; Group Centennial, *India 2039*; World Bank, *World Development Report 2010*; and Government of India (Planning Commission), *Mid Term Appraisal for the Eleventh Five Year Plan 2007–2012* (New Delhi: Government of India, Planning Commission, 2011).

If we assume (as we should) that official circles are aware of the usual Indian implementation gaps caused by official circles, the rhetorically sharp policy shift transforms into a more incremental change. We may nevertheless wonder why the government launched these multiple initiatives domestically and why it was showing at least some flexibility in international climate negotiations while it was long regarded as one of the principal deal-breakers. We may also ask why the new stand was not maintained consistently but put into question at the first resistance. These questions are not easy to answer.

6 Causes of the Climate Policy Shift

Summing up the position of the Indian government with regard to climate change, there was a shift from "a growth-first stance" to a "progressive realist" view.[45] It is not so easy to spot the root causes of India's shift in climate policies, as political actors like the national parties remain rather silent on this issue (see below) and as only few civil society associations publish their environmental positions. Additionally, their influence on the policies of the Indian government is difficult to assess. The slow erosion of the official position on climate policy can also be explained only with some difficulties with one of the traditional theoretical approaches to international relations in isolation. A neorealist position would have to clarify why India, which is becoming an economically and militarily stronger power, and which is at the same time not yet fully satisfied with its status in the international pecking order, should agree to compromises. More so when the strongest polluter and lone but slightly declining superpower (the United States) stood stubbornly aloof of the Kyoto Protocol and was therefore in a weak position to demand flexibility from other powers. An institutionalist approach with regard to India's climate policies would suffer from the fact that India has become far more integrated into the world market, that it has been granted a better position and greater say in international forums and institutions (most notably in the classical Bretton Woods institutions) and that it has been co-opted into the core negotiating circles of international climate, trade and finance policies, e.g. into the G8 and G20, as well as the core negotiating groups of the Doha Round. It could have been expected that India would have simultaneously developed a more pronounced global outlook and be prepared to shoulder a larger burden for the provision of global goods. So why did India not compromise earlier and more consistently in climate policies? A liberal approach has to struggle with the still modest (but growing) influence of

45 Dubash, *Toward a Progressive Indian and Global Climate Politics*, 15.

environmental and other interest groups on the technical aspects of India's climate policies and the opposition of most of them to any Indian compromise on the question of internationally binding commitments (see above). Last but not least, a constructivist approach would be justified by the acceptance of the core beliefs of international climate expertise (climate change is man-made, will have severe repercussions on agricultural production, poverty etc.), but not by any convergence of the Indian position to that of developed countries with regard to responsibility to move first or on burden sharing for mitigation and adaptation costs.

A common problem of these approaches is that they are dealing with only one part of the political equation, namely the international front, where the country has to deal with a more or less powerful coalition of other states backing or opposing its policies – a coalition, which, in addition, does not remain stable or always similarly united. The other and most often-neglected part of the equation is the domestic front, where the structure and relative power of coalition or veto powers against any change differs. It is obvious that on this front the government was able to implement a host of energy-saving and eco-friendly measures, although progressive rhetoric and planning went ahead of actual implementation and some difficult structural reforms (e.g. energy pricing and subsidies) were hardly tackled. It is nevertheless obvious that domestic actions were bolder than the international concessions the Indian government had to offer. The country's climate policies therefore have to be analysed by using a two-level approach[46] and by splitting climate-related reforms into more or less controversial ones.

Starting the explanation with the international level, the government's negotiating position was first affected because its traditional stance found fewer adherents, especially among its fellow developing partners. In view of the serious deterioration of the global climate situation (or at least the perception of it) and the shrinking global carbon space, emission reduction goals slowly took precedence over equity considerations in the international debate. Equity in the sharing of climate space means that the citizens of all countries have the right to more or less equal emissions of greenhouse gases. India was the main proponent of this position.[47] The government realised that its corresponding rhetoric lost credibility in view of India's growing prosperity (also in view of the unequal internal distribution of carbon space) and its rapidly expanding emissions, and only served as a pretext for Annex I (the developed) countries to hide

46 Putnam, Robert D., "Diplomacy and Domestic Politics: The Logic of Two-Level Games," *International Organizations* 42.3 (1988): 427–460.

47 Subramanian, Birdsall and Mattoo, "India and Climate Change."

behind 'deal-breaking' India and other emerging economies in their turn. This strategy was followed most prominently by the US government during the Copenhagen and Cancún summits. It became increasingly clear that by offering a pretext for recalcitrance to others, nothing at all would be achieved. As other emerging powers, especially China, came forward with voluntary reduction commitments for GHG emissions, and as partners in the BASIC Group (acronym for the group comprising Brazil, China, India and South Africa) left the common front – Brazil and South Africa became supportive of a legally binding agreement from the Cancún summit onwards – and finally, as India and the other emerging powers came under attack from the Alliance of Small Island States (AOSIS, comprising 43 member states), the group of African countries and even its own neighbours (Bangladesh, Bhutan, Maldives and Nepal) for their obstinacy, India could not maintain its traditional stand for fear of progressive international isolation: "We are not the bad guys, that's very important for India not to be isolated in any of these international forums and we have a larger stake in the world economy today than we had perhaps 25 years ago", the Minister of the Environment was quoted as saying.[48]

A second cause for flexibility at the international level is the desire of the Indian government to be perceived as a responsible member of the international community. This has been articulated with increasing frequency by its representatives in the last couple of years.[49] A special instance where responsibility was stressed was the conclusion of the US-India Civil Nuclear Cooperation Agreement (2005/08), going along with de facto admission to the nuclear club. Next, during India's participation in the creation and growing weight of the G20 during the global financial crisis in 2008/09 and on other occasions, India criticised the missing representativeness of the UN Security Council and of voting rights within the International Monetary Fund (IMF) and the World Bank. With power and admission to the inner power circle unavoidably comes responsibility, especially the preparedness to pay the price for admission.[50] Responsibility with regard to international climate policy comprises the admission that India is already a major emitter and therefore should shoulder a greater burden for mitigation of climate change, and, secondly, as the country is becoming more prosperous, without undue demand for financial support by

48 Cf. "On Commitments."

49 Ministry of External Affairs, "Remarks by PM."

50 Atteridge, Aaron, "Shifting Sands: India's New Approach to the Politics of Climate Change," Policy Brief, published by Stockholm Environment Institute, 2010; and Rastogi, Namrata Pastodia, "Winds of Change: India's Emerging Climate Strategy," *The International Spectator* 46.2 (2011): 127–141.

the international community for these activities. This implies that India is already a special case, not to be mixed up with poorer developing countries that also should be left with some remaining carbon space. This is, however, a conclusion the Indian government does not like to draw, as it always portrayed the country as being an integral part of the large group of poorer countries to be rightly privileged (i.e. entitled to ask for special, preferential treatment). The loss of this status and of alliance partners from the mass of other developing countries would most probably weaken its resistance power when put under pressure from the major developed countries in future climate rounds to make even more demanding commitments.

A part of the international puzzle is also explained by the fact that India did not promise too much in announcing its willingness to reduce the emission intensity of growth by 20 to 25 per cent until 2020 during the Copenhagen summit and afterwards. This would nevertheless imply, if actual growth rates in India can be maintained, that total emissions will double until 2020 and grow threefold until 2035.[51] This goal can be achieved if all the various departments and ministries simply continue their business-as-usual strategies.[52] All that is needed is carrying on with the existing efforts. If however the efficiency of power plants and fuel consumption of vehicles in India will be raised to the international average, if transmission losses can be curtailed to the best practice of Indian states, if the share of public transport be increased and some other (cost efficient) measures implying no international transfer of technologies can be implemented, emissions from India could be cut down by 35 per cent according to the calculations of the government.[53] Therefore, it would also have been possible to promise larger emission reductions in Copenhagen and afterwards without sacrificing growth unduly.

The gap between international commitments and domestic capabilities can also be demonstrated by (a) first looking at the historical development of emission intensity in India: It declined by 17.6 per cent from 1990 to 2005 *without* any special action programme operating; just by maintaining this ratio of emissions to growth, emission intensity will decline by a further 17.3 per cent by 2020 over the 2005 base.[54] Similarly, if only modest efforts are made to increase the en-

51 Rai, Varun, and David G. Victor, "Climate Change and the Energy Challenge: A Pragmatic Approach for India," *Economic and Political Weekly* 44.31 (2009): 78–85; and International Energy Agency, *World Energy Outlook 2010*.

52 Dutt, Gautam, "Reaching a Climate Agreement: Beyond the Copenhagen Accord," *Economic and Political Weekly* 45.17 (2010): 32–37.

53 Government of India, *National Mission for Enhanced Energy Efficiency*.

54 Government of India (Planning Commission), *Mid Term Appraisal*.

ergy efficiency of industrial production, of household appliances and waste management, emission intensity can be reduced by 25 per cent up to 2020 and by 35 per cent if more aggressive efforts are taken[55] – efforts whose scope and depth were termed as still rather unambitious by independent observers.[56] In conclusion, the Indian government is committing less in international negotiations than it is willing to push through domestically and far less what would be in reach, if a determined effort was made. This is, however, no special Indian idiosyncrasy. The other main emitters in the less developed world (China, Brazil, South Africa, Indonesia) are acting more or less likewise. The reason is very simple: By promising less than what could be achieved, leeway is kept for potential further demands internationally and leeway is kept for potential resistance against a major domestic backlash by affected interest groups or underachievements relative to the planning targets.

There is finally a dwindling hope (upheld by the government until the Copenhagen summit) that a solution to India's environmental and climate problems could be achieved by major concessions and commitments (for mitigation efforts of poorer countries plus transfers of concessionary capital and technology) from established economies. India already benefits disproportionally from the Clean Development Mechanism (financing emission-saving projects in developing countries) and cannot expect to be a major receiver of the Green Climate Fund agreed during the Cancún summit in 2010. This fund in addition will also not be operational before 2020 in a meaningful sense (e.g. staffed with adequate funds), after the date where decisive actions would keep global warming below two degrees centigrade and it is also still unclear who will provide the necessary means (of US$ 100 billion annually). Consequently major self-help actions are called for by India and other emerging economies.

Moving to the domestic level, the Indian government and major and/or influential social groups do acknowledge – at least in the past couple of years – that India will be among the countries most severely hurt by climate change. After long denying or at least playing down the impact of global warming on its own country (even the melting down of the Himalayan glaciers was disputed), the government today admits that

> India has a vital stake in the success of the negotiations [in Copenhagen] as we are among the countries most likely to be severely impacted by

55 Government of India (Planning Commission), *Low Carbon Strategies.*

56 Centre for Science and Environment, *Comments on Interim Report of Expert Group on Low Carbon Strategies by the Planning Commission of India* (New Delhi: Centre for Science and Environment, 2011).

> climate change. We have therefore adopted and started to implement a major National Action Plan on Climate Change, relying upon our own resources.[57]

India is indeed highly vulnerable to climate-related events. More droughts and flooding will affect agriculture, which supports the livelihood of around 60 per cent of the population. The melting of the Himalayan glaciers will affect water supply, and with the mounting water table of the oceans, parts of India's coastline will be submerged, leading to a massive loss of habitat.[58] Worst hit by climate change will be the poor, first of all because of the fall in agricultural productivity, only partly offset by higher cereal prices, which, however, will affect the urban poor. A scenario developed by some World Bank experts foresees an increase in the national poverty rate of 3 to 4 per cent up to 2040 compared to the counterfactual of zero warming.[59]

Next, India's energy supply is all but guaranteed. If economic growth follows the trajectory of the last few years, energy demand will increase fivefold until 2030–2031. Energy demand is met predominantly by domestic coal reserves but increasingly by oil and gas imports. Other rapidly growing emerging and also established economies are competing for the slowly increasing supply of oil and gas, which will have reached its peak in a decade, according to serious forecasts.[60] Therefore, India has (a) to save energy at any cost:

> Needless to say, even if there were no climate change arguments, considerations of energy security alone would require a medium-to-long term strategy of implementing a strategic shift from fossil fuels to non-fossil fuels [...]. It is because we recognize the linkages between climate change and energy security that we have adopted a National Action Plan on Climate Change [...].[61]

---

57 Ministry of External Affairs, "Remarks by PM."

58 Fujiwara, Noriko, and Christian Egenhofer, "Understanding India's Climate Agenda," CEPS Policy Brief No. 206 (Brussels: Centre for European Policy Studies, 2010).

59 Jacoby, Hanan, Mariano Rabassa, and Emmanuel Skoufias, *Distributional Implications of Climate Change in India*, Policy Research Working Paper No. 5623 (Washington, D.C.: World Bank, 2011).

60 International Energy Agency, *World Energy Outlook 2010*; and International Energy Agency, *World Energy Outlook 2011* (Paris: International Energy Agency, 2011).

61 Ministry of External Affairs, "India and Global Challenges: Climate Change and Energy Security" (address by Shri Pranab Mukherjee at The Asia Society, New York, September 30, 2008), accessed March 11, 2009. http://www.meaindia.nic.in/speech/2008/09/30ss03.htm.

India must also (b) develop other energy sources, although coal will remain the mainstay of the economy for the foreseeable future.[62] Hydro, wind, solar, nuclear power and the use of biomass are the most obvious candidates for replacing fossil fuels. Their larger use will necessarily lower emissions compared to business as usual. Climate-friendly policies will therefore be to a greater part the byproduct of policies trying to ensure energy security for India, and of developing an internationally competitive supplier industry for non-conventional energy appliances.[63]

This also explains a remarkable shift in the climate stance of Indian business associations, who, before long, were also staunch opponents of any sacrifice of growth for the benefit of the environment. There are still some associations (most notably Federation of Indian Chambers of Commerce and Industry (FICCI)) opposing any compromises (for example, binding international reduction commitments), but 85 per cent of companies interviewed by the more progressive business association (Confederation of Indian Industry) were in favour of absolute emission reductions by India, and two thirds of company representatives did not like the government's stance in international negotiations.[64] More than 90 per cent of companies surveyed by an American consulting firm were already engaged in green initiatives, moved by regulatory compulsions or the desire to build a stronger brand with consumers.[65] In addition to private companies asking for a more responsible attitude from the side of the government – certainly also motivated by realising attractive business opportunities, as India is becoming a more competitive supplier of technologies for renewable energies – a host of eco-friendly NGOs has sprung up in India, opposing environmentally unsound projects, deforestation, "land-grabbing" for infrastructural programmes, mining and industrial sites or damage to biodiversity. They are supported by a proactive judiciary enforcing environmental regulations even by administering the closure of factories.[66]

Regarding business opportunities, we have to bear in mind that India's wind power capacity is already ranking fourth globally and is expected to grow

62 Rastogi, "Winds of Change."

63 Wheeler, David, and Saubrah Shome, *Less Smoke, More Mirrors: Where India Really Stands on Solar Power and other Renewable*, Working Paper No. 204 (Washington, D.C.: Center for Global Development, 2010).

64 Federation of Indian Chambers of Commerce and Industry (FICCI), *FICCI'S 100 Day Action Agenda for the New Government* (New Delhi: FICCI, 2009).

65 Bhattacharya, Arindam, Rahul Jain, and Amar Choudhary, *Green Manufacturing: Energy Products and Processes* (Boston: The Boston Consulting Group/Confederation of Indian Industry, 2011).

66 Dubash, *Toward a Progressive Indian and Global Climate Politics.*

fourfold by 2030. India also has a sizeable capacity for photovoltaic cells and is endowed with vast solar energy potential.[67] The objective of the government is to establish India as a global leader in solar energy by creating the policy conditions for its diffusion across the country, although unit costs of solar energy are still very high compared to conventional sources.[68] To conclude with the words of the Confederation of Indian Industry: "Green has moved from being a 'necessary evil' to being seen as 'good business'."[69]

The shifting internal balance on environmental issues has already led to a modest impact on the programme of political parties. The environment was not an important issue for political parties formerly. It appeared more prominently in the election manifestos of national parties for the first time in 2009. The weakest effort was taken in the programme of the Congress Party, where the state of the environment and climate problems were dealt with on one half of a 21-page manifesto. In the manifesto of the Bharatiya Janata Party (BJP), they got a slightly more prominent place (1.25 of 48 pages); in this publication, and in the manifesto of the (Marxist) Communist Party, the reduction of GHG emissions was promised, as well as support for renewable energies, an increase in energy efficiency and even a reduction of fertiliser subsidies (BJP).

There is also a growing recognition within the government and Indian society that quite a few environmental problems are induced by unsound domestic policies. The following paragraph would have been unthinkable in earlier official publications:

> The proximate drivers of environmental degradation are population growth, inappropriate technology and consumption choices, and poverty [...] leading to development activities such as intensive agriculture pollution industry, and unplanned urbanisation. However, these factors give rise to environmental degradation only through deeper causal linkages, in particular, institutional failures, resulting in lack of clarity or enforcement of rights of access and use of environmental resources, policies which provide disincentives for environmental conservation (and which may have origins in the fiscal regime), market failures (which may be

67 Group Centennial, *India 2039*; and Confederation of Indian Industry, *Renewable Energy: The Next Wave* (New Delhi: Confederation of Indian Industry, 2010).

68 Government of India, *Jawaharlal Nehru National Solar Mission: Towards Building Solar India* (New Delhi: Government of India, 2009).

69 Confederation of Indian Industry, and The Boston Consulting Group, *Green Manufacturing: Energy, Products and Processes* (New Delhi: Confederation of Indian Industry, The Boston Consulting Group, 2011).

linked to shortcomings in regulatory regimes), and governance constraints.[70]

These shortcomings are later specified as subsidies for water, power, fertiliser and pesticides, overgenerous support prices for water-intensive crops, insufficient empowerment of local institutions, insufficient incentives for forest conservation etc.[71]

It is indeed no secret that transmission losses in electricity supply amount to 34 to 40 per cent of power generation,[72] and that a good part of these losses is caused by simple theft, deficient metering and defunct transmission lines. In addition tariffs for farmers are set so low that they have few incentives to save energy, but instead install always bigger pump sets on their fields, thereby depleting the water table, contributing to waterlogging and salination. Low tariffs for farmers and poor consumers are only partly compensated by much higher ones for private enterprises. Consequently, the State Electricity Boards are virtually insolvent, unable to finance maintenance or extension of grids and this in turn deters investments in power generation. India is the country with the highest net energy subsidies worldwide (US$ 21 billion per year). Slashing of subsidies would have a substantial effect on the country's CO_2 emissions[73] without having a negative effect on income distribution, as well-off farmers are the main beneficiaries. A similar story could be told about the effects of water and fertiliser subsidies on the environment.

Therefore, a good part of energy waste, ecological degradation and unnecessary GHG emissions is caused by populist policies, whose declared aim – protecting the poor – has more of an alibi function.[74] This does not mean that these policies are therefore easier to abandon. Quite to the contrary: Farmers in general and the more prosperous among them are a decisive 'vote bank'. Their political representation – the regional parties – have become indispensable partners in the coalition governments during the last one and a half decades at the centre. As small partners of a dominant party (either the Congress or the BJP) they have no real stake in any deal on global emission cuts, but a large one in maintaining entitlements for the groups they represent. We all know that it

70 Government of India, *National Environment Policy*, 24.

71 Ibid.

72 Government of India (Planning Commission), cf. fn. 54.

73 Cf. International Energy Agency, *Analysis of the Scope of Energy Subsidies and Suggestions for the G-20 Initiative* (Paris: International Energy Agency, 2010).

74 Cf. Government of India, *Jawaharlal Nehru National Solar Mission*; and Lal, Sumir, *Can Good Economics Ever Be Good Politics? Case Study of India's Power Sector*, World Bank Working Paper No. 83 (Washington, D.C.: World Bank, 2006).

is far easier to introduce subsidies than to abolish them – more so if benefits have already had a long tradition. This might partly explain why India's international climate stance was proceeding less than straightforwardly. A second reason for the meandering course may be the traditional foreign policy baggage of defending hard-won sovereignty, resistance to the pressure of established powers and aligning with the rest of poorer countries.[75] Although this position is finding successively fewer takers it is still of some use when it comes to international burden sharing. It remains to be seen, how long this position can be upheld in international climate negotiations.

7 Conclusion

There was indeed a shift in India's climate policies, far more at the domestic level than internationally. This testifies that only the simultaneous consideration of national as well as international political exigencies makes sense when trying to explain a country's international climate (and most probably also other international) policies. We can, of course, identify also a certain softening of India's position in international negotiations. While traditionally resisting the demand for any commitments by poorer countries, India approved the Copenhagen accord requiring all the major developing countries to implement nationally appropriate mitigation actions in 2009. It later moved even to the point of implicitly accepting binding reduction commitments in an appropriate form and monitored by joint bodies. This more compromising attitude was furthered by a mixture of motivations: The growing international isolation of India in climate negotiations coupled with the desire to be regarded as a responsible power, deserving more influence in global governance matters and the realisation that its traditional stance not only found fewer adherents but would present an alibi for other major emitters to do nothing and thereby render India even more vulnerable to climate change. There is hardly a chance for India's government to step back completely behind its newly-found pragmatism, unless the other emerging powers (especially China) do so and unless developed countries do not live up to their commitments any more. Unfortunately, this is, however, happening to some degree, as we witnessed during and after the Durban conference in late 2011. To explain India's position in international climate negotiations, the best candidate would therefore be some kind of rational

75 Narlikar, Amrita, *New Powers: How to Become One and How to Manage Them* (London: Hurst & Company, 2010).

choice realism, equipped with a long-time horizon and taking into account the repercussions of climate policies on India's overall standing.

This approach, however, has to be turned around when it comes to India's domestic energy and climate policies. We have first to notice that India was making less wide-ranging concessions on climate mitigation efforts in international conferences than would have been possible with regard to the likely results of its domestic actions for energy saving, afforestation etc. A possible explanation in accordance with the two-level analysis already mentioned[76] is that the government wanted to keep enough leeway for economic growth, the traditionally slow implementation of policies on the ground and potential resistance of interest groups affected by the changing climate policy course. In fact, the still somewhat rigid but in total more flexible international position is accompanied by unprecedented and serious internal efforts to reduce the energy intensity of industrial production, of home appliances and office building, to improve the efficiency of power generation and transmission, to push the supply of renewable energy and to make agriculture, consumption of the masses and individual traffic more eco-friendly. It would be not too difficult to explain this domestic activism by the threat of growing energy shortages, growing dependence on unstable foreign energy suppliers and the impact of global warming on India. We must, however, note that the bright picture of new climate initiatives within India is negatively affected by slow and intermittent progress in reducing energy, water and fertiliser subsidies with a considerable effect on the amount of India's GHG emissions and by significant underachievements in some climate-related policy fields. These deficiencies have less to do with the resistance of industry associations, who have become more enthusiastic about eco-friendly policies because of their profit potential for Indian companies, and more with the decisive weight of voter groups likely to be negatively affected by ecologically sound policies. Summing up, a mixture of a liberal and a rational choice (realist) approach would perhaps best explain the uneven picture of domestic climate policies.

The treatment of India's climate policies as a two-level game does not only make sense because of the differing strength and composition of forces at the tow levels opposing or backing a policy shift, but also because of the diverging payoffs on India's actions at these levels. Domestic initiatives can be started alone; their benefits are reaped directly by India, although there would be positive spillovers for other countries. Chances for free riding and the blaming of others for 'inaction' or costs imposed on citizens are, in addition, limited. Last but not least, and strengthening the case for splitting the climate game into a

76 Cf. Putnam, "Diplomacy and Domestic Politics."

domestic and an international part, the relative autonomy the government enjoys in international negotiations from undue interference by domestic interest groups does apply less to policies with a predominantly national reach. Taken together, this may explain the stronger ambition with regard to domestic climate-related actions but also the greater resistance when it comes to programmes with immediate negative effects on the welfare of large voter groups.

CHAPTER 7

New Politics, Old Paradigms: Urban Environmentalism and the Reshaping of New Delhi

Somnath Batabyal

Abstract

Taking two media-led environmental campaigns in India's capital, New Delhi, as case studies, this paper argues that climate change discourse has followed the patterns of development politics. While globalisation and economic shifts have altered the North/South power dynamics since the twentieth century, the post-development age has given rise to a transnational middle class that seeks to reemploy similar hegemonic paradigms and shape the world on its terms. This class has replaced the notional North; the world's poor replaces the notional South. This transnational middle class, the case studies highlights, moulds the environmental discourse in the public sphere to suit their own imaginations of urban space and the mainstream media are active partners in this articulation.

1 Introduction

Within ten years of the Brundtland report,[1] concerns around the environment, which had lurked for decades in the fringes of political discourse, entered the mainstream. Today, in both the global North and South, environmental rhetoric is built into political campaigns. It competes for attention with every other governmental concern, internationally and within nation states: issues of economic growth, development, poverty alleviation or even defence strategies. As the Tories in the UK are compelled to promise the "greenest government ever",[2] and Obama marked his 2008 victory speech as "the moment when the rise of

1 Known also as 'Our Common Future', the 1987 Brundtland Report published the findings of the Brundtland commission, an organisation commissioned by the UN secretary general in 1983 to raise awareness of the need for sustainable development and encourage nations to pursue united sustainable policies.

2 David Cameron's speech to civil servants at the Department of Energy and Climate Change, May 14, 2010.

 | DOI 10.1163/9789004273221_009

the oceans began to slow"[3] so in India, in battles between indigenous land owners in mineral-rich areas of Orissa and corporate entities like Vedanta, the courts are forced to legislate against indiscriminate mining. Brazil, like several other countries of Africa, uses land resources as a bargaining tool in international diplomacy, making Northern countries compensate her for not denuding the rainforest cover. Environment has replaced development as the overarching paradigm of 21st century politics.

This essay, however, seeks to culturally situate present-day environmental discourse and show it as a continuation of the development politics that emerged in the middle of the last century. It argues that despite a shifting of the goalposts – climate change instead of hunger or forest degradation instead of abject poverty – the rules essentially remain the same: the powerful decide, the marginalised negotiate, albeit with some crucial differences.

Towards the end of the last century, development policies, with their focus on comparative equity and the forces of globalisation gave rise to a new "transnational class of winners",[4] educated cosmopolitan people who share the same aspirations whether they live in New York or New Delhi. According to Sachs, their proportion today is roughly equal in both the global South and North.[5] With increased acceleration, this essay claims, these global citizens have started to impose their imagination on their habitation, especially cities, ignoring other claimants, their desires and needs. By moulding discourses in the public sphere through the use of media, especially mainstream media, this new class has managed to pursue its own agenda, presenting it as 'common sense', similar to the economic model of development which too has been perceived as the only possible alternative.[6]

Writing on New Delhi, where the two case studies are based, sociologist Amita Baviskar states:

3 Barack Obama's election victory speech, November 4, 2008, Grant Park, Chicago, Illinois.

4 Sachs, Wolfgang, "Preface to the New Edition," in *The Development Dictionary: Knowledge as Power*, ed. Wolfgang Sachs (London, New York: Zed Books, 2010), viii. The term is used by Sachs to refer to a new global middle class. This essay understands that to use the term uncritically, and for people across the globe, is problematic. Even in the Indian context, defining the middle class as a single set of people is not without its critics. For more on this, see: Lakha, Salim, "The State, Globalisation and Indian Middle-Class Identity," in *Culture and Privilege in Capitalist Asia*, ed. Michael Pinches (London: Routledge, 1999), 251–274; and Deshpande, Satish, *Contemporary India: A Sociological View* (New Delhi: Penguin, 2003).

5 Sachs, "Preface to the New Edition."

6 Esteva, Gustavo, "Development," in *The Development Dictionary: Knowledge as Power*, ed. Wolfgang Sachs (London, New York: Zed Books, 2010), 1–23.

> The dream of inhabiting a world-class city, enjoying a lifestyle of convenience and comfort, is fuelled by real-estate firms, manufacturers and retailers of consumer durables, the entertainment and hospitality industries, and, most of all, the media. It is the expansion of this corporate capital in India that provides the context in which urban projects are imagined and pursued.[7]

The first case study we examine, the Clean Air Campaign, was launched in the mid nineties. The Bus Rapid Transport (BRT) system, our second case study, was launched as a pilot project in April 2008. Both campaigns sought to reduce vehicular pollution in New Delhi but were divergently received by the media, which favoured the former and panned the latter. Critiquing the media representations around the campaigns, this essay will conclude that environmental concerns and their representations have been skewed by a powerful media, essentially middle class in its orientation, which has proved itself a most handy weapon for the globalised reimagination of our cities.

In trying to understand a global trend, it seems appropriate to situate the case studies in India, integral to the "resurgence" of the global South and a key member of the BRICS block.[8] This should help to contextualise how the more affluent Southern countries are shaping their urban environment and what the socio-political implications of this engineering are.

2 Methodology

The focal point of this essay, Delhi's media, arises out of the personal involvement of the author, who was both a journalist in the city for over a decade and is today a citizen concerned about the directions the metropolis is taking. This research, therefore, owes a particular debt to the standpoint theorists who helped develop a means of constructing an episteme that resonates intensely with the experiential.[9] Though standpoint theory has been critiqued for being insufficiently personal, the writer found that it helped to explain "why, I, the articulating academic, find a certain issue of value to explore but more importantly, how my subjectivity resonates with and through the subject matter as

7 Baviskar, Amita, "Between Violence and Desire: Space, Power and Identity in the Making of Metropolitan Delhi," *International Social Science Journal* 55.15 (2004): 89.

8 The BRICS countries include Brazil, Russia, India, China and the latest arrival, South Africa.

9 See Harding, Sandra, *The Feminist Standpoint Theory Reader: Intellectual and Political Controversies* (New York: Routledge, 2003).

I analyse it".[10] The choice of subject matter, the entering into research and the later analysis of the empirical evidence are influenced by this.

This personal interest is critically augmented with fieldwork in Delhi that includes a year-long ethnographic study of the Centre for Science and Environment (CSE), a prestigious knowledge-based NGO in New Delhi lobbying for environmental policies and one of the chief architects of the Clean Air Campaign. In-depth interviews were conducted with journalists, environmental campaigners and city planners, some of which find direct mention in this paper but all of which helped frame the author's ideas and thoughts. The paper also makes extensive use of secondary source material that ranges from critiques of development to political ecology and literature and media reports on cityscapes, which focus on Delhi. Besides these, the paper also makes use of the writer's previous ethnographic work conducted in India focusing on the media and its middle-class articulations.[11]

Theoretically, this paper relies on one of last century's most seminal thinkers, Michel Foucault. Foucault has changed our perceptions of knowledge, linking it intrinsically with power (power/knowledge complex) through discourse.[12] Foucault has argued that discourse constructs the topic, that "[d]iscourse is a group of statements which provide a language for talking about [...] a particular topic at a particular historical moment".[13] The study of discourses includes the rules, which prescribe certain ways of talking about some topics and excluding others and how knowledge about a topic acquires authority, a sense of embodying the 'truth' about it. This knowledge linked to power not only assumes the authority of 'the truth' but also has the power to make itself true, "not the 'Truth' of knowledge in the absolute sense but of a discursive formation sustaining a regime of truth".[14] Calling into question the overall category of 'reality', Foucault demystifies the "regimes of truth", (media, governments, agencies, world views, etc.) which claim that there is a singular version of reality that can be identified. This paper, through case studies, seeks to demystify the present day discourse of 'environmentalism' in urban spaces, arguing instead that it is a construct of

10 Sreberny, Annabelle, "Globalisation and Me," in *In Search of Boundaries: Communication, Nation States and Cultural Identities*, ed. Joseph Chan and Bryce T. Mcintyre (Westport, London: Ablex, 2002), 294.

11 See Batabyal, Somnath, *Making News in India: Star News and Star Ananda* (New Delhi: Routledge, 2012).

12 See Foucault, Michel, *The Order of Things: An Archaeology of the Human Sciences* (London: Tavistock, 1972).

13 Hall, Stuart, *Representation: Cultural Representations and Signifying Practices* (London: Sage, 1997), 44.

14 Hall, *Representation*, 49–50.

middle-class urban citizens, spearheaded in the public sphere by an activist media, middle class in its construction.

3 The Story of Development

Development as we know it – a continuous drive towards more and more economic prosperity – was literally invented post-Second World War and announced on June 9, 1949 by US President Truman.[15]

> We must embark on a bold new program for making the benefits of our scientific advances and industrial progress available for the improvement and growth of underdeveloped areas.
>
> The old imperialism – exploitation for foreign profit – has no place in our plans. What we envisage is a program of development based on the concepts of democratic fair dealing.[16]

No longer will the former colonies be exploited; they will receive "the benefits of our scientific advances and industrial progress" – in other worlds, be helped to become more like "us" – in this case, the United States. In this speech, Truman sets up the American lifestyle as the benchmark of aspirations, a phenomenon that is backed by Hollywood films depicting country ranches and farms, fast cars and trawlers, washing machines and dishwashers, and later through multimillion dollar advertisement campaigns and multinational brands as they enter the "underdeveloped" world.[17]

Development has its fair share of critics and cheerleaders and it is not the purpose of this essay to weigh in for either. But what the style of development championed by Truman definitely did was to speed up the extraction of earth's precious and limited resources. "It is the legacy of the twentieth century," Wolfgang Sachs points out, "that the desires of nations for a better tomorrow are predominantly directed towards development as growth. However, the multifaceted crisis of the biosphere turns this legacy into a tragic liability."[18] The last fifty years have seen an alarming depletion in ground water levels, greedy

15 Esteva, "Development," 1.

16 Truman, Harry S., "Inaugural Address," published by Gerhard Peters and John T. Woolley, *The American Presidency Project*, January 20, 1949, accessed December 3, 2012. http://www.presidency.ucsb.edu/ws/?pid=13282.

17 Ibid.

18 Sachs, "Preface," x.

exploitation of minerals, denuding of forest cover and a hole in the ozone layer; the human race is at the brink of a catastrophe from which there might be no recovery.

At this point, we arrive at one of the central dilemmas of the development ethos. The ethics of development is based on a founding principle: the idea of "comparative equity", that the earth's resources be divided and shared in equal measure by all. Into the second decade of the 21st century, there is almost no contesting that the earth has limited resources and a major share of it has been used up by the rich nations. Indeed, no more industrial revolutions can take place; the earth simply doesn't have enough in its coffers. If all of the world were suddenly to reach a Western level of living standards, we would need "five or six planets" to take the load.[19] There can be no relative equity.

Yet, the common refrain heard at every climate summit from every Southern corner is their "right to develop". Poor countries contend, and justly, that they must be given their due share of the resources and the atmosphere to develop.

> The longing for greater justice on the part of the South is one reason for the persistence of the development creed – even if, in this century, neither the planet nor the people of the world can any longer afford its predominance.[20]

The pursuit of 'economic justice', the realignment of the financial world and the insidious forces of globalisation have helped the rise of the Southern powers; China, Brazil and India being at the forefront of a GDP-driven revolution.

> For them, it is as if President Truman's promise at the birth of the development period in 1949 – that poor nations would catch up with the rich – had finally come true. [...]
>
> As a result of this shift, development came to mean the formation of a global middle class alongside the spread of the transnational economic complex, rather than a national middle class alongside the integration of a national economy. Seen from this perspective, it comes as no surprise that the age of globalisation has produced a transnational class of winners.[21]

19 Sachs, "Preface," xvii.

20 Ibid., ix.

21 Ibid., vii.

Today Indian tourists book out Thomas Cook charter flights to Paris and Beijing's cheap workforce keeps the global consumer industry afloat. But it is problematic to celebrate this as a 'revival of the South', when it represents a shift in and not an end to inequality. This, as Sachs and Esteva argue, is the legacy of development policies which have espoused a notion of comparative equity – where nations should have a fair share of the world's wealth – and not absolute equity – which recognises the individual's right to a fair share.

Indeed, Southern countries have managed to raise their economic profile significantly, but development policies, falling short of absolute justice, have left more than half of the world's population still without the right of access to food, shelter, clean air, water and basic health care. In Sachs' stark words:

> The globally oriented middle classes, although they push for development in the name of greater equality, largely disregard the plight of the poor. [...]. In just about all newly industrialising countries, social polarisation has been on the rise along with growth rates over the past thirty years.[22]

The earlier victims of inequality were the poor nations; today when money and influence are turning increasingly transnational, the dominated are the global poor. As Esteva explains in the same volume, the poor are not simply left poor – their means of resistance to hegemony is simultaneously undermined:

> Redevelopment implies the economic colonisation of the so-called informal sector. In the name of modernisation and under the banner of the war on poverty – pitting as always the waged against the poor, not a war on poverty itself – redeveloping the south involves launching the last and definitive assault against organised resistance to development and economy.[23]

Before this essay moves to New Delhi and the two particular case studies mentioned above, it is important to frame, if briefly, the 'emergence' of India as an economic power and along with it, the rise of its media which pushes an unashamedly middle-class, urban agenda.

22 Sachs, "Preface," x.

23 Esteva, "Development," 13.

4 India and Its New Middle Class

India is in the news; or rather, the country is making the news. Be it an Indian billionaire buying a British automobile or steel company or reports of Hollywood celebrities descending on a Bollywood theme wedding in Mumbai, the framing of 'India' from a poverty-stricken nation of famine, drought and hunger has changed drastically. Even the abject poverty of a *Slumdog Millionaire*[24] is glamourised.

It is by now accepted that the liberalisation of the economy in the early 1990's was a watershed moment which unleashed the growth potential of a slumbering country shackled by protective regimes. But even before the opening of the markets, between 1980 and 1989, India had witnessed a 47.5 per cent rise in consumption expenditure.[25] The much-vaunted Indian middle class, touted today as the world's largest had become tired of molly-coddled national brands, afraid of competition and producing inferior quality goods. A whole generation of consumers were waiting in the wings when the trade barriers were pulled down and Pepsi, Coca Cola, Nike and Adidas announced their arrival.

As the decade progressed, this new and confident middle class, animated by the vision of setting India on a liberated path of progress and economic prominence on the world stage,

> assumed for themselves the role of the makers of the nation in new ways. In the emerging middle class political vision, the nation is a community of citizens who are enfranchised by freedom of choice, consumption and material gratification and a lifestyle of enjoyment and pleasure.[26]

Fernandes notes that

> production of the new Indian middle class parallels comparative trends in the construction of the "new rich" as a social group that is the prime beneficiary of globalisation in contemporary Asia. The construction of

24 The 2008 Danny Boyle film won 8 Oscars.

25 Dubey, Salman, "The Middle Class," in *India Briefing*, ed. Leonard Gordon and Philip Oldenberg (Boulder, London: Westview Press, 1992), 150.

26 Chakravorty, Rangan, and Nandini Gooptu, "Imagi-Nation: The Media, Nation and Politics in Contemporary India," in *Cultural Encounters: Representing Otherness*, ed. Elisabeth Hallam and Brian V. Street (London: Routledge, 2000), 91.

> such a category, in effect, marks the potential benefits of globalisation for emerging market-oriented contexts in nations such as India.[27]

However, the discourse of economic prosperity aside, millions of Indians remain desperately poor.

> We [India] are a nation of nearly a billion people. In development terms we rank number 138 out of the 175 countries listed in the UNDP's Human Development Index. More than four hundred million lack even basic sanitation, and over two hundred million have no safe drinking water.[28]

How is a country where nearly 500 million live below the poverty line able to frame itself as an economic powerhouse? The average Indian family today consumes 100 kilograms less food than it did in 1991 when the process of economic reforms started to kick in. Incidents of farmers, unable to pay back loans, committing suicide have become so regular that they are relegated to the inside pages of local dailies. Female infanticide, famines and droughts, appalling health services and lack of educational infrastructure are glossed over for celebrity weddings and the arrival of Formula 1 events in New Delhi's suburbs, a trick which writer/activist Arundhati Roy in a 2012 BBC interview termed as a "false voice, a lie".[29]

Roy's comment on India's global image of a thriving democracy with a high growth output being manufactured by the media is crucial. It is often the media that becomes the "primary site for the exposition of different social and political discourses [...]"[30] and the discourse of India's well-being has been carried on the shoulders of a pliant media. Here we reach the crux of this essay: how a burgeoning, global middle class shapes their environments to their liking, ignoring other possibilities and imaginations, and how the media becomes a handy tool in this imposition of will by framing bourgeois desires as 'common sense' and reflecting a world of particular aspirations. If, in the last century, it was the North reflecting a way of life as the only possibility of development, today it is

27 Fernandes, Leela, "The Politics of Forgetting: Class Politics, State Power and the Restructuring of Urban Space in India," in *Urban Studies* 41.12 (2004): 2415.

28 Roy, Arundhati, "Introduction," in *India: A Mosaic*, ed. R. Silvers and B. Epstein (New York: New York Review of Books, 2000), xxiv.

29 BBC News, "Author Arundhati Roy: India's Economic Success a 'Lie'," last modified June 2, 2011, accessed May 23, 2011. www.bbc.co.uk/news/world-south-asia-13624077.

30 Chakravorty, and Gooptu, "Imagi-Nation," 93.

the affluent class in both halves of the globe who shape our cities according to their desires and imaginations.

To understand this collusion of media and middle-class desire in the Indian context, it is important to note that the upward curve in the economy, the liberalising of the markets and a sudden unprecedented growth in the media industry, especially television, happened simultaneously. Melissa Butcher states in the preface to her book, *Transnational Television, Cultural Identity and Change: When Star Came to India*:

> India was undergoing a sea-change as the impact of the state's deregulation began to visibly kick in at deeper and deeper levels within the country. The media landscape simultaneously embodied and reflected those changes.[31]

The media landscape was embodied in terms of spectacular growth: by 2006, the industry was estimated to be worth more than Rs 185 billion (approximately £230 million), a dramatic turnaround from its humble beginnings in 1959 as an educational project sponsored by the state.[32] Government-run Doordarshan, the only channel on Indian airwaves until the early 1990s, has now been eclipsed by an extraordinary explosion of private channels. With 360 channels currently on air, and 160 waiting for the government's permission to follow suit, the television viewer in India has more choice than anywhere else in the world. More than a hundred million households in India now own at least one television set. Seventy million of them are connected to cable or satellite TV. At their fingertips lie choices galore: movie channels, music channels, television soap operas and news broadcasts. CNN and BBC, the Rupert Murdoch-owned Star TV bouquet of channels, international sports channels and homegrown Zee Television all vie for their attention.

The media reflected the economic changes by screening more and more glitz and glamour of an India on the rise. The black and white days have been replaced with gloss, gossip, and money. As Hartley argues, news is "the sense-making practice of modernity"; it is an important medium through which we understand ourselves and the world around us; who we are, what we are, our

31 Butcher, Melissa, *Transnational Television, Cultural Identity and Change: When Star Came to India* (New Delhi: Sage, 2003), 9.

32 Kohli, Vineeta, *The Indian Media Business* (New Delhi: Sage, 2006), 62.

collective likes and dislikes.[33] Television created India in its own image: rich, booming and ready to compete on the world stage.

Within the medium of television, the news genre saw the most spectacular growth; prompting Thussu to ask "is India the world's biggest TV news bazaar?"[34] From just one news channel in 1998, today India has close to sixty 24-hour news channels spread across the country, most of which are "national, but many international in reach, some cater[ing] to the regional markets".[35]

This essay argues that an inherently undemocratic, advertisement-driven news media combined with a vocal middle class combined to give shape to our cities and our priorities. This has changed the nature of environmental movements and their reportage in India from an 'environmentalism of the poor' to a 'bourgeois environmentalism'[36]. Change of course presupposes history and this post-liberalisation phenomenon must therefore be understood in the light of what went on in the previous decades in Indian environmental movements and in its media.

5 Indian Media and Environmental Movements: A Very Brief Summary

The origins of the press in India are rooted in the country's Independence movement and it was therefore closely aligned with the main political party, the Congress. After the end of colonial rule, as the Congress came to power under Prime Minister Jawaharlal Nehru, the press easily fell into the role of backing the state in the task of nation-building. This cosy relationship continued till the 1970s. In 1974, when Nehru's daughter and then Prime Minister Indira Gandhi declared a state of emergency and curtailed constitutional freedom including gagging the press, the partnership ended. A new era of politically independent journalism was to emerge.

The Indian environmental movement that emerged around the same time was a direct result of the economic development prescribed by Nehruvian socialism. Within two decades of independence, "development had quickly been equated to channelising the nation's resources towards a narrow elite of

33 Hartley, John, *Popular Reality: Journalism, Modernity, Popular Culture* (London: Arnold, 1996), 32.

34 Thussu, Daya, *News as Entertainment: The Rise of Global Infotainment* (London: Sage, 2007), 96.

35 Ibid., 96–97.

36 The expression is coined in: Baviskar, "Between Violence and Desire."

omnivores – powerful landowners and urban people in the organised industries and service sectors".[37]

The environmental movement in the country emerged in response to this large-scale degradation of forest and agricultural soil, pollution of rivers and usurpation of land rights. Writing on the social attributes of the environmental movements, Karan states that

> [e]ven with limited resources the environmental movements have initiated a new political struggle for safeguarding the interests of the poor and the marginalized, among whom are women, tribal groups, and peasants.[38]

"Many of these groups," Anil Agarwal notes, "are old and their main concern is either civil rights, rural development, tribal welfare, appropriate technology, primary health care, or science demystification."[39] The 'red' and the 'green' were thus part of the same movement – each, with its own concerns, written into the other.

One of the earliest manifestations of the environmental movement was in "March 1974 when women of the Garhwal Himalaya hugged ash trees to save them from the axe of contractor's men, launching the Chipko Andolan".[40] Karan enumerates a number of other prominent agitations caused by the lopsided developmental programmes in post-Independence India:

> Save the Bhagirathi and Stop Tehri project committee in Uttar Pradesh; Save the Narmada Movement (Narmada Bachao Andolan) in Madhya Pradesh and Gujarat; youth organizations and tribal people in the Gandhamardan Hills whose survival is directly threatened by development of bauxite deposits; the opposition to the Baliapal and Bhogarai test range in Orissa, the Appiko Movement in the Western Ghats; groups opposing the Kaiga nuclear power plant in Karnataka; the campaign against the Silent Valley project; the Rural Women's Advancement Society (Gramin

37 Gadgil, Madhav, *Ecological Journeys* (New Delhi: Orient Blackswan, 2005), 27.

38 Karan, Pradyumna, "Environmental Movements in India," *Frontier Thinking in Sustainable Development and Human Well Being* 84 (1994): 32.

39 Agarwal, Anil, "An Indian Environmentalist's Credo," in *Social Ecology*, ed. Ramachandra Guha (Berkeley: Oxford University Press, 1994), 346.

40 Gadgil, *Ecological Journeys*, 26. Chipko Andolan was a non-violent movement protesting deforestation in Uttarkhand.

> Mahila Shramik Unnayam Samiti), formed to reclaim waste land in Bankura district; and the opposition to the Gumti Dam in Tripura.[41]

One of the unifying themes of these movements was that they were played out in remote parts of the country, away from the eyes of the urban Indian. Yet, movements like Chipko Andolan or NBA[42] attained iconic status, not only within India but globally, as showcase examples of grassroots movements. Without a doubt, the country's media, especially its newspapers, played a prominent role in keeping these movements alive in urban consciousness. In his 1994 work, *Social Ecology*, Ramachandra Guha duly notes that

> [...] there has been a spectacular growth in public consciousness about the forms of environmental degradation in India. With an amazing but welcome rapidity, this awareness is being translated [...] into substantial media coverage [...]. Continuous and sustained press coverage ensured that most of urban India, if not involved with the movements, was largely sympathetic towards this "environmentalism of the poor".[43]

A nascent environmental movement found strong support from a media emerging out of state shadow and taking its first dissident steps. What changed and how?

To answer this question, the essay now examines two case studies of vehicular pollution in India's capital New Delhi and the media campaigns surrounding them. In analysing the campaigns, this essay uses the experiential, interviews, court and media reports.

6 New Delhi and Indian Urban Environmentalism

The National Capital Territory of Delhi (NCT) became the capital of British India in 1919. At the time of Independence in 1947, its population was estimated to be 400,000 people.

41 Karan, "Environmental Movements in India," 32.

42 The Narmada Movement (Narmada Bachao Andolan) is a social movement campaigning against the Sardar Sarovar Dam construction in the Narmada river in Gujurat.

43 Guha, Ramachandra, *Social Ecology* (Berkeley: Oxford University Press, 1994), 3.

> Between 1951 and 2001, the population of the NCT grew from 1.7 million to 13.8 million (Census of India), and about half of the population now has to live in informal settlements and shanties (jhuggis).[44]

Today, 20 million are estimated to live in the metropolis with another three-million-strong floating population that comes in every day.

This exponential growth has led to varied struggles over the control of resources, especially over the use of land and water and increasingly over notions of 'beautification' and in efforts to 'clean up' the city and impose 'civic order'. The first of these battles, after the economic liberalisation of 1993, was over the right to clean air and this campaign, fought as much in courtrooms as in the front pages of newspapers, inadvertently charted the direction of urban environmentalism in India.

The beginnings of the campaign go back to 1985 when the environmental lawyer M. C. Mehta had filed a public interest litigation (PIL) against vehicular pollution in Delhi with the Supreme Court of India. This PIL was based on the constitutional obligation of the state to protect the health of its citizens, which case law implicitly extended to a right to a clean environment. In response to the PIL, the Court issued notices to the Union and Delhi government. The response was a Pollution Under Control Drive launched by the government in the early 1990's, which meant that owners were to get their vehicles tested at filling stations. Lax norms and rampant corruption in issuance of certificates, however, meant little overall improvement in Delhi's air quality. By the mid-nineties, Delhi's air quality deteriorated further and visibly. Black, acrid smoke hung in the air, buses and trucks belched out poisonous fumes and the addition of more and more diesel cars to the city's congested roads, a gift of economic liberalisation, added to the woes.

Somewhere around the end of 1997, a full-page advertisement was carried out in the *Times of India* (TOI) by a then little known NGO, the Centre for Science and Environment. The advert screamed, "51,779 Dead by Breathing".[45]

The CSE campaign, as their late director Anil Agarwal put it, "began without a premeditated design".[46] It was in the summer of 1995, while waiting to get the tail pipe of his car checked for pollution at a petrol pump that Agarwal felt vehicle owners were being unnecessarily penalised for a problem that did not start

44 Delhi Development Authority, 2000, cited in Sivam, Alpana, "Housing Supply in Delhi," *Cities* 20 (2003), 135.

45 Agarwal, Anil et al., *The Leap Frog Factor: Clearing the Air in Asian Cities* (New Delhi: Centre for Science and Environment: 2006), 5.

46 Loc. cit.

with them. The CSE launched an investigation into air pollution in our cities, which was published in the public domain as *Slow Murder* in November 1996. It brought to light not only the wide-ranging factors which had contributed to the "pollution mud pie", but also the politician–auto industry nexus, which turned a blind eye to the problem.

Just after the publication of *Slow Murder*, on November 18th, 1996, the Supreme Court of India joined hands with the activists and issued a *suo motto* notice to the Delhi Government to submit an action plan to control air pollution in the city in November 1996, merging the notice with M. C. Mehta's PIL and on which little action had been taken till now. To put pressure on the government to take action, the CSE also decided to launch a public interest campaign called the "Right to Clean Air" around the same time. And this is where the campaign came to life. Through advertisements in newspapers and lobbying journalists who were only too eager to comply, a forceful campaign was built up.

Anumita Roychowdhury, who heads the CSE's Air Pollution division, said in an interview:

> It was the Clean Air Campaign which marked a definite shift in the CSE policy. We became an advocacy group, a lobbying force for change. We understood how to use the media to pressurise the state through this campaign. Our later campaigns were shaped through this understanding.[47]

The Union Ministry of Environment and Forest, forced into action by a relentless media campaign, formed the Environment Pollution (Prevention and Control) Authority (EPCA) in January 1998, which would be responsible for the National Capital Region (NCR). It was headed by senior bureaucrat Bhure Lal, environmental activist Anil Agarwal, the then transport commissioner of Delhi, Kiran Dhingra, and Jagdish Khattar, former executive director (marketing and sales) of Maruti Udyog Ltd., who represented the automobile industry. The EPCA, therefore, saw for the first time the coming together of the state, activists and the corporate sector to give a definite shape to Delhi's physical environments.

Based on the recommendations of the EPCA, a comprehensive court order was issued in July 1998 that included the elimination of leaded petrol, the replacement of old autorickshaws (motorised three-wheel passenger vehicles) and taxis, the augmentation of the bus fleet in the city to 10,000 from 6600 and, most strikingly, the conversion of all buses, autorickshaws, and taxis from diesel

47 Interview conducted with the author on November 6, 2010 at the CSE office in New Delhi.

and petrol to compressed natural gas (CNG). After some back and forth on the CNG issue, the final phase-out period for diesel buses was from April to November 2002. Furthermore, new emission standards for private vehicles (Bharat-I and Bharat-II based on European standards) were introduced, and the driving of (diesel) trucks was banned during daytime in the NCT.[48]

The court order was one thing, implementation was another. "The orders could only be implemented if there was political will; unfortunately there was none. Politicians were not convinced about the problem."[49] The struggle to implement the court's orders, given the government's prevarication, was however dealt with severely by an activist media. For five years, the mainstream media kept pressure on the government along with civil society and environmental activists. When then Health Minister Harsh Vardhan claimed that air pollution was not a serious health problem, the media hounded him and forced him to retract. The Lieutenant Governor's remark at a press conference in late 1997 that Delhi pollution was simply "dust particles",[50] sparked off a media outcry. Public opinion, moulded by an activist press and judiciary, forced the Delhi government's hand and by mid-2003 the Clean Air Campaign launched by CSE had achieved most of its objectives. In the next few months, the air quality in Delhi improved visibly.

Within five years of the Supreme Court ruling however, the gains from the campaign seemed nullified. The almost double digit growth rates of the economy and easily available loan schemes meant a spurt in ownership of private vehicles and by August 2008, the average total suspended particulate (TSP) level in Delhi was 378 micrograms per cubic meter – approximately five times the World Health Organisation's (WHO) annual average standard. With approximately a thousand private vehicles being added every day in Delhi, vehicular pollution was at an all time high.

It is in this context that the mainstream media's hostile reception to the Delhi government sponsored Bus Rapid Transport (BRT) project in 2008 and the ensuing public outcry is surprising. Even before the project was fully operational, the media backlash had begun. Television channels, newspapers, even magazine reports, decried the initiative. So much so that despite the fact that more commuters were moving at a quicker pace in the BRT corridor, further expansion of the project was shelved by the government.

48 Veron, Rene, "Remaking Urban Environments: The Political Ecology of Air Pollution in Delhi," *Environment and Planning A* 38.11 (2006): 2099.

49 Agarwal et al., *The Leap Frog Factor*, 7.

50 Ibid., 8.

Simply put, the BRT is a dedicated corridor for public buses, which move more than 60 per cent of the daily commuters in Delhi. Despite the fact that they carry more people, the road space given to buses is disproportionately less than to private vehicles. By dedicating bus only corridors, travel time for the commuters was meant to be reduced and the use of private transport discouraged thereby reducing vehicular pollution.

Even if we leave aside for a moment the issue of clean air, the benefits of public transport in a metropolis like Delhi seem indisputable. The physical geography of India's capital has considerably changed in the past two decades, especially since the beginning of this century. Roads in Delhi increased from 22,487 km in 1991 to 31,183 km in 2008, an increase of 17 per cent. Flyovers and underpasses have come up and roads now account for nearly 21 per cent of Delhi's surface area, leaving little scope for further expansion. Yet, in the same period, the ownership of private vehicles has gone up more than 200 per cent and Delhi now has over 6.5 million vehicles, with approximately a thousand cars being added everyday. As cars owners supplement their fleet, four-wheelers replace two-wheelers and cyclists graduate to scooters and motorcycles, it is evident that no amount of road construction can keep pace with the pace of vehicle ownership.

The aims of the Clean Air Campaign and the BRT were similar in that both wanted to improve the cityscape and the health of the citizens by cleaning up the toxic nature of Delhi's air quality. One sought to do this by ushering in clean fuel and the other by incentivising the use of public transport. Yet both campaigns received markedly contradictory responses. In the next section, through interviews with journalists and city planners and analysis of both the media-led campaigns, the essay seeks to show how the bourgeois "strategies of exclusion" force a singular imagination in the public sphere of the cityscape.[51]

7 Analysis

The liberalisation process of the mid 1990s was, whichever side of the fence one sits on, a watershed moment, both for the economy and the wider socio-political sphere. Nowhere was this more marked than in the country's media where staid old Doordarshan[52] was giving way to Star's Bold and Beautiful and newspaper wars were hotting up as the *Times of India* and the *Hindustan Times* launched their one-upmanship campaigns to achieve dominance in the capital.

51 Baviskar, "Between Violence and Desire," 16.

52 This is a state-owned television channel.

But more importantly, the journalists – their aims, ambitions and desires for themselves – were changing. The particularities of the profession require a certain skill level, which in India, as in most other countries, is equated with the middle and more affluent classes. Media scholar Ginneken writes: "Most journalists are middle-class in social background, social position and social aspirations."[53] In an interview, Debashish Mukerji, senior Associate Editor of the Hindustan Times, put it succinctly:

> We suddenly had journalists who were travelling regularly, some on meaningless junkets, some on stories. Closer-to-home destinations like Singapore and Hong Kong were getting to be a regular fare. These journalists, dazzled by what they saw, started imagining the same for their city. Judges travelled, petty officials travelled and the glamour of the West was brought home. Now they wanted the same in their city.[54]

In the mid nineties, as the journalists started travelling, as Murdoch brought the 'dazzling' world of American television into Indian bedrooms, as the first multinational corporations started to take their baby steps on Delhi's soil, the air of the capital was dark. As a journalist on a daily city beat in a newspaper, this writer and other reporters around him knew how bad the air quality was but did not know what to do about it. The environment was not a beat and no one reported much on it. The government-launched *Pollution Under Control* (PUC) drive, which involved someone checking the exhaust of your car at petrol pumps, was assigned to the crime reporter to check for evasions. Journalists understood that things were bad but did not know whom to hold responsible, how to write the story. In a medium which likes simplicity above all else, the subject matter was complicated.

The first of the Clean Air Campaign advertisements, the CSE-sponsored "51,779 Dead by Breathing," in the *Times of India*, was a game changer. A simple narrative which linked air pollution to sickness and death and managed to bypass the complicated physics of harmful particulate matters in the air, their relationship to the petroleum and auto industries, the nexus between politicians and the transport lobby and so on, was picked up by the journalists and the media. The press community immediately got interested and headlines followed. Reporters understood through the experience of being out on the streets that bad air was harming them, their families and their friends. The story easily

53 Ginneken, Jack van, *Understanding Global News* (London, Thousand Oaks, New Delhi: Sage, 1998), 71.

54 Interview with the author, New Delhi, September 24, 2010.

became personal. And here perhaps lies the key to a campaign that was sustained for over six years, from 1997 till 2003, itself a miracle given the media's attention span: the story was personal, personal to the middle-class journalist and his/her audience.

K. S. Sachidananda Murthy, Resident Editor of the Malayala Manorama Group in Delhi said in an interview with the author:

> Of course the press is middle class with a middle class vision. We are so middle-class that we are also anti-super rich. We are happy when, say, the Delhi Development Authority (DDA) goes out to acquire land or the Noida Development Authority does so. We think we can get housing under their schemes, our readers can. But come Special Economic Zones[55] we turn against even the top corporate houses indiscriminately, be it an Indian company like Reliance or a foreign entity like the Salim group in Nandigram. But yes, it is the middle class interest we have in our hearts and that is very closely linked to our reader profile. One must keep that in mind.[56]

Murthy's link between the social status of the journalist and that of his/her reader is important; he emphasises that middle-class journalists write for a middle-class audience, and the success and longevity of the Clean Air Campaign were closely linked to this: a middle-class reimagination of Delhi as a "world class city".[57]

Along with the air quality of Delhi, therefore, a few other things changed which perhaps CSE's initial campaign had not bargained for. The campaign definitely helped in making urban environmental issues a legitimate concern in the city's mainstream media. But in attempting to push environment from the fringes to the centre of political discourse via the media, it probably did not envisage that essentially middle-class journalists writing for a largely middle-class audience would throw aside the earlier 'pro-poor' concerns of India's environmental movement.

---

55 A Special Economic Zone (SEZ) is a geographical region that has economic laws that are more liberal than the country's typical economic laws. Usually, the goal is an increase in foreign investment. In India, as the government has appropriated land for investment, several of these zones have turned into battlegrounds between peasants and the authorities.

56 Interview with K. S. Sachidananda Murthy, September 28, 2010, New Delhi.

57 Ibid.

As studies show, the poor had other, more immediate things to worry about than air pollution. While the media placed priority on clean air and beautification drives, it was "water supply and sanitation [that] formed the greatest environmental priorities for Delhi's poor inhabitants". In a comprehensive study done on Delhi's slums, inadequate and insufficient sanitation were found to be the most severe deficiency, while noise and air pollution are mentioned as merely "adding to the extremely poor conditions of environment".[58]

> Although Delhi's environmental policies centering on air quality have lessened exposure to environmental risks across socioeconomic classes, they have hardly addressed an environmental priority of the urban poor. I would argue that the intensity and effectiveness of environmental activism in the field of air pollution is reflective of a general middle-class bias, and that the larger (environmental) NGOs engaged in policy advocacy have paid little attention to the main environmental concerns of the urban poor.[59]

These concerns were however not the media's priority for Delhi, nor of the new kind of environmental awareness which was emerging in urban India. The 'environmentalism of the poor' vanished into the dustbins of history as an elite media and its readership sought to etch a 'bourgeois environmentalism' that privileged the concerns of the burgeoning middle class over any other.

Just a couple of years before the Clean Air Campaign was launched, academic Amita Baviskar wrote:

> [T]hrough a series of judicial orders, the Supreme Court of India has initiated the closure of all polluting and nonconforming industries in the city, throwing out of work an estimated 2 million people employed in and around 98,000 industrial units. At the same time, the Delhi High Court has ordered the removal and relocation of all jhuggi squatter settlements[60] on public lands, an order that will demolish the homes of more than 3 million people. In a city of 12 million people, the enormity of these changes is mind-boggling.[61]

58 Ali, Salim, *Environmental Situations of Slums in Delhi*, 2003, cited in Veron, "Remaking Urban Environments," 2101.

59 Veron, "Remaking Urban Environments," 2101.

60 Illegal temporary homes.

61 Baviskar, "Between Violence and Desire," 90.

What Baviskar perceives to be mind-boggling was nothing compared to what was to come.

> Between 1990 and 2003, 51,461 houses were demolished in Delhi under "slum clearance" schemes. Between 2004 and 2007 alone, however, at least 45,000 homes were demolished, and since the beginning of 2007, eviction notices have been served on at least three other large settlements.[62]

The fourth estate, with very minor exceptions, has been largely supportive of the drive.

It is in this light that one must look at the media campaign and furore over the BRT and the apparent public denouement of the scheme. Unlike the CNG, the BRT's public profile was doomed from the start. A residential colony of the city's influential journalists was situated close by and a host of them who make their way towards central Delhi for work every morning were inconvenienced. A senior editor in one of the city's biggest newspapers told this author that as his car was stuck in the traffic one morning, he decided to run a campaign against the corridor. Like the CNG campaign, it was also was personal. Other newspapers and television channels, especially in the English language, unequivocally panned the project. Even at its trial period in May 2008 (the height of summer is probably a bad time to start anything), the press coverage was unsparingly scathing. The project did not stand a chance.

"What the press missed," said Sam Miller of the BBC, "was that it was *supposed* to inconvenience cars. That was the whole point."[63] In contrast to the Metro, which started in the poorer parts of the capital, Miller pointed out that the BRT suffered from being situated in the heart of the affluent South Delhi: "The Metro came up quietly. People got to know about it, then they wanted to have a day out on it. They got used to it. The BRT inconvenienced the powerful."[64]

In another interview, Siddarth Pandey, a senior journalist covering environment in New Delhi Television (NDTV), one of India's most reputed news television organisations, started laughing when asked if he had ever travelled by bus on the BRT corridor: "Are you joking? I am just trying to get to my office in the morning. The bosses will lynch me if I try a stunt like that and say I am taking the bus." When asked how he knew, having never tried it, that the BRT was a

62 Bhan, Gautam, "'This Is No Longer the City I Once Knew': Evictions, the Urban Poor and the Right to the City in Millennial Delhi," *Environment and Urbanisation* 21 (2009): 127.

63 Interview with Siddarth Pandey, October 4, 2010, New Delhi.

64 Ibid.

disaster, he responded: "Well you should look at the cars stuck in the morning traffic. It is a long queue. I mean they [the government] should have tried something else."[65] When asked what sort of alternatives might exist, no clear answer emerges.

Writing in environment magazine *Down to Earth,* Sunita Narain, Director of CSE, writes "BRT is About Equity on the Road", stating that the logic of a bus corridor is irrefutable:

> The bus moves more than 50 per cent of the city but does not get proportionate space. In the first stretch of the Delhi BRT corridor, over this last year, buses have doubled to 3,000 – they now commute roughly 200,000 people each day.[66]

As Narain points out, the BRT "meant taking from a few and giving to many".[67] But amongst these few were the powerful articulators of our reality, the journalists and their audience. Together they created a public opinion against the BRT that was so strong that the Delhi Chief Minister Sheila Dixit was worried that her re-election in 2009 was doomed. The BJP's chief ministerial candidate, Vijay Kumar Malhotra, taking advantage of the media outcry, promised to make the BRT an 'election issue'.[68] A close aide in Shiela Dixit's office who monitors the press said in an interview that after much thought it was decided that the "song and dance"[69] about the BRT was being created by an elite media, which did not represent a vast section of the people. "The middle classes do not vote. The poor people who actually voted in the elections use buses. We decided to ignore the media's doom and gloom predictions," he said.[70] The predictions indeed turned out to be false and Sheila Dixit won a thumping victory. A media, divorced from ground realities, had egg on its face.

It would be simplistic to portray the media as 'lying' or 'misinformed'. Instead, it created a "regime of truth"[71] and started believing in it. Middle-class journal-

65 An opinion poll carried out by NDTV on May 1, 2008 reveals the class bias of the opposition to the project. While 65 per cent of the car drivers said that the BRT corridor made driving difficult, 75 per cent of the bus drivers disagreed.

66 Narain, Sunita, "Editorial," *Down to Earth,* March 1–15 (2010): 1.

67 Ibid.

68 "No Fund Crunch for BRT Project: Delhi Govt," *The Indian Express,* October 16, 2008, accessed December 5, 2012. http://www.indianexpress.com/news/no-fund-crunch-for-brt-project-delhi-govt/374191.

69 Private interview with the author, Novermber 4, 2010.

70 Ibid.

71 Hall, *Representation,* 49–50.

ists with middle-class visions writing for a middle-class audience found a convenient version of events, which suited 'everybody'. Democracy, or the 'republic of the streets', contested this version of events.[72]

8 Conclusion

In this chapter, it has been argued that present-day environmental debates follow the structures of the development paradigm, a legacy of 20th century politics. A globalised economy has created a new middle class, who today has cornered the debate around environment, especially in the urban arena. Just as the North forced its own notion of development on the South, this middle class forces its own version of urban environmentalism onto the landscape and politics of Indian cities.

In seeking to establish this middle-class environmentalism, this chapter briefly charted the course of development politics. Borrowing heavily from Wolfgang Sachs and Gustavo Esteva, it argued that development is a culturally situated concept that emerged in the aftermath of the Second World War from within the US. Resource intensive and exploitative of scant natural resources, this singular development trajectory has led the world to the brink of a crisis from where there might be no return; the ever-increasing hole in the ozone layer is leading to global warming, a rise in sea levels threatens to obliterate nations like Tuvalu and Maldives. Depleting forest cover, lack of agricultural land, scarcity of potable water have led to loss of habitats, stratospheric rises in food prices and famines. That things have to change, that development needs rethinking, was evident.

As the concern for our surroundings and the environment entered everyday discourse, it has been cornered by a transnational group of elites, themselves a product of economic development. This group drives the environmental discourse and directs it, marginalising other versions and voices. Taking urban environmentalism as a case in point and Delhi as the focus, two case studies in India were analysed. Through how a middle-class media viewed and reported the two cases, both intending to lower vehicular pollution, the essay argued that the public sphere is cordoned off from a pluralistic debate and Baviskar's 'bourgeois environmentalism' is shown as the only option. This is in stark contrast to the 'environmentalism of the poor' that the country's media had followed for two decades previous to the economic liberalisation.

72 Joyce, Patrick, *The Rule of Freedom: Liberalism and the Modern City* (London: Verso, 2003).

PART 4

Prospects

CHAPTER 8

Environmental Justice in Thailand in the Age of Climate Change

Susan M. Darlington

Abstract

While most research on climate change emphasises scientific impacts and responses, some researchers are beginning to examine critically its social dimensions, including economic and social justice. I draw from the approach of the environmental justice movement in the United States to shift the focus from the efforts by the government and international bodies to address climate change to the actions of some of the people most directly affected – poor farmers. Climate justice activists tend to look specifically at how to address the inequalities emerging from the burden of climate-related impacts. Environmental justice reminds us to consider the inequalities already in place, as well as how the various players involved understand and relate to their environments. While the issues involved range from the global to the local levels, I focus here on the local, presenting four cases of rural people in Nan Province, Northern Thailand, working to change their own and their neighbours' socio-economic situations through improving their agricultural methods. They demonstrate that those most affected by climate change are not mere victims. They are people who have agency, make choices, and often undertake innovative actions to help themselves.

1 Introduction

The world watched in shock through the summer and fall of 2011 as severe flooding spread from Northern Thailand to Bangkok. In July and August, agricultural fields in the north were flooded, followed shortly thereafter by the inundation of the ancient capital of Ayutthaya. By late October and November, much of Bangkok was underwater. People fought over floodwalls as the government struggled between saving economic resources and industrial factories or neighbourhoods. Over 800 people lost their lives.[1] The irony was that the year

1 Cf. "Death Toll from Months of Flooding in Thailand Rises Above 800," *Straits Times*, December 31, 2011, accessed January 14, 2012. http://www.straitstimes.com/BreakingNews/SEAsia/Story/

 | DOI 10.1163/9789004273221_010

before, Thailand suffered its worst drought in 20 years, followed by floods.[2] In both years, the events of which most observers blame in part on climate change, the impacts affected people differentially, hitting farmers and urban poor the hardest.

Poor farmers and urban poor are the most vulnerable to climate change in Thailand, and have the least say in the political decisions about how to adapt to and mitigate its effects. Because of the importance of rice in Thailand, both for food and export,[3] with 42.3 per cent of the employed population working in the agricultural sector,[4] the impact of climate change on farmers is of particular concern. Yet the official ways in which climate change is being addressed looks at the large picture – the impact on Thailand's economy, health sector, and infrastructure, for example – rather than the human impact. The people most affected are portrayed by the government as silent victims, people who must follow directions from above or, sometimes literally, get swept away by or bear a disproportionate amount of the effects of the changes.[5]

The title of a recent Bangkok Post article, "Flood Victims Were Not Faceless People," reframes this focus.[6] The author, Veera Prateepchaikul, makes the point that in dealing with and reporting on the floods of 2011, most officials (especially those from the Thai government and the World Bank) emphasise the economic costs (1.4 trillion baht, or ~44 billion US dollars), but not the human

STIStory_750551.html. Cf. also "Flood Toll 813: Three Missing," *Bangkok Post*, January 10, 2012, accessed January 14, 2012. http://www.bangkokpost.com/news/local/274506/flood-toll-813-three-missing; and Piyaporn Wongruang, "Flood Mitigation Plan 'Unlikely to Be in Place in Three Years'," *Bangkok Post*, January 15, 2012, accessed January 14, 2012. http://www.bangkokpost.com/news/local/275180/flood-mitigation-plan-unlikely-to-be-in-place-in-three-years.

2 Marks, Danny, "Climate Change and Thailand: Impact and Response," *Contemporary Southeast Asia* 33.2 (2011): 229.

3 Cf. Kawasaki, Jintana, "Thailand's Rice Farmers Adapt to Climate Change," published by Our World 2.0, United Nations University, last modified May 24, 2010, accessed January 29, 2012. http://ourworld.unu.edu/en/climate-change-adaptation-for-thailands-rice-farmers/; Asian Development Bank, *The Economics of Climate Change in Southeast Asia: A Regional Review* (Manila: Asian Development Bank, 2009), accessed January 28, 2012. http://www.adb.org/documents/books/economics-climate-change-sea/Economics-Climate-Change.pdf.

4 Ibid., 9.

5 Weida Lim makes a similar argument about how Singapore is addressing, or not addressing, climate change as "people are made to bear an unequal burden as the policy is class-blind and neglects the social reality that different classes contribute differently to environmental problems". Cf. Lim, Weida, "Climate Change Policies in Singapore: Whose 'Environments' Are We Talking About?," *Environmental Justice* 2.2 (2009): 83.

6 Cf. Veera Prateepchaikul, "Flood Victims Were Not Faceless People," *Bangkok Post*, January 10, 2012, accessed January 16, 2012. http://www.bangkokpost.com/opinion/opinion/274536/flood-victims-should-not-die-in-vain.

loss. I want to take Veera's perspective a step further.[7] Veera calls for compensation to the families of the flood victims, but he still portrays them as victims. Yet these are individual people, with hopes, dreams, and struggles of their own – and their own agency. For example, people in some of the lower-lying neighbourhoods surrounding Bangkok took it upon themselves to dismantle floodwalls that were diverting water into their neighbourhoods. The government erected the walls to prevent the flooding of the economically lucrative part of central Bangkok or industrial factories surrounding the capital.[8] My point here is not to criticise the government's difficult decision, but to highlight the actions of people affected by the floods – and other impacts of climate change. They do not merely sit around waiting for either the government to solve their problems (especially as the government's priorities may not always match their own) or weather changes that would ruin them.

In this paper, I draw from the approach of environmental justice to shift the focus from the efforts – underway, planned, and needed – by the government and international bodies to address climate change, to the actions of some of the people most directly affected. 'Environmental justice' emerged in the United States in response to the inequitable distribution of toxic waste dumping and other environmental hazards in poor communities of colour. Giovanna Di Chiro, Luke Cole, Sheila Foster, and Eileen McGurty all stress that environmental justice not only entails dealing with these inequities, but doing so through a reframing of underlying relationships.[9] In environmental justice, the 'environment' is defined as where people work, live, and play. It is not an 'othered' nature that remains separate from human life, but is intimately entwined with it. Environmental justice activists consequently deal with issues of social justice, a sustainable economy, health, livelihoods, community governance, and racial and social justice as much as explicit environmental issues.[10] While most

7 Thais go by their first names professionally. Following that convention, I will refer to Thais throughout this paper primarily by their first names.

8 Cf. "Khlong Hok Wa Barrier Torn Down," *Bangkok Post*, November 17, 2011, accessed January 16, 2012. http://www.bangkokpost.com/news/transport/266730/khlong-hok-wa-barrier-torn-down.

9 Cf. Di Chiro, Giovanna, "Nature as Community: The Convergence of Environment and Social Justice," in *Uncommon Ground: Rethinking the Human Place in Nature*, ed. William Cronon (New York: W.W. Norton & Co., 1996), 298–531. Cf. also Cole, Luke W., and Sheila R. Foster, *From the Ground Up: Environmental Racism and the Rise of the Environmental Justice Movement* (New York, London: New York University Press, 2001); and McGurty, Eileen, *Transforming Environmentalism: Warren County, PCBs, and the Origins of Environmental Justice* (New Brunswick, London: Rutgers University Press, 2009).

10 Di Chiro, "Nature as Community," 300; Cole, and Foster, *From the Ground Up*, 17.

work on climate change emphasises scientific impacts and responses, some researchers are beginning to examine critically the social dimensions, including economic and social justice.[11]

Cole and Foster point out that for the US communities they portray in their book about environmental racism and environmental justice the importance "is the self-representation and agency inherent in 'speaking for ourselves'".[12] These communities refused to accept explanations surrounding the condition of the environment in which they lived given to them by government and corporation officials, but undertook research and actions to be able to understand their situations and speak for themselves. The case of poor farmers in Thailand facing problems of environmental degradation from agricultural intensification, mono- and cash-cropping, all increased by climate change, parallels these aspects of the environmental justice movement in key ways. Not only do they face changes in the climate, but they struggle with socio-economic and political inequalities. As with US environmental justice cases, they have agency to begin to help themselves at the local level.

I agree with J. Timmons Roberts, who states that

> […] most of the adaptation to climate change will be done at the household and community levels, without outside assistance. People in the global South are not mere suffering victims, but have agency and autonomy, and are actively addressing this crisis every day.[13]

My own research focuses on Thai monks who undertake rural development and environmental conservation actions grounded in their interpretations of Buddhist teachings and the villagers with whom they work. I do not pretend to be an expert on climate change in Thailand. In working with these monks in the rural north of Thailand, however, I have witnessed both the struggles of farmers to make a living and their efforts, often in conjunction with non-governmental organisations or activist monks, to come up with solutions of their own. Many of these actions fit the recommendations of international bodies such as the Asian Development Bank and the IPCC and independent scholars for adapting to climate change. While the Thai government prioritises readiness to handle the next big flood,[14] rural farmers in many parts of the nation are already imple-

11 Cf. Mearns, Robin, and Andrew Norton, ed., *Social Dimensions of Climate Change: Equity and Vulnerability in a Warming World* (Washington, D.C.: The World Bank, 2010).

12 Cole, and Foster, *From the Ground Up*, 14.

13 Roberts, J. Timmons, "The International Dimension of Climate Justice and the Need for International Adaptation Funding," *Environmental Justice* 2.4 (2009): 186.

14 Cf. Piyaporn, "Flood Mitigation Plan."

menting activities to withstand and adapt to climate change through addressing broader underlying social inequalities that make them more vulnerable in the first place.

2 Inequities of Climate Change

Thailand is already experiencing the effects of climate change through irregular rainfall, prolonged droughts, severe flooding, sea-level rise impacting coastal areas and threatening to submerge Bangkok possibly within a couple of decades, and decreased agricultural and fishery yields.[15] Given the scale of these impacts, one would assume all Thais will be affected. But climate change does not occur on a blank slate – there exist social inequalities that are exacerbated by climate change and the ways in which societies respond to it. For example, climate change in Thailand will strongly affect the agricultural sector.[16] While large-scale, industrial agriculture will need to adapt, poor, small-scale farmers are far more vulnerable because their situations are already tenuous.[17]

Roberts demonstrates that these existing inequalities are not unique to Thailand, but can be found everywhere.[18] He argues that the main injustice within climate change "is that those who are least responsible for causing the problem are also those most likely to suffer directly its early impacts." The systemic inequalities that contribute to this injustice, Roberts says, are based on "profound and complex social inequalities of many types, notably between rich and poor

15 Marks, "Climate Change and Thailand," 231; Asian Development Bank, *Economics of Climate Change*; and Kisner, Corinne, "Climate Change in Thailand: Impacts and Adaptation Strategies," published by Climate Institute, Washington, D.C., last modified July, 2008, accessed January 14, 2012. http://www.climate.org/topics/international-action/thailand.htm.

16 Cf. Focus on the Global South, "Food, Livelihoods & Climate Change in the Mekong Region" (workshop report, Bangkok, August 9–11, 2010), accessed January 28, 2012. http://www.focusweb.org/sites/www.focusweb.org/files/Mekong%20Report.pdf.

17 Cf. Somchai Jitsuchon, "Adaptations to Climate Change of Poor Thai Farmers" (paper presented at the conference "Environments of the Poor", Asian Development Bank, November 24, 2010), accessed January 23, 2012. http://www.scribd.com/doc/43296343/Adaptations-to-Climate-Change-of-Poor-Thai-Farmers-paper; Lebel, Louis et al., "Adaptation to Climate Change and Social Justice: Challenges for Flood and Disaster Management in Thailand," in *Climate Change Adaptation in the Water Sector*, ed. Fulco Ludwig et al. (London: Earthscan, 2009), 127; and Roberts, "International Dimension."

18 Ibid. See also Adger, W. Neil, Jouni Paavola, and Sameemul Huq, "Towards Justice in Adaptation to Climate Change," in *Fairness in Adaptation to Climate Change*, ed. W. Neil Adger et al. (Cambridge, MA: MIT Press, 2006), 2.

nations, but also within nations between people of difference [sic] race, class, ethnicity, gender, and generation".[19] While all these kinds of injustices are at play in Thailand, I focus here on the rural poor. Having studied rural Thailand for over thirty years, I have witnessed the inequalities and challenges poor farmers face, as well as their innovative responses.

Most of the agriculture in Thailand outside of the central plan that surrounds Bangkok is rain-fed.[20] Irregular rainfall, combined with temperature increases of even a couple of degrees, negatively affects the production of the country's main crop, rice. Water management has become a major challenge to the whole country, but especially individual farmers. As Danny Marks, consultant for the Rockefeller Foundation's climate change program in Thailand, notes, "the impacts of climate change disproportionately hurt the poor and they will need the most help when natural disasters strike".[21]

The Thai government has not ignored the problems of climate change. In January 2008, the first Five-Year Strategy on Climate Change (2008–2012) began. The strategy focuses on capacity building, dealing with greenhouse gases through sustainable development, research and development to understand climate change, raising public awareness, strengthening relevant personnel and institutions, and engaging international efforts for mitigation and sustainable development.[22]

Marks points out the shortcomings of this strategy, particularly the limited opportunities for public participation. The plan prioritises the interests and views of government officials and the private sector. A gap exists between plans to deal with climate change issues and other environmental regulations.[23] Lebel et al. note that in disaster situations, such as floods, social inequalities play a key role in who suffers the most. Yet the government pays little attention to these inequalities in their responses or plans. Lebel et al. state that "securing the most affected and vulnerable people has not been a priority of [Thai] governments, and there is little reason to expect that emerging policies to adapt to climate change will change this".[24] Similarly, in a report produced in 2010 by several regional and national non-governmental organisations about the impact of climate change in the Mekong region, the authors stressed that

19 Roberts, "International Dimension," 185.

20 Marks, "Climate Change and Thailand," 235.

21 Ibid., 242. See also Lebel et al., "Adaptation to Climate Change," 134.

22 Lebel et al., "Adaptation to Climate Change," 131.

23 Marks, "Climate Change and Thailand," 244.

24 Lebel et al., "Adaptation to Climate Change," 135.

> [...] while governments claim to be formulating climate adaptation plans, they continue to follow development trajectories that undermine the capacities and resilience of local farming and fishing communities to withstand climate-related disasters.[25]

This report represents perspectives not often heard in climate discussions. The report emerged from a regional workshop held in Bangkok in August 2010, entitled, "Food, Livelihoods and Climate Change in the Mekong Region".[26] Throughout the report references are made to comments by farmers and fishers who shared their views. The conclusions that emerged take Marks's criticisms of the Thai government's plans a step further.[27] As the quote above indicates, the participants were concerned with "the dual challenges presented by a destructive development model and climate change".[28] They stressed the need for cooperation between people and the governments of the region to address the specific challenges faced in different areas of each country. A key point emerging from the workshop was the importance of incorporating climate justice principles in the process – not to ignore the experiences and inequalities faced by the most vulnerable, but to make them central to finding solutions.[29]

Lebel et al. discuss the importance of involving the most vulnerable people in the process of dealing with climate change:

> From mitigation through to rehabilitation there is a need to empower and enable the affected and disadvantaged [...]. They need places to articulate their needs and aspirations and space to build and develop their capacities. Access to, and control over, resources is needed, not just top-down allocations to agencies acting on "their behalf". [...] The links between knowledge and practice should be strengthened in both

25 Focus on the Global South, "Food, Livelihoods & Climate Change," 3. Adger, Paavola, and Huq point out this is a trend, especially in developing nations: "We live in a world where future climate injustices are likely to compound past injustices, such as underdevelopment and colonialism, that themselves have resulted in the uneven patterns of development in today's world." Cf. Adger, Paavola, and Huq, "Towards Justice," 2–3.

26 The workshop was organised by several non-governmental organisations, including Focus on the Global South, the Foundation for Ecological Recovery/TERRA, World Rainforest Movement, International Rivers, Bank Information Centre, and the Thai Working Group on Climate Justice. Cf. Focus on the Global South, "Food, Livelihoods & Climate Change," 1.

27 Cf. Marks, "Climate Change and Thailand."

28 Focus on the Global South, "Food, Livelihoods & Climate Change," 3.

29 Loc. cit.

> directions, recognising the valuable contributions which local experience and understanding of conditions can make to reducing risks [...]. A focus on building and maintaining resilience of affected and vulnerable groups rather than managing floods as generic hazards could also help address current injustices.[30]

Strengthening this argument, we should recognise that farmers have the ability to adapt to mild climate change already.[31]

In addition to the challenges arising from climate change itself, the situation of poor farmers is affected by the government's economic policies. As with other nations in the Mekong region, Thailand enacts a model of economic development based on natural resource extraction, including mining, dams for hydropower and irrigation, plantations, industrial agriculture, and industrialisation, dominated by multinational corporations and dependent on exports.[32] Such an emphasis on economic growth not only contributes to climate change, but undermines the ability of farmers to make a reasonable, debt-free living. Rural people are often pushed into economic agriculture, limiting their livelihood opportunities. Affected by the vagaries of the market economy, they fall into a cycle of debt and agricultural intensification. Many migrate to cities in search of jobs, contributing to the growth of the urban poor population.

A monk active in addressing issues of rural development and environmental degradation in Nan Province, Phrakhru Pitak Nanthakhun, has repeatedly told me over the past twenty years that the major problem facing the farmers of the province and the nation is debt. People expand their fields to grow more cash crops, shifting from subsistence farming to monocrop, cash cropping in order to buy more consumer goods. They are more likely to grow cash crops and buy their rice and other food. In the process, they get seeds, fertilisers, and pesticides from seed companies on advance for the crop. If the harvest is not enough to repay the output, the farmers go into debt, getting more seeds, fertilisers, and pesticides, and on into a vicious cycle from which many cannot escape.

---

30 Lebel et al., "Adaptation to Climate Change," 139.

31 Somchai Jitsuchon points out that farmers have difficulty countering the impacts of extreme climate change. Poor farmers in particular have limited coping potential. Cf. Somchai, "Adaptations to Climate Change," 8.

32 Focus on the Global South, "Food, Livelihoods & Climate Change," 6; cf. also Laird, John, *Money Politics, Globalisation, and Crisis: The Case of Thailand: Exploring New Paths Towards Sustainable Development* (Singapore: Graham Brash Pte Ltd., 2000); as well as Pasuk Phongpaichit, and Chris Baker, "Thai Capital After the Asian Crisis," *Economic & Political Weekly* 42.50 (2007): 63.

A rise in temperature, an irregular rainfall, severe drought, or flood can ruin such farmers. As Lebel et al. observe, "climate change could significantly exacerbate existing unfairness and inequalities".[33] Yet not all these poor farmers are merely awaiting their fate. Whether on their own initiative or with the assistance of grassroots organisations or local leaders, some farmers are seeking ways to take control of their situations, to get out of debt, and diversify their livelihoods so that they are both more self-sufficient and able to withstand at least mild changes in climate. They have agency, as Roberts notes,[34] and are not waiting for help from the government.

3 Farmers' Agency

To see how rural people are making efforts to help themselves, I present four examples of farmers or local leaders in Nan Province in Northern Thailand who have taken actions to change their own and their neighbours' situations. In each case, they instigated new approaches to earning a livelihood that reflects many of the principles called for to adapt to the impacts of climate change, while focusing on educating neighbouring farmers. The primary actions include diversification of livelihood and agriculture, particularly engaging in what the Thais call integrated agriculture: growing a variety of native crops and raising animals in ways that mimic nature's interconnections, such as planting nitrogen-producing legumes alongside plants needing nitrogen, or raising fish in flooded paddy.[35] The focus for these 'activists' – a title they probably would not give themselves, but one that I find appropriate for their efforts – emphasises taking care of families first, only selling surpluses to have cash for those expenses that cannot be covered internally. Their work does not idealise a past that was more 'simple', but recognises the challenges they face in the modern, changing world, both through evolving economics and climate change.

All four of these people demonstrate that poor farmers are not dependent on outside organisations to make positive changes in their lives and livelihoods. While Somchai Jitsuchon is correct that for extreme climate change, the most vulnerable group, which includes poor farmers, needs assistance and broader

33 Lebel et al., "Adaptation to Climate Change," 138.

34 Roberts, "International Dimension," 186.

35 Darlington, Susan M., "Translating Modernity: Buddhist Response to the Thai Environmental Crisis," in *TransBuddhism: Transmission, Translation, Transformation*, ed. Nalini Bhushan, Jay L. Garfield and Abraham Zablocki (Amherst, MA: University of Massachusetts Press, 2009), 199–201.

mitigation efforts, I also agree with his assessment of the potential of farmers to adapt:

> First, farmers should use scientific data on impacts of climate change as much as they can, and combine this with their own inherited knowledge derived from long history of combating with climate uncertainty. Second, a process of continuous adaptation is needed. Third, information and experience sharing among farmers can help screening and speeding up of appropriate adjustment measures.[36]

The four people I discuss below are all engaged in aspects of this potential. They all work from extensive experience farming, dealing with weather and climate uncertainties, and local conditions. They each share their experiences and learning with other farmers to expand the impact of their ideas and efforts to help as many others as possible.

Their cases present a positive outlook to dealing with climate change and climate justice. As with the environmental justice movement in the United States, much of the energy focuses on empowering local people to address their situations, tapping into outside support when needed, but primarily using their own agency to build social justice.

3.1 *Integrating Agriculture, Relieving Debt*

Dang[37] was the son of a farming family in Nan Province. His family had few resources, so he decided to join the army as a way to help his family and his country. After years in the military, he resigned and returned to Nan to take up farming again. He had land from his family, and only a small amount of cash from the military. To make ends meet, he began growing corn as a cash crop, selling his full harvest only to buy his family's food. He borrowed to buy the seeds and chemical fertilisers the seed company promoted on the promise of repaying them when the harvest came in. With the variable weather, each year his harvest decreased, and he was not able to repay his debts. He borrowed more, both to feed his family and to get new seeds and more chemicals for his crops. The cycle spiralled down, leading him into considerable debt with few options.

In the mid 1980s, Dang realised there had to be an alternative, one that was also less damaging to the forest. He began to teach himself about integrated agriculture, and approached the monk Phrakhru Pitak Nanthakhun asking for

36 Somchai, "Adaptations to Climate Change," 8.

37 I do not use the real names of these farmers to protect them.

help.[38] Phrakhru Pitak helped him attend a seminar on integrated agriculture, so he could gain further knowledge to break out of his cycle of debt. The monk gave him some fruit tree seedlings and seeds of native plants and rice, with the stipulation that he would return new seeds to be given to another farmer in the future. Dang began gradually to shift his fields to integrated agriculture.

Initially, Dang only converted a small portion of his land, planting a small plot of native rice along with pepper plants and fruit trees. He bought a few chickens, which he let run free in the fields. They ate the insects and refuse from the fruit trees, and naturally fertilised the fields. These few plants and animals provided some of Dang's family's needs, enabling him to save some funds from his cash crops to begin to repay his debts.

As he saw the progress he made with this first experiment, Dang realised that by continuing to diversify his crops, simplifying his needs and spending less, he could move himself out of debt. Eventually, he ended up with surplus crops, particularly peppers, which he sold at the local market. Over four years, he converted all his land to integrated agriculture, purchased more chickens and several pigs, paid off his debts, and secured his family's livelihood. Along the way, Dang noticed improvements in the quality of his crops – without chemical fertilisers. He realised he could withstand changes in the weather more easily because of both the methods used and the diversity of crops and animals. By the time I visited his farm in the early 1990s, Dang was not only thriving, but using his farm as a model for other farmers seeking similar security.

Dang's case occurred before the fears of climate change were as great as they are now. His situation illustrates instead the social injustices many farmers face, both because of his lack of options when he left the military (the military ironically often seen as a road to security and a way out of debt) and the huge debt in which he found himself through cash cropping. Dang was not alone, as many small farmers find themselves in similar situations. Phrakhru Pitak identifies the main problem of the farmers as being in debt, which is only exacerbated by climate change.

3.2 *Rice and Self-Sufficiency*

Around the same time Dang was working his way out of debt and discovering a more sufficient way to live, another young man in Nan became concerned about Green Revolution rice in the province. Having gone to an agricultural school nearby, Somchai began to study the impact of high-yield strains of rice and the methods used to grow them.

38 '*Phrakhru*' and '*Phra*' are ecclesiastic titles for Thai Buddhist monks. As the monks discussed here are well-known public figures, I use their real names with their permission.

The Green Revolution rice was genetically modified, and its supporters promoted a single strain across the region. Green Revolution rice, they touted, would increase yields, and help Thailand meet its aims of becoming a leading rice exporter. The problem with this rice, according to Somchai, was the destruction of native strains, forms of rice that had developed and adapted to the local region. He encouraged growing multiple varieties of rice, rather than placing all of one's resources in a single strain. If it failed, a farmer would lose everything. If the farmer planted multiple types, it would be easier to survive a difficult year.

Thailand had become a major supporter of Green Revolution rice, which was coming out of the International Rice Research Institute (IRRI) in the Philippines. When IRRI began in the 1960s, it developed new strains of rice aimed at higher outputs to help relieve the world's food crisis. More recently, many climate change researchers realised that the high-yield crops emerging from the Green Revolution contribute to the climate change in several ways. Besides forests being cut down to make way for expanded rice paddies, and the additional chemicals used to produce larger harvests, rice fields contribute to greenhouse gases.[39] The high-yield varieties are not native to most of Thailand, and require extra work and funds to maintain, especially in mountainous places like Nan Province.

Somchai began an experiment to compare the yields of the newer high-yield rice variety with native rice. He formed a small, three-person organisation to collect native rice seeds and grow them in the traditional methods, using only natural fertilisers. He monitored water usage and yields, and compared his results with his neighbours' paddies of high-yield rice. Over time it began apparent that his rice fared better, and provided more rice at less cost with fewer negative environmental consequences. As many strains of native rice are dry rice, not requiring flooded paddy fields, they can be grown at higher altitudes and on slopes, a plus in the mountainous terrain.

In addition to emphasising native rice, Somchai noticed the success of his mother-in-law's 'kitchen garden'. She grew a range of vegetables, fruits and herbs in her house compound, buying little to feed her family. He supplemented her yields with his native rice, and the family only needed to purchase meat. The yards in most compounds are relatively bare, as people find it easier to buy food than to care for a garden.

39 Cf. Frolking, Steve et al., "Short- and Long-term Greenhouse Gas and Radiative Forcing Impacts of Changing Water Management in Asian Rice Paddies," *Global Change Biology* 10 (2004): 1180–1196.

After severe flooding, when much of the village was inundated by the nearby Nan River, Somchai and his family provided food to help people who lost their fields. Situated higher in the hills, his rice fields were not flooded. His mother-in-law's garden produced enough surplus to share with several neighbours who could not get to the market, and had little to spare to buy food with the loss of their primary cash crop.

Observing Somchai's success with native rice, and his mother-in-law's thriving kitchen garden, villagers began to adopt these approaches. Somchai gave native rice seeds to many farmers, and demonstrated the organic methods he used. In the early 2000s, Somchai expanded his operations and founded an agricultural research organisation focused on native plants on a small plot of land in his village. The organisation is a good example of local initiative, and its impact has spread. He now employs several people, and hosts interns and volunteers who want to learn more about the methods the organisation uses. He emphasises organic, integrated farming, stressing that diversification is best for both the farmers and the environment.

3.3 *Pigs, Food, and Self-Reliance*

The third example is Somchai's wife, Pin. While there is little doubt that she got some of her ideas from her husband, Somchai is the first to acknowledge how much he learned from her. As a civil servant, Pin did not earn enough to send their two children to college, but believed that education would be their opportunity to create a stable livelihood.[40] Through her own initiative, Pin bought a small plot of land a few kilometres from the village on which to raise pigs. I visited her pig farm in the summer of 2011.

Climbing out of the truck, I heard the pigs right away. They knew it was feeding time, and squealed loudly in anticipation. Pin began the process of feeding her 100-some pigs with a mixture of grain and protein. I followed behind, watching her and the pigs. Then I realised – it didn't smell. Every other time I have been at a pig farm, the stench was horrendous, but Pin's farm smelled more of the feed than the pigs themselves. It was due, she told me, to the way she set up the pens, sunk into the ground to keep the pigs cool, with lots of air circulation.

Pin began raising pigs only a few years earlier, buying a few animals to start. She spent just a couple of hours a day caring for the pigs, before and after her job. The pigs thrived. Pin soon began breeding them, expanding her pens, and

40 One might argue that her example does not fit with that of the 'average' Thai farmer. I include her here because many of the recommendations for adaptation includes educating farmers. Pin demonstrates that again farmers do not need to wait for assistance to accomplish this goal on their own.

preparing the surrounding land to farm. She planted trees – bananas, mangos, and other fruits – along with native, local rice and bamboo. The roughage she fed to the pigs, reducing what she had to buy for them, while the crops provided her family with food. Soon she sold some of the animals, while continuing to breed and expand. Within a couple of years, she was making enough from the pigs to put her two children through school. She raised sufficient crops that she rarely needed to buy food, and could afford to quit her job.

Neighbours noticed. Initially they scoffed at Pin's plan, believing there was too much work involved, and it would take far too long to recoup her investments. After another flood inundated much of her village, Pin's family was again one of the only ones that had surplus food. As with the previous flood, they shared their food and seeds with neighbours, and encouraged them to diversify their own livelihoods.

Pin first invited neighbouring children to help her at the pig farm. Those who expressed the most interest and commitment received a couple of young pigs from her to raise for themselves. She taught them her methods, helped them get the animals immunised and tagged, and answered any questions they had. By the time I visited the farm, approximately five or six other families were raising pigs and growing much of the food for the pigs and themselves.

3.4 *Buddhism and Farmers*

Finally, I look briefly at a monk who, like Phrakhru Pitak Nanthakhun, promotes integrated agriculture. Rather than financially supporting individual farmers, Phra Somkit established a model integrated agricultural farm on land behind his temple. He is part of a nationwide movement of a small number of Buddhist monks who undertake environmental activism based on their interpretations of Buddhist teachings and values. The primary goal of Buddhism is to relieve suffering. These monks see environmental problems as contributing to the suffering of farmers. The monks therefore engage in environmental activism as part of their efforts to help poor farmers end their suffering in this world, in conjunction with teaching them more general Buddhist principles.[41] As with Somchai and Pin above, I have known Phra Somkit for over twenty years, and have frequently visited his temple and village. His story provides a good example of the environmental monks (*phra nak anuraksa*) who carry out similar projects across Thailand.

41 For more on this movement, see Darlington, Susan M., *The Ordination of a Tree: The Thai Buddhist Environmental Movement* (Albany, New York: State University of New York Press, 2012); and Darlington, "Translating Modernity."

At first, Phra Somkit's farm appears chaotic and overgrown, but through careful observation and discussion with the monk and his brother, who manages the farm work, the plan comes into focus. Central to the farm is a fish pond, which provides water for native dry rice, banana, papaya and mango trees, and other perennial crops. Water buffalo, pigs, and chickens roam freely, rooting out weeds and insects and fertilising the land with their excrement. One section of the farm is devoted to herbs, with signs identifying the plants and their uses.

The farm began when Phra Somkit's father donated land he had cleared to grow feed corn, a prevalent cash crop in the province. The land supported the corn for only a few years before erosion and decreased soil fertility inhibited production. Rather than continue to add fertilisers and other chemicals to keep the corn growing, Somkit's father decided to make religious merit through giving the land to the temple. Within ten years of leaving the land fallow, Somkit proudly informed me, trees again covered the land and a diversity of plants reemerged around them.

His father's donation of land adjacent to temple grounds motivated Phra Somkit to learn about sustainable or renewable forms of agriculture. His primary concern focused on being able to help the villagers deal with familiar problems: land worn out by overuse and chemicals; cycles of debt to seed and fertiliser companies; migration to cities in search of work; and the loss of local skills and handicrafts.

To expand and promote his farm as a centre of education, Phra Somkit adapted the Buddhist tradition of *bindabat*, the monks' daily alms rounds for food. In addition to allowing villagers to make merit through giving him food, he encouraged them to donate land for the integrated agricultural farm.

Phra Somkit's farm demonstrates the benefits of integrated agriculture to the local farmers. Under the care of his brother and a few villagers, the farm has done well. The monk invites children to play and work at the farm, giving them specific projects for which they each take responsibility. The children take their young chicks or yield from their fruit trees home to share with their parents. They plant trees, which are labelled with their names, teaching them the ongoing responsibility to care for them. He leads them on 'dharma walks' through the forest to instil a sense of the value of the forest and discourage cutting it down. Phra Somkit's aim is to educate the children, encourage them to think beyond cash crops and material goals, and to pull their parents into the process.

The high yields, in addition to the regeneration of degraded land, appealed to the farmers. Gradually, they followed their children to volunteer on the farm as a new form of religious merit making and to gain the knowledge and skills necessary for this agricultural approach. The monk claimed in 2011 that every family in the village practised at least some integrated agriculture, lessening their reliance on cash cropping.

In addition to integrated agriculture, Phra Somkit undertook a research project on the traditional weaving that had died out in the village. He worked closely with the village women's association to interview older women who remembered the techniques, and convinced them to teach other women. They agreed to use only natural materials, such as local threads and natural dyes. Within a few years, the women's weaving group created a cooperative, and began to sell their cloth to supplement their agricultural incomes.

Through his fastidious Buddhist practice and teaching, Phra Somkit receives the respect of the villagers. At the same time, he uses his position to filter ideas, concepts, and values entering the village, hoping to steer villagers away from models of consumerism and development that he believes will incur debt and suffering. Instead, he selects methods and concepts he hopes will help villagers adapt to the world in ways consistent with Buddhist values. While his primary motivation is to end suffering – the primary goal of Buddhism – Somkit believes that helping farmers deal with immediate problems such as debt, changing agricultural and economic conditions, and social injustices is the foundation for a better life. The degree to which the villagers have followed Phra Somkit's examples and taken his advice indicates, first, their respect for him, and second, that they understand that they can help themselves.

4 Responding to Climate Change

The actions these four people undertake illustrate ways in which poor farmers can take control of their lives and livelihoods in ways that make them less vulnerable to both socio-economic injustices and the impacts of climate change. While most of the literature on adapting to climate change discusses the multiple levels on which people must respond, from the global to the regional and state to the local levels, the place where the poor farmers have the most control and say over their situations is local. Most of the responses scholars, policy advisers, and scientists recommend the Thai state operate at the global or state levels. A few focus specifically on assisting the most vulnerable, such as the poor farmers. What is interesting about these four cases – and others similar to them across the country – is that the farmers themselves are enacting many of the recommendations to adapt to climate change without being directed to do so, or even supported in most cases, by the state.

These recommendations include building resilience through incorporating both knowledge and practice, educating vulnerable people about climate change and scientific approaches to adapting, expanding and diversifying their livelihoods, and restoring the ecosystems on which they depend,[42] through taking care of land in sustainable ways, such as organic agriculture and crop diversification.[43] Although the farmers discussed here do not articulate their actions in terms of climate change, what they are doing fits with these recommendations. In lowering their own and their communities' vulnerability to social and economic injustices, they are better preparing themselves to adapt to how changes in the climate are impacting agriculture.

These farmers and models enact these recommendations in several ways. They draw on local knowledge and practices, such as Somchai's learning how native rice was grown before the introduction of Green Revolution varieties and applying these techniques to the current conditions. Through integrated agriculture, raising livestock with new methods, and tree planting, they incorporate new knowledge within local knowledge to find ways of maintaining a livelihood with minimal impact on the natural environment. In the process, they are restoring their ecosystems. The organic, integrated farming methods not only cause less damage, but allow the land to rejuvenate. They use natural rather than chemical fertilisers that they no longer need to purchase. Phra Somkit recognised the value of time, allowing the degraded land his father donated to lie fallow for ten years, letting native plants and trees take hold and restore the land. In all cases, they diversified both their crops and their livelihoods, making themselves more resilient to weather fluctuations or disasters, such as drought or floods. Their motives aim more at addressing economic and social disparities than climate change, but their actions help with the latter in the process.

5 Enacting Environmental Justice for Climate Justice

All four people discussed above, and the communities to which they belong, are part of a broader, albeit undefined movement that parallels the environmental justice movement in the US. As with the environmental justice movement,

42 Lebel et al., "Adaptation to Climate Change," 135; Agrawal, Arun, "Local Institutions and Adaptation to Climate Change," in *Social Dimensions of Climate Change: Equity and Vulnerability in a Warming World*, ed. Robin Mearns and Andrew Norton (Washington, D.C.: The World Bank, 2010), 180.

43 Marks, "Climate Change and Thailand," 242; Somchai, "Adaptations," 7, 9; Mearns, Robin, and Andrew Norton, "Equity and Vulnerability in a Warming World: Introduction and Overview," in *Social Dimensions of Climate Change: Equity and Vulnerability in a Warming World*, ed. Robin Mearns and Andrew Norton (Washington, D.C.: The World Bank, 2010), 10.

community is central for these people, and they are acting in creative ways to redefine how to live their lives in ways that simultaneously lessen the negative impacts of social and economic injustices and strengthen their own agency. Di Chiro remarked that what is significant about the environmental justice movement is that it is "transforming the possibilities for fundamental social and environmental change through processes of redefinition, reinvention, and construction of innovative political and cultural discourses and practices".[44]

Dang, Somchai, Pin, and Phra Somkit all engage, consciously or not, in similar processes. They act against the pressures from the dominant society to participate in a consumerist economy that emphasises cash cropping and materialism. They have found innovative ways to tackle the debt and inequalities that result from cash cropping, economic intensification, and environmental degradation, shifting the values on which villagers base their livelihoods from material accumulation to self-sufficiency. They are not trying to turn back the clock to an earlier, 'simpler' age, but to find ways within the modern world to live comfortably while caring for the environment of which they are a part.

Community is central to making these efforts work. These activists do not only work for themselves, but share their ideas, efforts, and even materials with neighbours and other farming families facing similar struggles of debt and challenging agricultural conditions. Again, Di Chiro's concept of community in the environmental justice movement fits the Thai farmers whom I have profiled when she writes, "[c]ommunity becomes at once the idea, the place, and the relations and practices that generate what these activists consider more socially just and ecologically sound human/environment configurations".[45] Pin's efforts to share the benefits that she has gained from raising pigs with her neighbours demonstrate how central this village community is to her. She is not simply working for herself. Phra Somkit embodies this dedication to his community through the ways in which he pulls families into his integrated farming. He gives seedlings, fish, chickens, and pigs to children and their families to support their own efforts at integrated farming.[46] He invites any villager to use the crops and livestock on the temple farm. Dang similarly uses his farm to teach others about his methods of integrated farming and to demonstrate the success he has had. He makes a particular point of informing people about the large

44 Di Chiro, "Nature as Community," 303.

45 Ibid., 310.

46 Phra Somkit gives away livestock and fish in particular because, as a Buddhist monk, he cannot harm any sentient being. The farm is on temple land, and while he raises fish and livestock, he cannot kill them. If he did not give them away, the farm would soon be overrun.

debt he was in when he started this approach. As with Phra Somkit, he stresses the patience and time it takes to restore the land and shift one's livelihood from cash-cropping to sufficiency.

None of these farmers is entirely removed from the market economic system. They participate actively, selling their surplus crops. In Pin's case, selling her pigs is how she gets the funds to send her children to school. The values they emphasise shift the focus from accumulation to sufficiency. None of these farmers is solely reliant on the market in order to meet their daily needs.

Besides reflecting individual actions that can help Thai farmers adapt to the impacts of climate change, these cases illustrate people acting to counter the social and economic injustices that many poor farmers face. Climate change does affect people already dealing with social, economic, and political inequalities. How their actions help them adapt to climate change reflects how they are dealing with social injustices.

Climate justice tends to focus on the inequalities between developed and developing nations, greenhouse-gas-emitting countries and resource-poor countries, elite and urban and rural poor, dominant citizens and indigenous or ethnic minorities, and intergenerational actions and aspirations. Mearns and Norton point out that these dichotomies oversimplify the realities and complexities within climate justice issues, stating, "[v]ulnerability to climate change, viewed first and foremost as a development challenge, cuts across national borders".[47] The focus remains, however, on adaptation and mitigation, responding to the immediacies and dangers of climate changes. Incorporating an environmental justice perspective places the focus on the people themselves, especially the most vulnerable. Di Chiro stresses the agency of these people, and their awareness of the larger forces against which they need to act:

> [...] environmental justice activists explicitly undertake a critique of modernist and colonial philosophies of unlimited progress, unchecked development, the privileging of Western scientific notions of objective truth and control of nature, and the hierarchical separation between nature and human culture.[48]

The distinction to be made is that climate justice activists tend to look specifically at how to address the inequalities emerging from the burden of climate-related impacts. Environmental justice activists remind us to consider the inequalities already in place, as well as how the various players involved under-

47 Mearns, and Norton, "Equity and Vulnerability," 7.

48 Di Chiro, "Nature as Community," 310.

stand and relate to their environments. While the justice issues included in both range from the global to the local levels, I focus here on the local. This is where the most vulnerable have the opportunity to act on behalf of themselves. Other levels are crucial in the long run for addressing the inequalities of both the socio-economic systems in place and the burdens of climate change, and must be considered for overall climate-related plans. As Agrawal states, however,

> [...] to strengthen the adaptive capacity of the rural poor population, governments and other external actors also need to understand, take advantage of, and strengthen already existing strategies that many households and social groups use singly or collectively.[49]

Examining the efforts of these rural activists through the lens of the environmental justice movement help us to do just this. In the process, we are reminded that these are real people, facing challenging situations, who have the capacity to act and speak for themselves.

---

49 Agrawal, "Local Institutions," 181.

CHAPTER 9

Mitigating Climate Change: An Additional Role for Technology and Policy in India as well as International Arenas

Rüdiger Haum

Abstract

The following chapter will argue that in order to integrate climate protection and economic development in India, both goals should be explicitly addressed in climate policy and supported with adequate resources. The argument is supported by a theory-based comparison of two policy programmes supporting photovoltaic technology. Although there is no linear or causal connection between the articulation of policy goals and policy outcome, the argument underlines that achieving both climate protection and economic development is unlikely to be realised if only the former is supported through policy. This finding stands contrary to politically articulated ideas of so called win-win situations, which are often suggested in international climate policy discourse to solve conflicts over climate protection and development.

1 Introduction

Many scientists and policy makers consider anthropogenic climate change the most pressing global environmental problem today, as they consider its solution the prerequisite for any form of sustainable development. The effects of anthropogenic climate change on the earth's ecosystem, human livelihoods and societies are, from their perspective, so severe that if not averted with fast and ambitious political means they will make any other efforts to protect the environment or support human development futile.[1] Responses to anthropogenic climate change are generally categorised under adaptation and mitigation. While the former refers to efforts to react to current or anticipated effects of anthropogenic climate change, the latter refers to efforts to reduce greenhouse gas emissions.

1 Rockström, Johan et al., "A Safe Operating Space for Humanity," *Nature* 461 (2009): 472–475.

 | DOI 10.1163/9789004273221_011

It is beyond doubt that rural as well as urban communities around the globe will be affected by climate change. Communities in developing countries including South Asia are in general particularly vulnerable due to limited capacities for adaption.[2] This chapter will draw attention to the fact that South Asian communities will also be affected by the way mitigation policy is formulated. The research discussed here suggests that if development aspects are not explicitly addressed in policy and adequately supported in the context of (international) climate policy, action taken to mitigate climate change might be at the expense of economic development. It will do so by comparing a Global Environment Facility (GEF) programme seeking to reduce carbon dioxide emissions by supporting photovoltaic technology in India with a rural electrification programme supported by the Indian government.

I will argue that climate protection adds additional requirements to technology and subsequently to policy and that both climate protection and development goals will have to be explicitly integrated and supported into mitigation policy in order to ensure that climate protection is not achieved at the expense of economic development. This argument stands counter to the political rhetoric of what are called win-win situations, which suggest that merely supporting climate change mitigation through the application of climate friendly technology might automatically entail economic benefits.

In order to support my argument I will proceed as follows: I will firstly sketch some developing countries' concerns in relation to anthropogenic climate change. I will secondly look into the role of technology and policy achieving economic development through industrial development and innovation as well as mitigation. I will thirdly compare the GEF Photovoltaic Market Transformation Initiative (PVMTI) in India with the Indian government's Solar Photovoltaic Programme (SPP) to empirically support my argument. Both projects support photovoltaic (PV) technology. I will fourthly discuss the effects of both programmes on Indian communities and draw some conclusions on international policy making.

2 Developing Countries' Concerns with Regard to Anthropogenic Climate Change

From the very beginning of international climate policy in the late 1980s, developing countries have been concerned that efforts to mitigate climate change

2 Parry, M. L. et al., ed., *Climate Change 2007: Impacts, Adaptation and Vulnerability* (Oxford: Oxford University Press, 2007).

might be undertaken at the expense of their own development goals. Although individual developing country governments' positions on international climate policy differ, numerous studies have revealed a broad consensus with regard to possible trade-offs between climate protection and development and linked common positions in international climate negotiations.[3]

The preamble of the United Nations Framework Convention on Climate Change (UNFCCC) states that:

> The global nature of climate change calls for the widest possible cooperation by all countries and their participation in an effective and appropriate international response, in accordance with their common but differentiated responsibilities and respective capabilities and their social and economic conditions.[4]

From the perspective of many developing countries, the principle of shared responsibilities entails that developed countries have to bear the main share of emission reductions and associated costs for three reasons. First, developed countries are responsible for the majority of historic greenhouse gas emissions that resulted from burning fossil fuel and land use change and laid the foundation for their high economic growth rates since the industrial revolution. They additionally possess better financial, economic, and technical capacities to mitigate emissions. Second, emission reductions within developing countries are perceived in the self-understanding of many developing country governments as a strain on economic growth and as unfair, as per capita, emissions in developing countries are much lower than in developed countries. Emission reductions are, in the interpretation of many developing countries, only justified after adequate levels of economic development have been achieved. A third very much related concern for developing countries is the aspect of additionality. Any form of mitigation support for developing countries shall be additional to and not at the expense of conventional development aid.[5]

3 Depledge, Joanna, "The Opposite of Learning: Ossification in the Climate Change Regime," *Global Environmental Politics* 6.1 (2006): 1–22; William, Marc, "The Third World and Global Environmental Negotiations: Interest, Institutions, and Ideas," *Global Environmental Politics* 5.3 (2005): 48–69; as well as Kasa, Sjur, Anne Gullberg, and Goril Heggelund, "The Group of 77 in the International Climate Negotiations: Recent Developments and Future Directions," *International Environmental Agreements* 8 (2008): 113–127.

4 *The United Nations Framework Convention on Climate Change* (Berlin: UNFCCC Secretariat, 1992).

5 Najam, Adil, Saleemul Huq, and Youba Sokona, "Climate Negotiations Beyond Kyoto: Developing Countries Concerns and Interests," *Climate Policy* 3 (2003): 221–231; Missbach,

3 Local Realities and Policy in India

India is one of the fastest growing economies in the world with a real growth in the gross domestic product (GDP) of 9.2 per cent in 2009, 8.8 per cent in 2010 and 7.6 per cent in 2011.[6] In absolute terms, India is the ninth largest economy in the world, with a GDP of US$ 1,727,111,096,363 in 2010.[7] India's per capita GDP is, however, considerably smaller. It was around US$ 1,400 compared to US$ 4,354 in China.[8] The country has a population of 1.1 billion, of which almost 42 per cent earn less than US$ 1.25 a day (at present purchasing power parity (PPP)) and almost 44 per cent of all children under five are considered malnourished.[9]

In 2005, India was the fifth largest emitter of greenhouse gases (excluding land use changes) with 1,865 metric tons of carbon dioxide equivalent ($MtCO_{2e}$). In terms of per capita emissions, India ranked 148th with 1.7t CO_{2e} per person in the same year.[10] As India is undergoing strong economic growth and shows high population growth rates, the country's emissions are expected to grow in the future, the absolute number depending on the growth rates of the economy and the population and efforts to decouple economic activities from greenhouse gas emissions.[11]

At the same time, India is vulnerable to effects potentially resulting from climate change. The most dramatic effects would result from changes in temperatures and rainfall patterns affecting agriculture that is necessary to support

---

Andreas, *Das Klima zwischen Nord und Süd: Eine regulationstheoretische Untersuchung des Nord-Süd-Konflikts in der Klimapolitik der Vereinten Nationen* (Münster: Westfälisches Dampfboot, 1999); Roberts, J. Timmons, and Bradley C. Parks, *A Climate of Injustice: Global Inequality, North-South Politics, and Climate Policy* (Cambridge: The MIT Press, 2007); German Advisory Council on Global Change (WBGU), *Solving the Climate Dilemma: The Budget Approach* (Berlin: WBGU 2009); Gong, Gloria Jean, "What China Wants: China's Climate Change Priorities in a Post-Copenhagen World," *Global Change, Peace & Security* 23.2 (2011): 159–175.

6 Organisation for Economic Co-operation and Development (OECD), *OECD Science, Technology and Industry Scoreboard 2007* (Paris: OECD, 2007).

7 The World Bank, "Gross Development Product Data (Current US$)," accessed January 16, 2012. http://data.worldbank.org/indicator/NY.GDP.MKTP.CD.

8 United Nations Development Progam (UNDP), "Per Capita GDP at Current Prices – US Dollars," accessed May 21, 2012. http://data.un.org/Data.aspx?d=SNAAMA&f=grID%3A101%3BcurrID%3AUSD%3BpcFlag%3A1.

9 International Monetary Fund (IMF), *India Sustainability Report 2011* (Washington: IMF, 2011).

10 Climate Analysis Indicator Tool (CAIT), "Per Capita Emissions," accessed May 19, 2012. http://cait.wri.org/.

11 International Energy Agency, *World Energy Outlook* (Paris: IEA, 2011).

the livelihood of two thirds of all Indians. Also, the melting of the Himalayan glaciers would have consequences for water security in the northern part of the country.[12]

In 2011, about 833 million Indians (72.2 per cent of the Indian population) lived in rural areas.[13] It is estimated that 70 to 80 million households, of which 92 per cent are located in rural areas, still depend on kerosene for lighting purposes. Furthermore, it is estimated that 124 million households in India rely on traditional biomass for cooking purposes.[14]

These few figures already indicate quite clearly that India needs, despite its impressive economic achievements, further economic development to reduce poverty, further rural electrification to increase the quality of life in rural areas, but also mitigation efforts to reduce its growing greenhouse gas emissions.

Since gaining independence from Great Britain in 1947, the Indian government aimed for industrialisation as part of the country's economic development strategy to substitute for imports. Initially, the focus was placed on large-scale capital-intensive industries such as steel, petrochemicals, engineering and machinery in the form of public sector companies. Soon, rural, small-scale industries producing consumer goods were also supported.[15] Likewise, the Indian government aimed at the increased electrification of rural areas.

On the international level, for example within UNFCCC negotiations, the Indian government has historically held the position that the country's most important goals are poverty alleviation, economic growth and food security. Greenhouse gas emission reduction is considered to be the responsibility of developed countries due to their responsibility for climate change resulting from historic emissions. This position is slowly changing and India has taken first steps to reduce domestic emissions with explicit reference to climate change. Many of the policy measures taken historically that unintentionally led to the reduction of greenhouse gas emissions, like the promotion of energy conservation or the support of renewable energy, were undertaken for reasons

12 Fujiwara, Noriko, *The Political Economy of India's Climate Agenda*, CEPS Working Document 35 (Brussels: Centre for European Policy Studies, 2010).

13 Census of India, "Census 2011," accessed May 18, 2011. http://censusindia.gov.in.

14 Srivastava, Leena, and I. H. Rehman, "Energy for Sustainable Development in India: Linkages and Strategic Directions," *Energy Policy* 34 (2006): 643–654; Bhattacharya, Subhes C., and Prasanta K. Dey, "Risk Management in a Large Rural Electrification Programme," *CEPMLP Internet Journal* 16.13 (2006): 1–24.

15 Mukherjee, Anit, and Xiaobo Zhang, "Rural Industrialization in China and India: Role of Policies and Institutions," *World Development* 35.10 (2007): 1621–1634.

of energy security or rural electrification.[16] With regard to low-carbon technology, the Indian government stresses on the international (UNFCCC) level that India needs to receive those technologies via international technology transfer, as it does not hold the technological capacities to develop them itself.[17]

4 The Dual Role of Technology with Regard to Climate Change

One possible solution to link greenhouse gas emission mitigation and economic growth, which would also address the concerns of developing countries, is the creation and realisation of what are known as win-win situations. Win-win situations are opportunities in which environmental protection through related policy also leads to economic benefits rather then additional costs. An often-mentioned example of a win-win situation is energy efficiency. Policy supporting or even mandating the application of energy efficient technology, for example, leads to decreased carbon dioxide emissions resulting from decreased fossil fuel consumption as well as economic gains through decreased spending for fuel (after the investment for the energy efficiency measure has been paid off). While win-win situations conventionally express the mutual benefit of two parties in a given situation, they often get a slightly different twist in the context of environmental and climate policy. The way win-win situations are articulated in climate policy discourse often suggests that the effort (including the financial means) in achieving a certain environmental (e.g. climate) goal will also lead to some sort of economic benefit without additional efforts or commitment of resources.[18] They are often associated with the application of

16 Wagner, Christian, "India: A Difficult Partner in International Climate Policy," in *International Climate Policy: Priorities of Key Neogiating Parties*, ed. Susanne Dröge (Berlin Stiftung Wissenschaft und Politik, 2010), 67–73; Ghosh, Prodipto, "Climate Change: Is India a Solution to the Problem or a Problem to the Solution?," in *Climate Change Perspectives from India*, published by United Nations Development Programme, India (New Delhi: UNDP India, 2009), 17–36. http://www.in.undp.org/content/dam/india/docs/undp_climate_change.pdf.

17 Haum, Rüdiger, *Transfer of Low-Carbon Technology under the United Nations Framework Convention on Climate Change: The Case of the Global Environment Facility and Its Market Transformation Approach in India* (Brighton: University of Sussex, 2011).

18 Muchie, Mamo, "Old Wine in New Bottles: A Critical Exploration of the UN's Conceptions and Mechanisms for the Transfer of Environmentally Sound Technologies to Industry," *Technology in Society* 22 (2000): 201–220; Bäckstrand, Karin, and Eva Lövbrand, "Planting Trees to Mitigate Climate Change: Contested Discourses of Ecological Modernization,

economic instruments in environmental policy.[19] References to possible win-win situations abound in discussions on climate change and climate policy.[20]

The GEF also aims to create win-win situations through their climate change projects, although it is very cautious about the usage of the term and has admitted that win-win situations are hard to actually create and implement through their projects.[21] However, the logic of the GEF PVMTI project discussed later follows a typical win-win argumentation.

4.1 *Theoretical Underpinnings of Win-win Situations and Policy Implications*

The discursive suggestions of win-win situations gain their attractiveness from an inherent, seemingly intuitive plausibility as well as the apparent possibility to reconcile goals that have for long been considered mutually exclusive: environmental protection and economic growth. Understanding the nature of win-win situations and being able to evaluate their chances of exploitation beyond political rhetoric requires some theoretical elaboration. Only by gaining a deeper understanding of the processes addressed one is able to derive suggestions about what kind of policy is needed for the manipulations of the processes in question.

Central to win-win situations is technological change, which in economics is often referred to as 'innovation'.[22] In the following, I will shortly introduce the

---

Green Governmentality and Civic Environmentalism," *Global Environmental Politics* 6.1 (2006): 50–75.

19 Economic instruments aim to change the polluting behaviour of actors through setting economic incentives, e.g. imposing a tax (additional cost) on water consumption rather then regulating water consumption through strict regulation (e.g. non-tradeable quotas).

20 E.g. Ockwell, David et. al., *UK–India Collaboration to Identify Barriers to the Transfer of Low Carbon Energy Technology*, Final Report (Falmer: SPRU-Science and Technology Policy, 2007); Bierbaum, Rosina et al., *Confronting Climate Change: Avoiding the Unmanageable and Managing the Unavoidable*, Scientific Expert Group Report on Climate Change and Sustainable Development (New York: United Nations Foundation, 2007); and El-Ashry, Mohamed, "Adaptation to Climate Change: Building Resilience and Reducing Vulnerability," in *Climate Change Policy: Recommendations to Reach Consensus*, ed. Brookings Institution (Washington: Brookings, 2009), 58–65.

21 Global Environment Facility, *GEF Contribution to Agenda 21 – the First Decade* (Washington: Global Environment Facility, 2000); Global Environment Facility, *Report of the GEF to the Twelfth Session of the Conference of the Parties to the United Nations Framework Convention on Climate Change* (Washington: Global Environment Facility, 2006).

22 Jänicke, Martin, "Ecological Modernisation: New Perspectives," in *Environmental Governance in Global Perspective-New Approaches to Ecological Modernisation*, ed. Martin Jänicke and Klaus Jacob (Berlin: FFU Report 01, 2006), 2–24; Jänicke, Martin, and Stefan

relationship between innovation and economic benefits and climate protections as well as the specific role of policy. This elaboration will serve as a framework for the discussion of the GEF programme and the Indian government's policy on PV technology.

In an economic sense, technology is understood as the knowledge needed to produce goods and services. The term 'innovation' refers to the commercially successful application of new knowledge, mostly within firms. The Organisation for Economic Co-operation and Development (OECD)[23] defines innovation as:

> The implementation of a new or significantly improved product (goods or service), or process, a new marketing method, or a new organisational method in business practices, workplace organisation or external relations.[24]

The concept of 'new' relates to new in the world, new to a country or new to a firm.

In economics, innovation is associated with economic benefits. From the perspective of a business firm, innovations offer revenues from new, improved or cheaper products. From the perspective of a national economy, primary innovation, the widespread diffusion of new technologies and related further innovation offers the possibility of growth of new industries and associated benefits like employment, increased household incomes and tax revenues. Innovation and related industrial development have been major sources of economic growth in Europe, the US and Japan, in several 'latecomer' economies (South Korea, Taiwan, Hong Kong, and Singapore), as well as in the emerging economies (China, India, Brazil). Given the contributions that innovation and industrial development might make to economic growth, it is no surprise that many developing countries consider industrial development as part of their

---

Lindemann, "Governing Environmental Innovations," *Environmental Politics* 19.1 (2010): 127–141.

23 The OECD currently comprises of 34 member states: Australia, Austria, Belgium, Canada, Chile, Czech Republic, Denmark, Estonia, Finland, France, Germany, Greece, Hungary, Iceland, Ireland, Israel, Italy, Japan, South Korea, Luxembourg, Mexico, the Netherlands, New Zealand, Norway, Poland, Portugal, the Slovak Republic, Slovenia, Spain, Sweden. Switzerland, Turkey, the United Kingdom and the United States.

24 OECD, *OECD Science, Technology and Industry Scoreboard*, 95.

overall development strategy and actively seek to support innovation via domestic policy and foreign support.[25]

The ability to use, develop and improve technology on a firm or country level may be captured in the notion of technological capabilities.[26] Technological capabilities refer to the ability of a firm to use and improve technology for their interest.[27] Bell and Pavitt define technological capabilities as "the resources needed to generate and manage technical change, including skills, knowledge and experience, and institutional structures and linkages"[28].

How do firms build technological capabilities for innovation? The formation of technological capabilities is a learning process. What can be learned depends firstly on the existing capabilities. Deciding and learning to do one thing will to a certain extent predispose the direction of future innovation as the learning and complementary strengths developed in the former effort provide a basis for the next round. Secondly, technological learning is an active process that requires "purpose change generating activities" and "setting in motion a process of technological and organisational change". Thirdly, the formation of technological capabilities is guided by access to relevant knowledge. Firms may generate knowledge through (re)searching in the form of their own research and

---

25 Soete, Luc, "International Diffusion of Technology, Industrial Development and Technological Leapfrogging," *World Development* 13.3 (1985): 409–422; Fu, Xiaolan, Carlo Pietrobelli, and Luc Soete, "The Role of Foreign Technology and Indigenous Innovation in the Emerging Economies: Technological Change and Catching-up," *World Development* 39 (2011): 1204–1212; Hobday, Mike, "East Asian Latecomer Firms: Learning the Technology of Electronics," *World Development* 23.7 (1995): 1171–1193; Hobday, Mike, and Howard Rush, "Upgrading the Technological Capabilities of Foreign Transnational Subsidiaries in Developing Countries: The Case of Electronics in Thailand," *Research Policy* 36.9 (2007): 1335–1356.

26 Archibugi, Daniele, and Alberto Coco, "Measuring Technological Capabilities at the Country Level: A Survey and a Menu for Choice," *Research Policy* 34 (2005): 175–194.

27 Technological capabilities are one prerequisite for industrial development and the result of technological learning, for which there are a number of sources: investments in different knowledge inputs, research and development, prototyping, design activities, taking advantage of tacit knowledge embodied in single persons or teams, the hiring of staff, interactions in inter-firm linkages (like supply chains and networks), technology imports, past learning efforts, investments in human capital, learning by doing, and linkages to public knowledge sources (research and development labs, technology intermediaries). Cf. Lall, Sanjaya, "Promoting Technology Development: The Role of Technology Transfer and Indigenous Effort," *Third World Quarterly* 14.1 (1993): 95–108; Bell, Martin, and Keith Pavitt, "Technological Accumulation and Industrial Growth: Contrasts between Developed and Developing Countries," *Industrial and Corporate Change* 2.2 (1993): 157–210.

28 Bell, and Pavitt, "Technological Accumulation," 163.

development activities, learning by doing, or through learning through linkages with other sources of relevant knowledge. Linkages might include research laboratories, consultants, customers, suppliers, trade shows, etc.[29]

The important point here is that the acquisition of technological capabilities is an active learning process and not all firms in all countries may innovate in all technological sectors easily. In fact, learning activities, especially if they should lead to the continuous improvement of the technology, might need public support. This is especially the case for developing countries, where firms on average show lower technological capabilities than those in developed countries.[30]

At the same time, technology also holds the potential to contribute to the solution of environmental problems including climate change. Environmental technologies are defined by the EU, for example, as technologies which, when applied, are less harmful than relevant alternatives.[31] Technologies reducing GHG emissions have been labelled as 'climate-friendly' or 'low-carbon' technologies (in those cases in which they aim at the reduction or substitution of carbon dioxide (CO_2) emissions).[32]

Pacala and Socolow have identified 15 sets of technologies to reduce carbon emissions, which can be grouped into four technology groups (see table 9.1).[33]

29 Bell, Martin, "Technology Transfer to Transition Countries: Are there Lessons from the Experience of the Post-War Industrialising Countries?," in *Technology in Transition*, ed. David Dyker (Budapest: Central European University Press, 1997), 68; Nelson, Richard, "Why Do Firms Differ?," *Strategic Management Journal* 12.2 (1991): 61–74.

30 Freeman, Christopher, "The Learning Economy and International Inequality," in *The Globalizing Learning Economy: Major Socio-Economic Trends and European Innovation Policy*, ed. Daniele Archibugi and Bengt-Åke Lundvall (Oxford: Oxford University Press, 2001), 320–334; Freeman, Christopher, and Luc Soete, "Developing Science, Technology and Innovation Indicators: What We Can Learn from the Past," *UNU-MERIT Working Paper #2007-001* (Maastricht, UNU-MERIT, 2007). There are of course exceptions.

31 European Commission, *Stimulating Technologies for Sustainable Development: An Environmental Technologies Action Plan for the European Union* (Brussels, EU Comission, 2004).

32 German Advisory Council on Global Change (WBGU), *World in Transition: A Social Contract for Sustainability* (Berlin: WBGU, 2011); Ockwell et al., "UK-India Collaboration."

33 Pacala and Socolow also list examples of changes in technology-user behaviour (less car use) and changes in land use practices (carbon sinks) that will be neglected in this context. Cf. Pacala, Stephen, and Robert Socolow, "Stabilization Wedges: Solving the Climate Problem for the Next 50 Years with Current Technologies," *Science* 305.13 (2004): 986–972. I have also excluded carbon capture and storage technologies as they do not reduce the carbon dioxide emissions but rather enable alternate disposition.

TABLE 9.1 *Climate change mitigation technologies*[34]

	Group	Example
1	(More) energy-efficient industrial technologies (capital goods and production processes)	Steel production technology requiring less energy input than comparable processes with similar results
2	(More) energy-efficient end-use products	More energy-efficient cars, insulation technologies in buildings
3	Less carbon-intensive energy generation technology	More efficient coal-based power stations, more gas-based power plants
4	Low-carbon energy generation technologies	Renewable energy technologies (large scale and small scale)

From a climate change perspective, rural electrification through PV technology opens up the possibility to displace or avoid carbon dioxide emissions, as rural households typically use kerosene lanterns, batteries charged through diesel generators or paraffin lamps to satisfy their lighting needs. If solar photovoltaic applications are used as an alternative to a grid connection, they also avoid carbon dioxide emissions as almost all electricity supplied through the grid is based on fossil fuel power stations in India.[35]

Emissions are reduced through the diffusion of an environmentally benign technology. The overall amount of emissions reduced is a function of the level of the operating effectiveness of a specific installation of technology and the number of applications relative to emissions from the technology that has been replaced.[36] How does climate friendly technology diffuse? Diffusion is the

34 Chart created by the author after figures provided in Pacala, and Socolow, "Stabilization Wedges."

35 Drennen, Thomas E., Jon D. Erickson, and Duane Chapman, "Solar Power and Climate Change Policies in Developing Countries," *Energy Policy* 24.1 (1996): 9–16; Kaufman, Steven et al., *Rural Electrification with Solar Energy As a Climate Protection Strategy*, REPP Research Report No. 9 (Washington: REPP, 2002); Taele, B. M., K. K. Gopinathan, and L. Mokhuts'oane, "Renewable Energy Technologies for Rural Development in Lesotho," *Renewable Energy* 35 (2007): 609–622.

36 Kemp, Rene, *Environmental Policy and Technical Change* (Cheltenham: Edward Elgar, 1997); and Jänicke, Martin, "Ecological Modernisation: New Perspectives," *Journal of Cleaner Production* 16 (2008): 557–565. Effectiveness in this context relates to the technical efficiency of a technology (e.g. the efficiency of a photovoltaic cell) and the way it is operated (e.g. properly maintained etc.).

aggregate outcome of multiple adoption decisions of technology users. Whether or not a technology is adopted depends on whether adopters know about the technology, whether adopters associate particular benefits with the technology relative to alternatives, technology cost, and whether the technology may be adapted to specific user needs.

Adoption is also influenced by whether and when adopters receive information. Diffusion rates differ because some adopters learn faster about certain technologies than others. In addition, the quality of the information is relevant as it must be sufficient to allow adopters to consider the risks and benefits of adoption. Firms have different internal characteristics (available capital, willingness to pay, size, competencies, vintage technology stock, search cost, input prices, factor productivity and regulatory cost) that will lead to different perceived benefits in relation to the cost of adopting a technology. Adopters perceiving benefits to be high will adopt the technology while others will do so later or not at all. If costs fall over time and perceived benefits change, the number of adopters is likely to increase. Cost reduction of technologies may result from five factors: technological progress, input price changes, internal efficiency improvements, learning by doing, and economies of scale.[37] The next aspect relates to learning in order to productively apply the acquired technology and learning to change and to adapt the technology to different user needs.

The diffusion of low-carbon technologies usually requires some policy intervention. Environmental economists argue that polluters do not invest in environmental technologies due to a double market failure. The first market failure is that polluting firms cannot pass additional costs resulting from investments in environmental technologies on to customers. Customers are considered unwilling to pay additional costs for environmentally friendly goods. The second market failure is due to the public nature of the environmental firms investing in environmental technology, which, as a result, cannot fully claim the benefits of their investments. Competitors not investing in clean technologies also benefit from the reduced pollution as the need to invest further decreases with investments taken by others.[38] Industry is only likely to adopt environmentally friendly technologies if induced by the state.

37 Ibenholt, Karin, "Explaining Learning Curves for Wind Power," *Energy Policy* 30 (2002): 1181–1189.

38 Cf. Jaffe, Adam B., Richard G. Newell, and Robert N. Stavins, "A Tale of Two Market Failures: Technology and Environmental Policy," *Ecological Economics* 54.2/3 (2005): 164–174; and Rennings, Klaus, "Redefining Innovation – Eco-innovation Research and the Contribution from Ecological Economics," *Ecological Economics* 32 (2000): 319–332.

In addition, political science, most notably in the form of the concept of ecological modernisation, has argued that policy intervention is a necessary condition for the diffusion of low-carbon technologies, as users have little incentive to adopt them without this.[39] Supply, demand and markets for environmental technologies are primarily created through governments.[40]

In conclusion, the reduction of greenhouse gas emissions gives technology another role and an additional objective. From an economic perspective, technology understood as the knowledge to use, produce and improve goods and services serves industrial development. Hence, one goal of policy is to increase the technological capabilities of firms for production and innovation (regardless of where the produced goods and services are consumed). Policy aiming at the increase of technological capabilities should therefore support research and development activities, the education of the workforce, knowledge networks, supplier-customer relationships with relevant policy instruments etc. if it aims to develop technological capabilities and innovation.

In order to reduce emissions, technology must be effectively applied as end-use technology as often as possible. The main goal of policy must therefore be to support the diffusion of such technology (regardless of where climate-friendly technologies are produced). In order to support diffusion of climate technologies, policy might hence promote information about the technology, its distribution, moderate the extra cost for adoption, or directly support adoption.

There is a long and complex discussion within political science as well as the interdisciplinary field of innovation studies on how to achieve mitigation and the formation of technological capabilities. These cannot be discussed here at length. However, the recapitulation of some basic insights from each of the two serve as the basis for the main argument of the chapter, namely, that in order to integrate environmental and developmental aspects, the explicit support of both goals is necessary and favouring one might lead to neglecting the other.

Environmental policy analysis has underlined, on the basis of numerous empirical studies, that explicit, clear formulation of the various goals of a policy instrument is a major prerequisite for its successful implementation. If the policy aims to achieve a number of goals, they should not only be explicitly addressed and if possible quantified, but should also not be in conflict with each

39 Jacob, Klaus et al., *Lead Markets for Environmental Innovations* (Mannheim: Physica, 2005).

40 Lanjouw, J. O., and A. Mody, "Innovation and the International Diffusion of Environmentally Responsive Technology," *Research Policy* 25 (1996): 549–557.

other.[41] In line with that insight, policy analysis concerned with achieving environmental protection (including climate protection) through innovation needs to address not only effective diffusion, but also technological innovation. If, for example, incentives for the diffusion of a certain technology are given, the technology in question would possibly need to be adapted through innovation. As stated before, corresponding innovation is not necessarily undertaken by private actors, and must be supported by policy as well. The underlying instruments, however, differ from those supporting diffusion. In other words, supporting diffusion might not necessarily lead to any innovation. Likewise, support for technology development and innovation alone does not guarantee widespread application of a certain technology.[42]

The insights from environmental policy mirror insights from innovation studies. Innovation studies have long been pointing out that demand for a certain technology is insufficient to trigger adequate innovation. At the same time emerging innovation, e.g. as a result of research (supply), is insufficient to automatically stimulate demand and diffusion.[43]

What does this mean for win-win situations? As seen above, achieving environmental as well as economic benefits relates to two different processes and requires the application of different policy instruments. It seems unlikely that policy support for one goal automatically supports achieving a different goal.

41 Jänicke, Martin, Philip Kunig, and Michael Stitzel, *Lern- und Arbeitsbuch Umweltpolitik: Politik, Recht und Management des Umweltschutzes in Staat und Unternehmen* (Bonn: Diez, 2002); Haum, Rüdiger, Klaus Jacob, and Julia Hertin, "Integration von Umwelt-, Innovations- und Industriepolitiken: Eine Evaluation von europäischen Strategien und Maßnahmen," in *Nachhaltig regieren: Eine Bilanz zu Governance Prinzipien und Praktiken*, ed. Reinhard Steurer and Rita Trattnig (München: Oekom, 2010), 97–122.

42 Jacob et al., *Lead Markets*; Jänicke, and Lindemann, "Governing Environmental Innovations."

43 Dosi, Giovanni, "Technological Paradigms and Technological Trajectories: A Suggested Interpretation of the Determinants and Directions of Technical Change," *Research Policy* 11.3 (1982): 147–162; Dosi, Giovanni, L. Marengo, and C. Pasquali, "How Much Should Society Fuel the Greed of Innovators? On the Relations between Appropriability, Opportunities and Rates of Innovation," *Research Policy* 11.3 (2006): 1110–1121; Freeman, and Soete, "Developing Science, Technology and Innovation Indicators;" Mowery, David, and Nathan Rosenberg, "The Influence of Market Demand Upon Innovation: A Critical Review of Some Recent Empirical Studies," *Research Policy* 8.2 (1979): 102–153.

5 The GEF PVMTI Project in India and the Solar Photovoltaic Programme of the Indian Government

The simple theoretical elaborations introduced in the preceding paragraph will serve as a discussion point for two very different efforts related to technology and climate change in India in the following. In general, both innovation and diffusion may be supported through a number of very different policy instruments. The choice of policy instrument depends, generally, on the characteristics of the technology, its development stage, the characteristics of the target group of the policy, and the goals of the policy. Within the context of this chapter, we cannot go into discussing the appropriateness of different policy instruments in different settings. The following comparison will, however, highlight how the formulation of goals is reflected in the choice of policy instruments and how this affects policy results. Of course, policy outcomes are influenced by many factors during policy implementation and policy analysis has long warned against a mechanistic interpretation (goal determines → instrument → determines outcome) of the policy process. However, within international policy negotiations, emphasising the importance of articulating and striving for support of different goals is of crucial importance. Keeping in line with the theoretical framework, after a short introduction of each programme, I will discuss its goals, the chosen policy instruments and the results. The discussion is based on an in depth case study including document analysis, expert interviews and secondary literature review.[44]

5.1 *Overview*

The Global Environment Facility (GEF) was founded in 1991 and currently serves as the financial mechanism for four international environmental conventions (The Convention on Biological Diversity, The United Nations Framework Convention on Climate Change, The United Nations Convention to Combat Desertification, and the Stockholm Convention on Persistent Organic Pollutants). In its role as a financial mechanism to the UNFCCC, the GEF aims with regard to mitigation to "support developing countries and economies in transition towards a low-carbon development path".[45] In order to achieve such goals, the GEF operates "the largest and most comprehensive global portfolio of investments in energy efficiency, renewable energy and other climate-friendly

44 A more detailed analysis is found in Haum, *Transfer of Low-Carbon Technology*.

45 Global Environment Facility, *GEF 5 Focal Area Strategies* (Washington: GEF, 2011), 17.

projects".[46] According to the last publicly available GEF internal evaluation, the climate programme consisted of 659 projects from 1991 to 2009, for which it has allocated US$2.74 billion.

The GEF has financed projects supporting off-grid solar photovoltaic technology since 1991, aiming primarily at applications for rural households.[47]

The Photovoltaic Market Transformation Initiative (PVMTI) project is one of the earliest projects financed through the GEF and follows the market transformation approach. PVMTI took place from 1998 to 2009 and was implemented in Kenya, Morocco and India. In the following, the results of PVMTI India will be presented as of 2007. The Global Environmental Facility bore the US$30 million cost of the project and the International Finance Corporation (IFC) served as the implementing agency.

The Indian government started considering renewable energy sources as early as 1971. It started with the formation of the Fuel Policy Committee (FPC) in 1971, followed by the formation of the Working Group on Energy Policy (WGEP) in 1977, the Commission for Additional Sources of Energy (CASE) in 1981 in the Department of Science and Technology (DST), the Department of Non-conventional Energy Sources (DNES) in 1982, and the Advisory Body on Energy (ABE) in 1983. In 1992, the Ministry of Non-conventional Energy Sources (MNES) was founded, which was renamed the Ministry of New and Renewable Energy in 2006.

Being a planning economy and relying strongly on central government control at that time, support for renewable energy was included in the government's five-year plans. Significant financial support started in 1980 within the sixth five-year plan, in which the government allocated almost 1.1 billion Rupees (0.1 per cent of the overall budget).[48] The Solar Photovoltaic Programme (SPP) is part of the Indian government's wider renewable energy programme, which was launched primarily as a response to the 1970s energy crisis. The SPP started in 1975 and is still running today.

46 Eberhard, Anton et al., *GEF Climate Change Program Study* (Washington: Global Environmental Facility Office of Monitoring & Evaluation, 2004), 1.

47 Martinot, Eric, "Renewable Energy Markets and the Global Environment Facility," *Financial Times*, February 12, 2000.

48 Panda, Haribandhu, "Review of Renewable Energy Policy in India," in *Biomass Based Decentralised Power Generation*, ed. B. S. Pathak and N. Srivastava (Gujarat: Sardar Patel Renewable Energy Research Institute, 2005); Sastry, E. V. R., "The Photovoltaic Programme in India" (paper presented at the IEA PVPS International Conference, Osaka, 2003); and Sastry, E. V. R., "The Photovoltaic Programme in India: An Overview," *Solar Energy Materials and Solar Cells* 47 (1997): 63–69.

5.2 *Goals and Policy Instruments*

The main goal of PVMTI has been to:

> Stimulate PV business activity in selected countries and to demonstrate that quasi-commercial financing can accelerate its sustainable commercialisation and financial viability in the developing world.[49]

If private sector sales companies were to receive public funding, these companies would succeed in increasing sales of small-scale PV systems (solar home systems, solar lanterns, solar streetlights etc.).

The GEF expected that 11 subprojects (four of them with a low probability) would receive funding in response to the PVMTI request for proposals. The number of possible investments was actually expected to exceed PVMTI funds.[50] The IFC further expected that PVMTI-funded projects (plus those companies imitating PVMTI-funded business models) would install a total of 10 MW PV generation capacity from 1998 to 2003 in addition to all other Indian PV installations (which were projected to amount to 8 megawatts (MW) installed capacity).[51] The targeted users are private end-users able to afford electricity but unlikely to receive a grid connection and commercial users in need of non-grid energy sources. Secondary goals were the displacement of carbon dioxide. The GEF further assumes that the total avoided carbon emissions would be 1,207,800 tonnes of carbon (C). This aggregate figure for expected environmental benefits was not broken down for India, Kenya and Morocco.

To achieve both goals, PVMTI lowered the cost and risk for private businesses to develop the necessary market infrastructure as a basis for increased sales through subsidised finance in the form of equity, soft loans, and bank default guarantees. Offered finance should be used as working capital for the development of sales and distribution systems or to subsidise down-stream, end-user finance. It was further hoped that the subprojects would be commercially viable by the end of the project period in 2009 and in the position to repay parts of the IFC/GEF capital. By supporting the distribution of PV application and reducing cost for individual PV systems, the PVMTI project supports instruments for the diffusion.

---

49 International Finance Corporation, *India, Kenya, and Morocco: Photovoltaic Market Transformation Initiative (PVMTI)*, Project Document (Washington: International Finance Corporation, 1998), 5.

50 Ibid.

51 Ibid.

The strengthening of technological capabilities and further innovation is not directly supported. However, the project excludes neither as being a result of increased demand created through PVMTI. The GEF expected "additional investment in manufacturing" as a result of growing demand created through the PVMTI supported sales companies.[52] This expectation mirrors a classical win-win situation. Aiming at climate protection through the support of diffusion of PV applications shall lead to investment in manufacturing and technological capabilities (not supported), promising wider economic benefits.

Let us now look at the second programme under discussion. The SPP of the Indian government aims at developing commercially viable PV applications, creating a strong manufacturing base as well as energy access through the diffusion of PV applications to remote and rural areas of India for electrification. It covers different PV applications like solar streetlights, solar home systems, solar water pumping, solar lanterns, and PV power plants. While the goals of the programme have remained the same over the past forty years, the emphasis on each of the goals as well as the policy instruments changed over time. The initial focus was on industrial development and the production of technically feasible PV applications. The main policy instruments were the establishment of pilot production plants for cells and modules, financing research and development, as well as setting up demonstration projects. Other project components included the field-testing of applications, the training of engineers for field installation, and awareness-raising for PV technology among government agencies and ministries.[53] In the mid-1980s, the Indian government spent 13 times more money on the research and development (R&D) programme than in the demonstration programme (65,000 lakhs[54] of rupees on R&D, 5,000 lakhs on demonstration programmes).

From the mid-1980s onwards, the aim of widespread application of PV systems gained more importance. The applied policy instruments consisted of large government procurement programmes through different ministries and state agencies (e.g. 25,000 PV systems for powering rural radio communication systems) and increased rural electrification programmes with an emphasis on PV powered streetlights. The Ministry for Renewable Energy bought PV modules and allocated them to the state renewable energy agencies. The state renewable energy agencies organised system integration and developed and

52 International Finance Corporation, "India, Kenya, and Morocco," 16.

53 Department of Non-Conventional Energy Sources (DNES), *Annual Report 1985–1986* (New Delhi: Ministry of Energy, 1986); The Energy and Resources Institute (TERI), *Report on Solar Photovoltaic Sector Review and Market Potential in India* (Delhi: TERI, 2006).

54 One lakh stands for an amount of 100,000.

implemented deployment projects through which systems were installed.[55] From the 1990s, the emphasis changed yet again towards increased application. Spending on research and development were further decreased but private investment in PV manufacturing were permitted and supported through tax breaks etc. Government procurement of PV application was stopped in the mid-1990s, but the rural electrification programme expanded and was combined with an end-user subsidy scheme. It contributes a certain sum to end users willing to buy PV applications like solar home systems (SHS). Private companies were also allowed to supply PV systems to the government programmes and to directly sell to end-users.[56] The subsidy scheme was modified several times but its core principle remains unchanged.

5.3 *Results*

An overview of both programmes' results is given in table 9.2. With regards to diffusion, PVMTI made investments in four companies. Their activities resulted in the sale of approximately 70,000 solar home systems in four Indian states (West Bengal, Andhra Pradesh, Karnataka, Kerala). The contribution of each of the four companies varies significantly and the overall results are well below initial GEF expectations. With regard to the formation of technological capabilities, no PV manufacturing company in India had made any investment in production technology due to the demand created through PVMTI. Their first and foremost explanation was that the demand created through PVMTI sales was too small to stimulate further investment in production technology or any other form of learning to innovate. Furthermore, they explained that even if demand was greater, companies might have invested in increased production, but even then would have seen no reason for increased technological capabilities and innovation because the kind of demand created could be met with existing technological capabilities.

The Solar Photovoltaic Programme of the Indian government (SPP) achieved the sale of approximately 65,000 SHS during roughly the same period of time in the same regions of PVMTI activity. With regard to the formation of technological capabilities, the SPP supported (over a long period of time) the creation of PV industry that included 20 PV cell and module manufacturers and held a world market share of almost two per cent in 2009.[57]

55 DNES, *Annual Report.*

56 Ministry of Non-Conventional Energy Sources (MNES), *Annual Report 1993–1994* (New Delhi: MNES, 1994).

57 European Photovoltaic Industry Association (EPIA), *Global Market Outlook for Photovoltaics until 2015* (Brussels: EPIA, 2009); and Haum, *Transfer of Low-Carbon Technology.*

TABLE 9.2 *Overview of project results*[58]

	PVMTI			SPP	
Direct goals	Diffusion: Sales activities		Technological Capabilities	Diffusion: Energy access	Technological Capabilities: industrial development
Indirect goals		CO_2 displacement	Industrial development		
Instruments	Company finance, end-user finance		None	State programmes support private companies, end user finance	R&D, state production, support private investment, testing, demonstration
Results	Four companies funded, 70,000 SHS sold in four Regions (2001 to 2007)	92,503.2 t CO_{2e} substituted	None	Approx. 65,000 SHS sold between 2001 and 2006 in four regions. Approx. 450,000 SHS systems sold all over India (1986–2009).	20 PV cell and module manufactures, 90 MW production capacity -modules, 50 MW production capacity cells, cell efficiency between 14 and 16 per cent.

Much is to be said about the individual project results and their explanations. PVMTI did not go as expected and has suffered from various conceptual flaws and implementation difficulties.[59] Most relevant in this context are the relation

58 Source: Haum, *Transfer of Low-Carbon Technology*, 169.

59 Results for both projects are based on official government data, GEF publications and own research. For a more detailed discussion, see Haum, *Transfer of Low-Carbon Technology.* For a detailed discussion of results, their reliability and in-depth explanations see Haum, Rüdiger, "Energy Security, Rural Electrification, and Market-Based Climate Change

of project goals and project outcomes as well as a project comparison. PVMTI aimed for the diffusion of PV technology by supporting private sales activities. PV sales companies received subsidised finance to develop PV distribution and to subsidise end-consumer loans. Very much related was the goal of CO_2 substitution. Both goals were directly supported and achieved, although there are indications that many of the PV systems were not sold in rural markets but to urban customers. These usually use PV systems not to substitute kerosene lamps or other fossil energy sources, but as back up systems for power failures. In this case, no CO_2 is substituted.

Investment in manufacturing, which possibly would have led to increased technological capabilities, was expected as a reaction to growing demand. This (secondary) goal was hence not directly supported and not achieved. Companies explained that the PVMTI project gave them no incentives to do so. All investments in increasing their technological capabilities were made as reactions to demand abroad and competing firms abroad that were supplying their products to the same markets. PVMTI had effectively left the choice of investing in technological capabilities to third parties and relinquished control over it by not making it an explicit project goal or directly supporting it with adequate instruments.

The Indian government's SPP aimed for diffusion of PV systems for energy access and the development of a PV industry and supported both goals with various instruments. Both goals were achieved, but again, strong qualifications must be made. The list of criticisms of the government programmes supporting PV for rural electrification is, despite its achievements, long. The SPP has not been aiming directly at mitigating CO_2 emissions, so the extent to which the programme contributed to emission substitution is unknown. However, we can assume that some substitution has taken place. And although diffusion was supported with energy access as a priority, supporting diffusion with mitigation in mind might have led to different results and as yet unknown trade-offs.

Also, although the Indian PV industry might have grown substantially over the past twenty years, it is certainly not the most competitive in the world.

Intervention by the Global Environment Facility in India: The Limits of Transformation," *International Energy Journal* 11.4 (2010): 115–125; Haum, *Transfer of Low-Carbon Technology*; and Haum, Rüdiger, "Project Based Market Transformation in Developing Countries and International Technology Transfer: The Case of the Global Environment Facility and Solar PV," in *Low-Carbon Technology Transfer to Developing Countries: From Rhetoric to Reality*, ed. David Ockwell and Alexandra Mallet (London: Earthscan, 2012), 185–210. Assumptions on sales had to be made for a number of years in SPP (on a national and regional level), as annual data is incomplete.

Interviewees were quite frank to admit that they were neither cost competitive, e.g. with PV manufacturers in China, nor did they possess the technological capabilities to produce 'state of the art' PV technology, like cells with 20 per cent efficiency or thin film cells. The Indian government, has it seems, despite the initial success in industrial development, neglected to support the further formation of technological capabilities of its domestic PV industry. However, in comparison to PVMTI, industrial development was directly supported and directly achieved.

6 Conclusion

Developing countries are concerned about the neglect of development issues within international climate policy. India is still in a position where any effort to reduce greenhouse gas emissions must not be made at the expense of its own economic development ambitions. More importantly, environmental protection and growth must both be achieved. The open and hotly debated question is whether and how both may be combined.

The policy discourse of win-win situations often suggests that both goals can be combined. The comparison of PVMTI and SPP suggest that combining environmental protection in the form of climate protection and economic benefits in the form of industrial development through innovation is certainly possible. However, both goals must be formulated and addressed individually as well as supported with adequate resources through domestic government or other institutions like international organisations. This finding is in line with the theoretical framework developed above. However, it runs counter to the political rhetoric of win-win situations, which suggest that addressing emission reduction might lead to positive economic benefits.

Local realities for Indian communities are most certainly affected by the way climate policy is framed and articulated. The diffusion of PV technology and associated electrification brings many benefits like improved lighting or averting climate change (although emission reduction potential is small when compared with industrial production). Both PVMTI and SPP underline that rural electrification may be combined with mitigation. However, both programmes did not include mitigation as direct goals (PVMTI only as a secondary goal). A stronger integration of the goal would have possibly led to a different design of the programme, which could have in turn led to reduced CO_2 emissions.

The development of an Indian PV manufacturing industry and continuous innovation opens the possibility of economic benefits, either directly through employment in the industry's value chain or indirectly through increased governmental public welfare spending through tax revenue (provided the Indian government makes the effort). Investments in manufacturing, innovation and technological capabilities are most certainly not a guarantee for successful, competitive firms since it takes more to effectively sell products in the world markets. They are however a crucial contribution in many industries.

Indian climate policy should be aware of the important role technology can play in mitigating domestic greenhouse gas emissions. At the same time, Indian policy makers should not succumb to the temptation of win-win situations and leave issues of green industrial development purely to demand, markets, and market actors. At the same time, it is important for Indian policy makers to aim for the explicit support of both climate and development goals on an international policy level.

The analysis undertaken here is not so much a critique of the GEF, as its primary task is the protection of the global environment and it certainly does not possess the financial means (resulting from the contributions of nation states) to support greenhouse gas mitigation as well as industrial development. It is more a critique of the international climate policy communities of developed countries, which are often at pains to support climate mitigation as well as necessary economic development in developing countries. Although changing the agenda of the GEF and increasing its resources could be a step in the right direction, the decision is to be made by the leaders of the powerful, yet strongly polluting, industrialised nations.

CHAPTER 10

Institutional Dynamics of Climate Change Adaptation in Southeast Asia: The Role of ASEAN

Dennis Eucker

Abstract

There is an increased interest in institutions and the role they play in the context of adaptation to climate change. The present chapter contributes to the growing debate by providing an analysis of the Association of Southeast Asian Nations (ASEAN). The chapter shows that while member states originally made use of the institutional framework of ASEAN to generate a common understanding of climate change mitigation and to apply it externally, the association is now becoming increasingly accepted for internal framing of adaptation processes. Yet while ASEAN has undoubtedly 'come to terms' with climate change, the notion of sustainability, which is a key concept for any adaptation strategy, remains on a broad rhetoric level only. Results thus show that without elaborating more fundamental linkages between sustainability and the overarching topic of economic development, any adaptation strategy will be prone to fail.

1 Introduction

The rapid rise in warming of the Earth's surface over the last fifty years is well accepted among scientists. Arguably, there still are uncertainties about the degree to which the climate is changing and some debates whether the experienced warming of is entirely due to the 'greenhouse effect'. But the fact that the climate is changing is not in doubt.[1] Two of the central messages of the Intergovernmental Panel on Climate Change (IPCC)'s Third Assessment Report

1 Levine, Simone, Eva Ludi, and Lindsey Jones, *Rethinking Support for Adaptive Capacity to Climate Change: The Role of Development Interventions* (London: ODI, 2011); Schelling, Thomas, "Climate Change: The Uncertainties, the Certainties, and what they Imply About Action," published by The Berkeley Electronic Press, 2007, accessed May 31, 2012. http://are.berkeley.edu/courses/EEP131/fall2007/ClimateChange/Climate%20Change%20The%20Uncertainties...%20%28Schelling%29.pdf; Oreskes, Naomi, "The Scientific Consensus on Climate Change," *Science* 306.3 (2004): 1686.

 | DOI 10.1163/9789004273221_012

(TAR) in 2001 – and which were reinforced in the Fourth Assessment Report (AR4) in 2007 – were that, first, climate change is already having observable and adverse impacts and, second, that due to time lags in natural and physical systems, mitigation measures alone cannot avoid climate change entirely any more.

The TAR has demonstrated the importance of adaptation,[2] a topic that then received extensive coverage and was demonstrated in its complexity also in the AR4.[3] Over the past decade, the discussion on adaptation to climate change has therefore truly started.[4]

While the projected consequences of climate change remain highly uncertain, they have been spreading unequally across regions so far, and will continue to do so. Evidently, the impacts of climate change will affect developing countries to a disproportionately high degree.[5] But response options are said to be limited as they possess only a limited degree of human, institutional, and financial capacity to anticipate and respond to the changing conditions.[6] Climate change, therefore, as a risk, is particularly seen as exogenous to 'peripheral' countries, and to those "that have little capacity for control and much greater potentiality for catastrophe".[7] Climate change will add another layer of complexity to existing development challenges and to economic growth. Hence, there is a strong need for starting now to carefully think about, and to respond adequately to, the challenges posed by a changing climate.

The adverse consequences of climate change are increasingly observed and witnessed in many places in South East Asia, particularly among many if not most of the member countries of the Association of Southeast Asian Nations

2 Smit, I. Barry, and Olga Pilifosova, "Adaptation to Climate Change in the Context of Sustainable Development and Equity," in *Climate Change 2001: Impacts, Adaptation and Vulnerability*, ed. IPCC (Cambridge: Cambridge University Press, 2001), 877–912.

3 Working Group II, "Impacts, Adaptation and Vulnerability," in *Fourth Assessment Report of the IPCC Technical Summary*, ed. Martin L. Parry et al. (Cambridge: Cambridge University Press, 2007), 23–78.

4 Berkhout, Frans, Dave Huitema, and Katrien Termeer, "The Governance of Adaptation" (paper presented on an international symposium, Amsterdam, March 22–23, 2012), accessed May 31, 2012. http://www.ivm.vu.nl/en/Images/The%20Governance%20of%20Adaptation%20Call%20for%20Abstracts_tcm53-227922.pdf.

5 See, for example, Stern, Nicholas, *The Economics of Climate Change: The Stern Review* (Oxford: Cambridge University Press, 2006); and Sachs, Jeffrey D., *The End of Poverty: How We Can Make It Happen in Our Lifetime* (London: Penguin Books, 2005).

6 Sperling, Frank, ed., *Poverty and Climate Change – Reducing the Vulnerability of the Poor through Adaptation* (Washington, D.C.: AfDB et al., 2003).

7 Beck, Ulrich, *World Risk Society* (Cambridge: Polity Press, 1998), 42.

(ASEAN).[8] Forecasted climate changes in the region will include increases in the average temperature, decrease in rainfall, and increased frequency, duration and intensity of extreme weather events, such as heavy precipitation, droughts, storms, floods, typhoons, and heat waves. Moreover, sea level rise poses a significant risk to many of the coastal areas, which are mostly heavily populated, resource-rich,[9] and drivers of economic change in most of the Southeast Asian countries.

Surprisingly, while there is a broad range of actors involved in developing climate change response strategies, ASEAN and its role for fostering the agenda has not much been in the focus of research yet. Governments in the region, while having started to take measures for reducing emissions of climate-hostile greenhouse gases some few years ago, are increasingly concerned now about adapting to the adverse impacts of climate change.[10] Yet it has been argued that since change is a constant, interventions can only be sustainable if people can adapt them in the future to a changing context.[11] Facilitating a more sustainable development in the face of climate change is an important topic that needs to be carefully linked with economic development agendas in order to make adaptation work.

Therefore, this chapter analyses the role of ASEAN – an organisation that originally was built for peaceful regional integration and for fostering economic cooperation and development throughout the Southeast Asian countries – for dealing with the topic of climate change. As a matter of fact, ASEAN has recently taken a stronger position around the communication of climate change. Yet, in many facets, including adaptation, its role still remains uncertain.

8 Eucker, Dennis, and Jonas Hein, *Klimawandel in Südostasien: Die ASEAN als Wegbereiter einer regionalen Klimapolitik? GIGA Focus Asien* 4 (Hamburg: GIGA, 2010); Lian, Koh Kheng, and Lovleen Bhullar, "Adaptation to Climate Change in the ASEAN Region" (draft dated October 4, 2010), accessed May 31, 2012. http://www.ucl.ac.uk/laws/environment/docs/hong-kong/Adaptation%20to%20CC%20ASEAN%20%28KL%20Koh%20AND%20Lovleen%20Bhullar%29.pdf.

9 Cruz, Rex Victor et al., "Asia," in *Climate Change 2007: Impacts, Adaptation and Vulnerability: Working Group II Contribution to the Fourth Assessment Report of the Intergovernmental Panel on Climate Change*, ed. Martin Parry et al. (Cambridge: Cambridge University Press, 2007), 469–506.

10 Eucker, and Hein, *Klimawandel in Südostasien.*

11 Levine, Ludi, and Jones, *Rethinking Support.*

A variety of research found that institutions are at the heart of the lack of sustainability of adaptation-related interventions.[12] The present chapter is therefore based on two overall considerations:

- First, adaptation policies and strategies need to be based on broader considerations of sustainable development;
- Second, adaptation policies and strategies will only be of use if there are institutions that ensure effective implementation.

The chapter aims to provide answers to three questions. First, is ASEAN based on an institutional framework that can be perceived as conducive to the promotion of sustainability in the face of adapting to a changing climate? Second, how is the topic of climate change dealt with in ASEAN and what are the potential leverages for the development of sustainability-driven approaches to deal with climate change? And, third, is the ASEAN Strategy to Deal with Climate Change an indication of an increased awareness towards sustainability in the region?

Based on these underlying questions, the chapter contributes to the increased interest in institutions and the role they play in the context of adaptation to climate change. By dealing explicitly with ASEAN, it examines institutional awareness from the perspective of one of the major political players in the Southeast Asian region. The chapter employs a human development-based understanding of sustainability which also reaches into the adaptation agenda, and which is opposed to another major direction of adaptation, i.e. the 'risk-based' approach.[13]

---

12 For example, Levine, Ludi, and Jones, *Rethinking Support*.

13 In a 'risk-based' understanding, climate change is perceived a risk that exacerbates development and macroeconomic growth and which, as a consequence, mainly demands for scientific forecasts and the elaboration of climate change models in order to provide meaningful adaptation in terms of 'structural' solutions. The aim is to respond to impacts directly associated with climate change, based on technical adaptation to future climate change. Conversely, in the 'human development' based understanding, climate change means an elusive challenge not only for disadvantaged population groups, but also for the prosperity of broader societies. The aim is to tackle vulnerability to a range of stressors, including current climatic stimuli. Even though these two perspectives have much in common in what they may offer, a closer look reveals striking differences – and leads to different understandings of the underlying approaches. See Georg, C., "Climate Change and Social Vulnerability" (paper presented at IHDP Open Meeting, World Conference Center, Bonn, April 26–30, 2009).

2 Institutions, Sustainability and Adaptation to Climate Change

There is an increased interest in institutions and the role they play in the context of adaptation to climate change. The following part of this chapter gives an overview of current understandings of institutions and the role they play under conditions and processes that can promote features of adaptation. Some useful understandings and definitions of 'institutions' and institutional analysis will be presented, and ideas are presented on what is needed to make institutions 'work' for adaptation. To this end, a context-specific understanding of institutions and institutional theory will be applied. It will be argued that institutions and their role for adaptation can only be understood usefully within the broader sustainability framework, which is closely linked to adaptation, and which therefore merits detailed attention and analysis.

The role that institutions play in sustaining levels of development, economic growth, prosperity and well-being has been the topic of lively debates in political science in recent years.[14] Despite researchers putting increased interest in governance, i.e. the ways that political and societal power is exercised under 'new' forms of governing,[15] many authors still consider public institutions to be a major catalyst if not the major agents of change in the direction of these fields.[16]

---

14 Jackson, in this instance, identifies an almost undisputed role for the state in maintaining macro-economic stability. "For better or worse," he argues, "government [...] 'co-creates' the culture of consumption, shaping the structures and signals that influence people's behaviour. At the same time, of course, government has an essential role to play in protecting the 'commitment devices' that prevent myopic choice and support long-term social goals." See Jackson, Tom, "Prosperity without Growth? The Transition to a Sustainable Economy," Sustainable Development Commission, 2009, accessed May 31, 2012. http://www.sd-commission.org.uk/data/files/publications/prosperity_without_growth_report.pdf.

15 Evidently, governance does also play an important role in the science of adaptation. In a recently held international symposium, it was argued that "while governors and societal actors at a range of governance levels have started to develop adaptation policies, it is still not clear what is the extent to which deliberate interventions need to be made [...]. This means that the concept of governance is highly relevant for adaptation". See Berkhout, Huitema, and Termeer, "The Governance of Adaptation," 1.

16 Sonderforschungsbereich 700, ed., *Grundbegriffe der Governance-Forschung: Ein Beitrag aus dem Teilprojekt A1*, SFB Governance Working Paper Series 8, June 2009 (Berlin: Freie Universität, 2009; Bruyninckx, Hans, "Sustainable Development: The Institutionalization of a Contested Policy Concept," in *Advances In International Environmental Politics*, ed. Michele Betsill, Kathryn Hochstetler and Dimitris Stevis (Hampshire: Palgrave Macmillan, 2005), 265–298; Eckersley, Robyn, *The Green State: Rethinking Democracy and*

Generally, institutions are important to consider in policy making, but the roles they play are complex and hard to decipher.[17] Given its wide range of consequences on and interactions with wider aspects of development, climate change will inevitably have considerable implications for the way institutions work in a wide array of domains and how their role may vary according to them. This includes industrial and infrastructure development, agriculture, health and tourism, as well as poverty reduction, social development and environmental protection. Policies related to adaptation are therefore neither necessarily attached to only one level of government nor to just one ministry or sector.[18] Actions based on domain logics may not necessarily relate to each other, and each domain will bring a set of own specifics with itself. Accordingly, there is a need to consider how public institutions can help to elaborate strategies that support societies to adapt to a changing climate or, at the very least, prevent actions that lead to further adverse consequences.[19]

Political science has come up with a multitude of ways for analysing institutional dynamics over time,[20] which only recently have found entrance to the

---

Sovereignty (Boston: MIT Press, 2004); Frickel, Scott, and Debra J. Davidson, "Understanding Environmental Governance: A Critical Review," *Organisation & Environment* 17.4 (2004): 471–492.

17 Young, Oran R., Leslie A. King, and Heike Schroeder, ed., *Institutions and Environmental Change: Principal Findings, Applications, and Research Frontiers* (Cambridge, London: The MIT Press, 2008), xiv.

18 Jones, Lindsey, Eva Ludi, and Simone Levine, *Towards a Characterisation of Adaptive Capacity: A Framework for Analysing Adaptive Capacity at the Local Level*, Background Note, December 2010 (London: ODI, 2010).

19 Adger et al. have provided useful concepts and understandings of adaptation practices, options, constraints and capacity, showing that adaptation to climate change needs to consider the various scales of actors, institutions, and actions involved. See Adger, W. Neil, Nigel W. Arnell, and Emma L. Tompkins, "Successful Adaptation to Climate Change Across Scales," *Global Environmental Change* 15.2 (2005): 77–86; as well as Adger, W. Neil, Shardul Agrawala, and M. Monirul Qader Mirza, *Assessment of Adaptation Practices, Options, Constraints and Capacity: Climate Change 2007: Impacts, Adaptation and Vulnerability*, contribution of Working Group II to the Fourth Assessment Report of the Intergovernmental Panel on Climate Change (Geneva: IPCC, 2007).

20 Shanks et al. point out that the connection between political studies and development studies "has waxed and waned over time". See Shanks, Edwin et al., *Understanding Pro-poor Political Change: The Policy Process* (London: ODI, 2004), 1. Some approaches have looked in detail at the nature of the state and its embeddedness in society, sometimes in highly context-specific matters, approaching politics through a focus on political culture and discourse. Another level of analysis used by, for example, Bratton and van de Walle (1997) – that of political regimes – can be described as the sets of procedures that determine the distribution of power. See Bratton, Michael, and Nicholas van de Walle,

domain of climate change-related actions and domains. A central part of the problem for defining a theory of institutional adaptation lies in the understanding of what an institution is. Existing theories are mostly contested ones, and "there is no single theory of institutional adaptation in the face of climate [...] change".[21] Institutions, in a broad understanding, can be defined as the humanly created formal and informal mechanisms that shape social and individual expectations, interactions, and behaviour.[22] For political analysis, there is a commonly accepted distinction between institutions and organisations.[23] In daily usage, however, the term 'institution' is often used interchangeably with 'organisation', which can be both of a public and non-public nature. Bratton and de Walle[24] hence provide the term 'political institutions' in order to make the analytical focus clear, so that a ministry (or a government, a regional organisation, etc.) can be clearly described as a political institution.[25] It is in this context that Adger[26] deals with two manifestations of institutions: as structures of political power and legitimacy, and as predetermined social commitments and worldviews, in which diverse characteristics and motivations of a given

Democratic Experiments in Africa (Cambridge: Cambridge University Press, 1997). Various attempts have been made to define criteria with which to categorise different types of regimes and their institutional effectiveness. Other approaches, then, aim at the contours of power relations within a society and locate political regimes along two axes, according to the degree of political competition and by the degree of political participation.

21 Adger, W. Neil, "Institutional Adaptation to Environmental Risk under the Transition in Vietnam," *Annals of the Association of American Geographers* 90.4 (2000): 740.

22 Uphoff, Norman T., and Louise Buck, *Strengthening Rural Local Institution Capabilities for Sustainable Livelihoods and Equitable Development*, paper for Social Development Department (Washington, D.C.: The World Bank, 2007).

23 North, Douglass C., *Institutions, Institutional Change and Economic Performance* (Cambridge: Cambridge University Press, 1990).

24 Bratton, and de Walle, *Democratic Experiments.*

25 Shanks et al., *Understanding Pro-poor Political Change*. Tompkins et al. assume that operational arrangements of political institutions generally exist at smaller spatial scales, whereas constitutional arrangements are rather observed at higher levels. See Tompkins, Emma L., W. Neil Adger, and Katrina Brown, "Institutional Networks for Inclusive Coastal Zone Management in Trinidad and Tobago," *Environment and Planning A* 34.6 (2002): 1099. Firmin-Sellers (1995, cited in Tompkins, Adger, and Brown, "Institutional Networks,") argues that political institutions even operate at three levels: constitutional, institutional arrangement, and operational. Finally, Huntington (cited in Shanks et al., *Understanding Pro-poor Political Change*, 3) links institutions and organisations in an interesting manner, defining institutions as "expected patterns of behavior", which helps to establish the distinction between organisations that are politically institutionalised (i.e. predictable in their procedures) and those that are not.

26 Adger, "Institutional Adaptation," 740.

institutional framework can both overlap and conflict. In line with the prevailing academic debate, we therefore should explicitly accept that the term 'political institution' provides for at least three understandings: First, it gives a description of the underlying framework for formal mechanisms that shape political (or even social and individual) expectations, interactions, and behaviour; second, it shows that formal arrangements that affect institutionalised problem-solving measures are mostly decided upon national or even higher levels of decision-making; third, the term makes clear that at these levels of political organisation, patterns of behaviour are based on some particular levels of predetermination – both for external observers and for the individual stakeholders involved. Against this background, institutionalist perspectives analyse policy choices and outcomes through the lens of political institutions, and often see them either as actors on their own terms, or at least semi-autonomous of society.

Political institutions play a crucial part in the generation and dissemination of knowledge,[27] but also in constraining knowledge and information[28] related to climate risks, as well as information and communication of opportunities and possibilities to adapt to them in a pre-determined way. We can therefore conclude that institutional frameworks create a strong impetus for political-societal action in which "actors maximize their goals by changing their strategies or by changing the institutional setting that transforms their strategies into outcomes".[29]

Climate change is a transnational issue, but adaptation measures are usually developed at the regional, national and local levels.[30] Doubtlessly, regional organisations can play a critical role in creating mechanisms that help nations and societies to adapt. It has been widely accepted, therefore, that it is mostly institutions from the regional level, and from there down to the local and even to the household level,[31] that build the front lines to adapt to the changing climate, and that policies in these areas can create an enabling environment for

---

27 Adger, W. Neil et al., "Governance for Sustainability: Towards a 'Thick' Analysis of Environmental Decision-making," *Environment and Planning A* 35 (2003): 1095–1110.

28 Jones, Lindsey et al., *Responding to a Changing Climate: Exploring how Disaster Risk Reduction, Social Protection and Livelihoods Approaches Promote Features of Adaptive Capacity*, Working Paper 319 (London: ODI, 2010).

29 Tsebelis, George, *Veto Players: How Political Institutions Work* (Princeton: Princeton University Press, 2002), 96f.

30 Lian, and Bhullar, "Adaptation to Climate Change."

31 Climate Change Agriculture and Food Security (CCAFS), *Open Call: A Discussion Paper Analyzing the Role of Rural Institutions in Enabling Adaptation to Climate Change in the Agricultural Sector* (Nairobi: CCAFS, 2011).

supporting long-term social goals and human development.[32] These may also reach beyond the logics of specific domains. Yet, in spite of these assumptions, current research suggests that existing institutional frameworks are rather unlikely to be able to cope efficiently and equitable with climate change outside their range of experience.[33]

In addition to climate change, the future risk and exposure of developing countries will also depend on their development path.[34] Many sectors providing basic livelihood services to the poor in developing countries are not able to cope even with today's climate variability and stresses in a sustainable manner, and the way many state institutions operate in developing countries is unfavourable to disadvantaged population groups.[35] In order to understand gaps and potentials for adaptation to climate change, the analysis of economic sectors as well as broader socio-economic factors is important, but great significance should be given to institutional factors.

These considerations directly take us to the question of how public institutions frame sustainability in the face of adaptation to climate change. While 'adaptation' refers to actions taken to deal with the consequences of climate change, 'sustainable adaptation' is based on a continuous stream of activities, actions, decisions and attitudes, which reflect both existing as well as changing social norms and processes. The emphasis lies on the 'balancing' of environmental, social and economic goals: this understanding is closely linked to the broader sustainability debate in which a stable economy is perceived as one which is able to produce enough welfare for everybody, and to distribute the benefits and the cost in an equitable way, without endangering the environment.[36] Therefore, while the notion of 'sustainability' has always put a strong focus on environmental concerns that are embedded into broader conceptualisations of society and economy,[37] sustainable adaptation should not restrict

---

32 Watkins, Kevin, *Human Development Report 2007/2008: Fighting Climate Change: Human Solidarity in a Divided World* (New York: UNDP, 2007), 89.

33 Tompkins, Emma L., and W. Neil Adger, "Defining Response to Enhance Climate Change Policy," *Environmental Science and Policy* 8.6 (2005): 562–571.

34 Sonderforschungsbereich 700 ed., *Grundbegriffe der Governance-Forschung.*

35 World Bank, *World Development Report 1999–2000: Entering the 21st Century* (Washington, D.C.: The World Bank, 2000).

36 Fisher, Julie, *The Road from Rio: Sustainable Development and the Nongovernmental Movement in the Third World* (Westport: Praeger, 1993).

37 In order to build a closer link to the notion of sustainability, it is worthwhile to take a closer look at the idea behind the concept. Bruyninckx argues that only very few concepts have made such a fast and pervasive career in academic and policy discourses as sustainable development. See Bruyninckx, "Sustainable Development." In the 1980s and 1990s, the discussions held in the name of sustainability created an important momentum for

itself on normative concepts for a more equitable use of environmental resources. The discussion is rather based on the vision of how a more inclusive and 'just', i.e. human, development can be facilitated and fostered.

Naturally, the practicability of sustainable adaptation raises some concerns, and the question remains of 'how to do' sustainable development that promotes adaptation. In here, it has been argued that dealing with change and unpredictability are central challenges for human development all over the world.[38] In terms of promoting collective security, success in both sustainable and human development initiatives should be redefined by how they promote "legitimate, broad-based development that allows individuals and societies to cope with risk and adapt to changing circumstances over time".[39] Even though many human activities that accompany the processes of development may have destructive consequences, it is positively assumed that it is also within human power to resist and reverse many of these bad consequences if timely action is taken.[40] Hence sustainable adaptation practices will require an awareness of the dynamic interactions between economic, social and ecological processes, and integrating aspects of human development into all of the approaches for adaptation.[41]

---

innovation in ideas, political mobilisation, and policy change – particularly in connection with the UN Conference on Environment and Development (UNCED) in Rio de Janeiro in 1992. Since its introduction as a guiding policy principle, sustainable development has been accepted as a framework for policy agendas. Yet at the same time, the concept remains contested at different levels. Critics point, for example, to the vagueness of the concept, to a level of aggregation that is not suitable for pragmatic policy-making, or to the normative 'Western bias'. Yet sustainable development has become important only since the publication of the Brundtland report on 'Our Common Future', where sustainable development was broadly introduced and defined as "development that meets the needs of the present without compromising the ability of future generations to meet their own needs". See World Commission on Environment and Development (WCED), *Our Common Future, Report of the World Commission on Environment and Development* (Oxford: Oxford University Press, 1987), 8. This definition is clearly embedded in a number of currents which comprise, for example, the Club of Rome's 'Limits to Growth' report (1971), the Brandt Commission Report (1977), and the United Nations Convention on the Human Environment held in Stockholm 1972.

38 Beck, *World Risk Society.*

39 Adger, W. Neil, "Building Resilience to Promote Sustainability," *IHDP Update* 2 (2003), 3.

40 Sen, Amartya, "Climate Policy as Human Development, Special Contribution to the Human Development Report 2007/2008," in *Human Development Report 2007/2008*, ed. UNDP (Hampshire, New York: Palgrave Macmillan, 2007), 29.

41 Drawing on wider political concerns about the relationships between economy, well-being, and the environment, Dobson delineates political theories that incorporate a 'green' politics perspective, placing sustainability concerns at the centre of a normative

In terms of institutional analysis, science should not only concern itself with the way adaptation options can be facilitated, but should also include the question of how institutions themselves are able to adapt to changing circumstances. In sum, the issue once again is whether public interventions and institutions work well – and according to the interests of society – in the way they address climate change. Adaptation interventions themselves can therefore be approached as an integral part of 'good development'. Addressing the underlying drivers that constrain sustainability will not only help societies to respond to climatic changes, but also to adverse trends more generally.[42] While this idea underlies most of the academic literature on public institutions, it has not necessarily been reflected enough in the process of rule-making for sustainable adaptation so far.

3 Sustainable vs. Economic Development? An Overview of Institutional Dynamics in the ASEAN

After having provided the above insights into the conceptual understanding of institutions and what roles they play in adaptation to climate change, the objective of the following is to provide an overview of ASEAN. Against the background of the prevailing development and economic growth agenda, the question is whether ASEAN is based on institutional dynamics that can be perceived as conducive for the promotion of sustainability in the face of adapting to a changing climate. This opens the space for a more detailed analysis in the chapters to follow.

Before we start with the analysis, though, it will be important to clarify the application of the term 'development' and what is meant by it. Basically, one important view is provided by Hart, who introduces a fundamental distinction between the "geographically uneven, profoundly contradictory set of historical

understanding of social and political change. See Dobson, Andrew, *Justice and the Environment: Conceptions of Environmental Sustainability and Dimensions of Social Justice* (Oxford: Oxford University Press, 1999). Kates et al. in turn offer an integrative synthesis, linking the economic, environmental, and socio-political dimensions of sustainability into what they have dubbed a 'sustainability science'. See Kates, Robert W. et al., "Sustainability Science," *Science* 292 (2001): 641–642.

42 Riché, Beatrice et al., *Climate-related Vulnerability and Adaptive Capacity in Ethiopia's Borana and Somali Communities* (Manitoba: International Institute for Sustainable Development, 2009).

[developmental] processes", and the "project of [development] intervention".[43] This is supported by Cowen and Shenton, who distinguish between "development as an immanent and unintentional process [...] and development as an intentional activity".[44] These notions provide an explanation to the reasons why the word 'development' started buzzing in dominant parlance: the term

> [...] rested on a mere – albeit unquestioned – assumption, and no one cared to define it properly. Both elements characterize a buzzword: an absence of real definition and a strong belief in what the notion is supposed to bring about.[45]

However, Bebbington et al. add that this approach

> [...] offers a means of clarifying the relationship between development policy and practice and the underlying processes of uneven development that create exclusion and inequality for many just as they lead to enhanced opportunities for others.[46]

These initial considerations are of crucial importance not only for providing us with an entrance point for where planning for sustainable development is about to start; they also provide us with some initial hindsight about the institutional dynamics upon which ASEAN is based. Built in 1967, ASEAN is marked by high levels of political and economic heterogeneity amongst its ten member countries.[47] While, in political terms, countries such as Indonesia have *de facto* undertaken a transition to more democratic structures during the past, there are others, including Vietnam and Laos, where communist parties build the

43 Hart, Gillian, "Development Critiques in the 1990s: culs de sac and Promising Paths," *Progress in Human Geography* 25.4 (2001): 649–658.

44 Cowen, Michael, and Robert W. Shenton, *Doctrines of Development* (London: Routledge, 1996), 50.

45 Rist, Gilbert, "Development as a Buzzword," in *Deconstructing Development Discourse: Buzzwords and Fuzzwords*, ed. Andrea Cornwall and Deborah Eade (Oxford: Practical Action Publishing, 2010), 20.

46 Bebbington, Anthony J., Samuel Hickey, and Diana C. Mitlin, "Introduction: Can NGOs Make a Difference? The Challenge of Development Alternatives," in *Can NGOs Make a Difference? The Challenge of Development Alternatives*, ed. Anthony J. Bebbington, Samuel Hickey and Diana C. Mitlin (London, New York: Zed Books, 2008), 5.

47 Since 1999, the list of member countries comprises Thailand, Indonesia, Malaysia, the Philippines, Singapore (founding states), Brunei (1984), Vietnam (1995), Myanmar (1997), Laos (1997), and Cambodia (1999).

government and largely control the policies of the state. Moreover, while countries such as Singapore, Malaysia and Thailand have reached enormous levels of economic progress during the past decades, other ASEAN states such as Laos and Cambodia still belong to the world's poorest countries.

Against this background, ASEAN's main aim is to accelerate economic growth based on better integration between the ASEAN economies, social progress, and cultural development in the region. In 2010, nominal GDP in the region was estimated to have grown to US$ 1.8 trillion.[48] Its member countries have a total population of 600 million, which is more than almost any other regional organisation across the globe – theoretically, at least, this provides for an enormous potential for sustained economic growth.

The question then is what framework for action is needed to promote 'sustainable development' in all its dimensions? Generally, sustainable development requires action "beyond the economic domain".[49] But as Perch argues,

> We are still struggling to find the right balance between human and sustainable development. Reconciling the imperatives of society and the environment remains one of the weakest areas of development policy.[50]

As the world's economic history has told, social and ecological differentiation is a typical implication of the development process. In many instances, environmental problems have been aggravated through accelerated growth under the conditions of social poverty as well as through misery.[51] To put it differently, economic growth – widely hailed as a prerequisite to prosperity – takes place more often than not at the expense of either the environment or human beings.[52]

Doubtlessly, one direct implication of the economic direction of many, if not most, of the ASEAN countries is that economic achievements have led to widening inequalities and vulnerabilities among the population. Evidently, the Gini index – which describes and compares income inequalities on an annual basis

---

48 Delegation of the European Union to Indonesia, Brunei Darussalam and ASEAN, accessed May 31, 2012. http://eeas.europa.eu/delegations/indonesia/index_en.htm.

49 World Bank, *World Development Report*, 33.

50 Perch, Leisa, "Inclusive and Sustainable Development: For Whom?" One Pager 126 (Brasilia: UNDP International Policy Centre for Inclusive Growth (IPC-IG), 2011), accessed May 31, 2012. http://www.ipc-undp.org/pub/IPCOnePager126.pdf.

51 Hein, Wolfgang, "Globalisierung und nachhaltige Entwicklung in den Ländern des Südens," in *Nachhaltigkeit*, ed. K. Rennings and O. Hohmeyer (Baden-Baden: Nomos, 1997), 165–218.

52 Rist, "Development as a Buzzword."

and in countries throughout the world – has shown that the majority of ASEAN countries have been affected by ever-growing income disparities over the past decade. In this context, integration into the globalised market economy, shaped by a liberalisation of domestic and regional markets and agriculture has led to great income inequalities and to an undermining of livelihoods of disadvantaged groups, especially in rural areas.[53] In Rist's view, the undeniable success of 'development', linked to its undeniable failures in improving the condition of disadvantaged population groups, therefore needs to be called into question, and that not only will development fail to narrow the gap between rich and poor, but

> [...] in fact it both widens and deepens this division and ultimately creates poverty, as natural resources and human beings alike are increasingly harnessed to the pursuit of consumption and profit.[54]

Tim Jackson, in his much-cited 2009 report for the Sustainable Development Commission (SDC), an independent executive non-departmental public body acting as the UK Government's independent adviser on sustainable development, brings the notion of 'prosperity' into the sustainable development debate and asks whether economic development can be achieved without prosperity (or, vice versa, to achieve prosperity without economic growth). In his report, by questioning the underlying vision of a prosperity built on economic growth, he proposes to search for alternative visions, "in which humans can still flourish and yet reduce their material impact on the environment".[55] Yet, he admits that this is not easy to achieve – since in many instances growth is a necessary precondition for granting prosperity, by maintaining economic and social stability – but that the overarching vision of human prosperity based on sustainability should not be given up lightly. To this end, it has become rather clear from the ongoing debate that 'enhancing security' is a central means to foster sustainable development. This finding helps to link the debate with the more detailed focus on adaptation to climate change. However, as we have learned, it might be clear that this is mainly an institutional task in which clear commitment among all public institutions involved is needed.

Based on the ASEAN Charter, which was signed by all member countries in 2009, it is formally agreed that ASEAN is

53 Taylor, Philip, "Poor Policies, Wealthy Peasants: Alternative Trajectories of Rural Development in Vietnam," *Journal of Vietnamese Studies* 2.2 (2007): 3–56.

54 Rist, "Development as a Buzzword," 19.

55 Jackson, "Prosperity without Growth?," 7.

> [...] resolved to ensure sustainable development for the benefit of present and future generations and to place the well-being, livelihood and welfare of the peoples at the centre of the ASEAN community building process.

Yet the formal agreements made in the Charter to strive for more sustainable forms of development successes registered on the institutional front of the ASEAN must not mask difficulties on the way to realisation of this task. Sustainable development, even though being highlighted in the Charter, has not progressed fast enough. A major cause of this observed challenge may lie in a lack of mainstream sustainable development principles in the overarching principle of economic development – mainly on the regional level, but also on the level of many of the countries in the region. A general difficulty to overcoming this challenge might be that, as argued by LeBlanc,[56] most of the challenges at the local and national level pertain to the mainstreaming of sustainable development thinking and principles into the daily routine of sectoral institutions. Another challenge is noted by Keskitalo and Liljenfeldt,[57] who make note of the creation of institutions dedicated to sustainable development that fail to find a durable 'niche' in the existing institutional setting, and of the relegation of sustainability to the background when new priorities driven by immediate political concerns emerge.

In order to understand the existing challenges to bridge the gap between formal document rhetoric and day-to-day reality of policy planning and implementation, one has to delve deeper into the institutional dynamics by which ASEAN is characterised and the role it plays for its members. At the institutional level, ASEAN is based on a general principle of non-interference into any of its members' internal policies and affairs. Policymaking is characterised by consensus-based decision-making, a principle which is also enshrined in the Charter.[58] On a more informal level of decision-making, processes have often been managed in the past through close interpersonal contacts among the governing elite. Against this background, Stoker makes an important point – which is that analysing institutions can just be "a simplifying lens to a complex reality",

56 LeBlanc, David, "Special Issue on Institutions for Sustainable Development," *Natural Resources Forum, Special Issue: Institutions for Sustainable Development* 36.1 (2012): 1–2.

57 Keskitalo, E. Carina H., and Johanna Liljenfeldt, "Working with Sustainability: Experiences of Sustainability Processes in Swedish Municipalities," *Natural Resources Forum, Special Issue: Institutions for Sustainable Development* 36.1 (2012): 16–27.

58 Goh, Gillian, "The 'ASEAN Way': Non-Intervention and ASEAN's Role in Conflict Management," *Stanford Journal of East Asian Affairs* 3.1 (2003): 113–118.

but that such an analysis is crucial "for understanding changing processes of governing, characterised by processes of adaptation, learning and experiment".[59] Any analysis must therefore look both at the 'technical' capacity of institutions and the power relations behind decision-making.[60] However, defining a 'well-developed' institution is problematic and subjective. Institutional arrangements dictate the social and political acceptability of different policies, they structure worldviews, and they determine the provision of resources. The ability to respond to new conditions is therefore both enabled and constrained by institutional conditions. The 'dominant worldview' of an institution, as Adger shows, is itself a product of power, and is not fixed.[61]

Moreover, the economic, administrative, and overall development orientation of ASEAN, as well as of several of its member countries, has often been described as being 'technocratic'[62] in style,[63] or as a 'government by experts',[64] in which a governing elite puts its focus on development and nation building. Against this background, ASEAN's development models have been criticised in a variety of ways. In the particular case of the term, the difficulty across many examples in the ASEAN region lies in the a priori positive meaning of the word 'development', which derives both from its supposedly 'natural' existence and from its inclusion in a cluster of unquestionable shared beliefs.[65] Despite its widespread usage, the meaning of the term 'development' remains vague, and particularly for the ASEAN case tends to refer to a set of beliefs about accelerating economic growth, social progress, and cultural development in the region.

---

59 Stoker, Gerry, "Governance as a Theory: Five Propositions," *International Social Science Journal* 155 (1998): 26.

60 Jones, Lindsey, "Overcoming Social Barriers to Adaptation," published by Overseas Development Institute, 2010, accessed May 31, 2012. http://www.odi.org.uk/resources/download/4945.pdf.

61 Adger, "Institutional Adaptation."

62 Milne thoroughly works out that the word 'technocracy' "was coined after World War I to describe a system of economic thought inspired by the kind of rational analysis at the base of the physical sciences". See Milne, R. S., "Technocrats and Politics in the ASEAN Countries," *Pacific Affairs: An International Review of Asia and the Pacific* 55.3 (1982): 405.

63 Schmidt, Guido, *Die Governance-Strukturen der ASEAN* (Norderstedt: Grin Verlag, 2009); Shiraishi, Takashi, and Patricio N. Abinales, ed., *After the Crisis: Hegemony, Technocracy and Governance in Southeast Asia* (Kyoto: Kyoto University Press, 2005); Heinrich Boell Foundation, *Towards Good Society? Civil Society Actors, the State, and the Business Class in Southeast Asia – Facilitators of or Impediments to a Strong, Democratic, and Fair Society?* (Berlin: HBS, 2004); Milne, "Technocrats and Politics."

64 "Technocrats – Minds like Machines," *The Economist*, November 19, 2011, accessed May 31, 2012. http://www.economist.com/node/21538698.

65 Rist, "Development as a Buzzword."

Lee has recently brought in a new vision and understanding of decision-making in ASEAN politics. By referring to 'intra-institutional balancing',[66] which refers to the strategy that competing states cooperate to create an institution when the demand for public goods is high, he observes that member countries "seek various cooperative as well as balancing behaviours within institutions such as bargaining, struggle, and confrontation".[67] Lee shows that the proliferation of these agreements is driven not by pure economic interests but by political and strategic domino effects, and concludes that an overemphasis on solely economic factors would fail to shed light on institutional balancing behaviours.[68]

In sum, what becomes fairly clear is that impacts of economic growth are unsustainable, which put into question the conventional formula of development in the region. Economic growth has delivered its benefit unequally across the ASEAN region. While this chapter shall by no means deny the ASEAN's countries' need for economic development, there is a great point for arguing that the Association should play a stronger role in facilitating growth economies that are based on sustainability principles. Hence this demands an agreement among all its member countries, which goes beyond mere intentional agreements, even more so against the background of the institutional and the socio-economic context.

4 The Institutional Context for Dealing with Climate Change in the ASEAN

In order to understand existing approaches for dealing with climate change, it is important to recognise the interactions between climate change and wider development pressures, since adaptation is driven by a range of different pressures acting together.[69] Only this will help us to understand how the topic of climate change is dealt with, and what the margins for further refinement of appropriate strategies are. Based on the institutional context of ASEAN, we have

66 On using the 'logic of institutional balancing' to properly capture institutional dynamics upon which his analysis is based, Lee explains that "the logic of institutional balancing in particular makes sense given that East Asian countries do not display hard balancing behaviours despite the rapidly shifting structural environments in the region". See Lee, Seungjoo, *The Evolutionary Dynamics of Institutional Balancing in East Asia*, EAI Asia Security Initiative Working Paper (Seoul: EAI, 2012), 18.

67 Lee, *The Evolutionary Dynamics*, 6.

68 Ibid., 12.

69 Levine, Ludi, and Jones, *Rethinking Support for Adaptive Capacity to Climate Change.*

to ask how underlying processes of dealing with climate change are characterised, where they are located, and which specific challenges they aim to address. While the first two aspects will be dealt with in the following, the latter will be approached in the chapter below.

For an understanding of how processes for dealing with climate change are characterised and where they are located in the institutional context, again, the prevailing political and economic background of ASEAN member states needs to be taken into account. Statements and policies among ASEAN leaders have often been said to be decided upon on the basis of non-interference and under conditions where none of the member countries sees their own interests at stake. Yet it has been widely agreed among the ASEAN community that, with agriculture and natural resources building the economic mainstay of the overwhelming part of their populations, the majority of countries are now aware of the potential highly adverse consequences that climate change will bring about.

Having that said, it should be taken into consideration that it is especially the drastic economic differences between the ASEAN countries that build the frame for their overall political orientations. One vivid example lies in the emission of greenhouse gases. In terms of greenhouse gases produced by the individual countries, emission is distributed quite differently across the ASEAN region. Above all, in terms of total emissions (which is due to high rates of deforestation), Indonesia is already today one of the world's largest emitters. At the same time, Singapore reaches per capita emissions at western European levels. Other countries contribute little to climate change.

This fundamental heterogeneity within ASEAN makes the emergence of a far-reaching joint climate policy difficult. For example, while the highly industrialised Singapore – as part of its National Climate Change Strategy[70] – aims at a reduction by 16 per cent below the business-as-usual (BAU) scenario[71] till 2020,[72] and while Indonesia aims to introduce a carbon tax and far-reaching emission reduction – mainly through the expansion of protected forest areas,[73]

70 Singapore Government, *Singapore's National Climate Change Strategy* (Singapore: Singapore Government, 2008).

71 A business-as-usual scenario describes the development of emissions without the influence of policy instruments to avoid further increase of the emission base.

72 "Fifty-five Countries Pledge to Cut Greenhouse Emissions, UN Hails 'Important Step', but 137 Nations yet to Make Promise," *The Guardian*, February 2, 2010, accessed May 31, 2012. http://www.guardian.co.uk/enviroronment/2010/feb/02/55-countries-greenhouse-emissions-pledge.

73 Ministry of Finance of Indonesia, *Ministry of Finance Green Paper: Economic and Fiscal Policy for Climate Change Mitigation in Indonesia* (Jakarta: Ministry of Finance and Australia Indonesia Partnership, 2009).

the efforts of other members have focused mainly on adaptation to climate change impacts, and not so much on the reduction of greenhouse gases.

These are not the only facts that need to be kept in mind when judgments are made about the role of ASEAN in connection to responding to climate change. It also needs to be asked how institutional dynamics in the ASEAN countries are themselves responsive to the challenges posed by climate change. Berkhout et al.[74] pose a straightforward question in this context. While they attribute governing to a set of purposive activities, they ask: Which are the actual problems that governing seeks to address? "Problems are not simply out there waiting to be tackled", as they put it so aptly. On the contrary, as they explain, actors often engage in struggles to present (or 'frame') particular phenomena into problems that suit their pre-existing political interests or policy competencies.

Against this background, the climate-policy agenda of ASEAN is closely linked mainly to its environmental efforts. Since the ASEAN summit in Kuala Lumpur (1997), environmental policy and sustainable development have been attributed with a certain degree of significance and importance. In the ASEAN Vision 2020, member states – apart from developing a political and economic community – have committed themselves to establish a

> [...] clean and green ASEAN with fully-developed mechanisms for implementing sustainable development, safeguarding the environment in the region, the sustainability of natural resources and quality of life of its people.[75]

In the preamble of the ASEAN Charter of 2007, sustainable development is also defined a fundamental objective of the Community.[76] However, the area of environmental policy remains subordinated to other thematic contents, as reflected in the Roadmap for an ASEAN Community 2009–2015. In this institutional context, climate change remains a topic that is subordinated to the topic of environmental sustainability and that is on par with five other topics, i.e. human development, social welfare and protection, ASEAN identity, and

74 Berkhout, Frans, Dave Huitema, and Katrien Termeer, "The Governance of Adaptation: An International Symposium" (paper presented in Amsterdam, March 22–23, 2012), 1, accessed May 31, 2012. http://www.ivm.vu.nl/en/Images/The%20Governance%20of%20Adaptation%20Call%20for%20Abstracts_tcm53-227922.pdf.

75 Association of Southeast Asian Nations (ASEAN), "ASEAN Vision 2020," accessed May 31, 2012. http://www.aseansec.org/1814.htm.

76 Association of Southeast Asian Nations (ASEAN), *The ASEAN Charter* (Jakarta: ASEAN, 2008).

narrowing the development gap. This range of topics then belongs to the 'Socio-Cultural Community' that is one of three pillars of the ASEAN community.

Even though the thematic subordinance of climate change is not prevalent only in the ASEAN context, it deserves special mentioning. Often, institutions dedicated to sustainable development even fail to find a durable 'niche' in existing institutional settings. Therefore, the institutionalisation of climate change, even though playing a small role in the ASEAN framework, is a noteworthy step in ASEAN.

Indeed, most of the challenges pertain to the mainstreaming of sustainable development thinking and principles into the daily routine of sectoral institutions.[77] In this context, Karlsson-Vinkhuyzen[78] explores how the final articulation of the allocation of responsibilities for implementation among the different levels is the result of compromise rather than of an articulated, coherent vision. There is good reason to support this argument from what we can track in ASEAN's institutional framework.

5 ASEAN's Strategy to Deal with Climate Change

The pivotal question of the following is whether the ASEAN Strategy to Deal with Climate Change builds an indication of an increased awareness towards sustainability in the region. Since "adaptation to climate change can take multiple forms, occur at different governance levels, and in different policy sectors, presenting governors with a range of possible actions",[79] we have to ask which specific challenges the underlying processes of dealing with climate change aim to address, in order to make clear then how climate change is being dealt with and which strategies are applied.

In this context, let us start with the observation of Lian and Bhullar who point out that, in the past, "regional organizations in Southeast Asia have played a relatively small role only in international climate change policymaking".[80] Overall, however, this view suffers from some misperception, and it should be noted that the Community has indeed endeavoured to establish a joint position towards the international community for several years now. As we have seen,

77 LeBlanc, "Special Issue."

78 Karlsson-Vinkhuyzen, Sylvie I., "From Rio to Rio via Johannesburg: Integrating Institutions across Governance Levels in Sustainable Development Deliberations," *Natural Resources Forum, Special Issue: Institutions for Sustainable Development* 36.1 (2012): 3–15.

79 Berkhout, Huitema, and Termeer, "The Governance of Adaptation."

80 Lian, and Bhullar, "Adaptation to Climate Change," 5.

dealing with climate change in the emergence of the ASEAN Road Map is a significant topic. Yet it is subordinated to more prominent topics. However, it is striking that the topic has recently played an important role on various Joint Meetings of Environment Ministers and among Heads of State and Government.

Interestingly, since 2007 ASEAN joins international climate conferences by formulating Joint Statements.[81] These were formulated at the Conference of Parties (COP 15) in Copenhagen in 2009, as well as beforehand over the course of the Bali Roadmap (COP13) in 2007. The Joint Statements are clearly meant to show that ASEAN is making proactive progress towards climate change, and progress in which climate change is understood as a fundamentally important risk. The main message is that even though ASEAN has contributed little to anthropogenic climate change, all members of the Community show commitment to face the challenges within the range of given possibilities. Yet a central argument included in this message has been that "all countries should do their part based on the principle of common but differentiated responsibilities".[82]

Therefore, ASEAN serves as an important platform for articulating a particular climate policy-related position on the international level and for getting heard in a more lasting way than would be possible based on individual countries' initiatives. In particular, and directed towards the western industrialised nations, the ASEAN position on facing climate change comprises the following:

- reduction of greenhouse gases as recommended under existing IPCC guidelines;
- financial support for adaptation measures; and
- enhanced levels of knowledge and technology transfer.

In summary, therefore, it can be stated that the Joint Statements and the Roadmap for an ASEAN Community 2009–2012 builds a framework that all member countries unanimously agree to and which aims at the guidance and practical implementation of joint climate change-related policies and activities.

81 See, for example, Association of Southeast Asian Nations (ASEAN), *ASEAN Declaration on the 13th Session of the Conference of the Parties to the UNFCCC and the 3rd Session of the CMP to the Kyoto Protocol* (Singapore: ASEAN, 2007); and Association of Southeast Asian Nations (ASEAN), *ASEAN Joint Statement on Climate Change to the 15th Session of the Conference of the Parties to the United Nations Framework Convention on Climate Change and the 5th Session of the Conference of Parties Serving as the Meeting of the Parties to the Kyoto Protocol* (Cha-am Hua Hin: ASEAN, 2009).

82 Association of Southeast Asian Nations (ASEAN), *Fourth ASEAN State of the Environment Report* (Jakarta: ASEAN, 2009).

However, the existing framework does not comprise binding emission reduction targets yet, and it remains highly doubtful whether this will happen one day, given the differences across the ASEAN economies.

Apart from the external significance that the Joint Statements have and the role they play for creating common understanding, there is a range of initiatives tractable within the ASEAN in order to face the challenges coming along with climate change. In the Joint Statements, member states agree to enhance scientific collaboration to downscale climate change effects and incorporate mitigation and adaptation strategies into national development strategies and policies.

The ASEAN Roadmap aims at enhancing regional and international cooperation for dealing with climate change and its impacts. Member states are encouraged to develop strategies for mitigation and adaptation, and to implement them. In order to facilitate a development that is as much based on economic as well as on sustainable and environmental goals, the community has now started to envisage ways to build synergies between climate change and sustainable development.[83] In this context, some of the existing Blueprints[84] play a stronger and more direct role for sustainability and adaptation than others: The ASCC Blueprint, for example, lists some key concerns for environmental sustainability, such as transboundary environmental pollution. The APSC Blueprint lists 'transboundary challenges', and identifies climate change as a 'non-traditional security issue'. This latter emphasis is elaborated in more detail in the 2005 ASEAN Agreement on Disaster Management and Emergency Response. The topic of climate change, then, is again referred to in the ASCC Blueprint where the demand for an ASEAN Climate Change Initiative is expressed.

Apart from that, joint efforts among environmental ministers and ministries have led to a variety of initiatives. Liana Bratasida, Chairperson of the ASEAN Working Group on Environmentally Sustainable Cities and Special Assistant to the Minister of Environment of Indonesia, said that while ASEAN cities have varying capabilities and capacities in handling climate change, the discussions show that they are all "taking innovative approaches to counter the effects of climate change through initiatives in adaptation, mitigation and climate proofing".[85]

---

83 Eucker, and Hein, *Klimawandel in Südostasien.*

84 Blueprints build the basis upon which the ASEAN Roadmap is built, and which comprise the ASEAN Socio-Cultural Community (ASSC) Blueprint, the ASEAN Economic Community (AEC) Blueprint and the ASEAN Political Security Community (APSC) Blueprint.

85 Mass Communication of Thailand (MCOT), "ASEAN Explores Tools to Address Climate Change," accessed May 31, 2012. http://www.mcot.net/cfcustom/cache_page/159435.html.

Above all, the establishment of a joint Working Group on Climate Change that shall aim at closer cooperation within the framework of adaptation and prevention strategies is a priority in the institutional context. Three components are of prime importance:

1. joint research on the effects of climate change in the ASEAN region and to the sectors of agriculture, fisheries and forestry as well as an integration of prevention and adaptation strategies into national development strategies of ASEAN member states;
2. enhanced collaboration on pilot projects to improve land use practices, and to build knowledge and experience, particularly on agricultural and forestry production systems; and
3. the strengthening of regional networks in the field of climate change and food security. In addition, a common implementation of the UNFCCC is sought. Here, the population shall participate at the local level, particularly in the implementation of joint programs.

There also exist guiding principles for future dealings with the UN-REDD (Reducing Emissions from Deforestation and Forest Degradation) mechanism. These tools will play an important role especially for Indonesia in the future, and were therefore introduced by the country in the international climate negotiation as part of the COP13.

However, all these components are still in an early phase, and implementation and success are still at an early stage. The planned activities in the field of energy efficiency and energy security are not yet at the level of formulating a joint action plan,[86] and far-reaching steps towards implementation are not detectable so far.

In sum, this assessment shows that while mitigation was the initial focus of ASEAN's climate change initiatives, the importance of adaptation is now increasingly recognised. While member states over the past years used the institutional framework of ASEAN to generate a common understanding to how to communicate their understanding of climate change and responsibilities, ASEAN as a platform is now increasingly used for internal framing of adaptation processes.

It is clear that some caution has to be applied to the above results. The need for sustainability remains a rhetorical key concept in ASEAN documents, but

86 Association of Southeast Asian Nations (ASEAN), *Annual Report 2008–2009: Implementing the Roadmap for an ASEAN Community 2015* (Jakarta: ASEAN, 2009).

challenges persist in the way forward. A recent paper by Da Fonseca et al.[87] focuses on the gaps

> [...] between rhetoric and reality, and specifically on how the rhetoric of sustainable development is sometimes used by international and national development organizations to create and drive processes [...] that, while following the letter of sustainable development practice as codified by different agencies, may squarely conflict with local values, power relationships, and interests. [In such cases, the authors assume, ...] while the rhetoric of sustainable development is upheld and principles such as participation are applied on paper, the reality is a disconnect, which translates into a lack of buy-in by the supposed "beneficiaries" of the process being put in place, a possible capture by elites, and ultimately the mitigated success or even failure of the whole process.[88]

6 Conclusion

Based on the initial consideration that there is an increased interest in institutions and the role they play in the context of adaptation to climate change, this chapter provided an analysis of ASEAN and how, in one of the biggest political institutions in Southeast Asia, the topic of climate change and climate change adaptation is framed. Based on three underlying questions, the chapter reflects on ideas generated amongst the ASEAN community to deal with the adverse impacts of climate change and how to face them. The chapter showed that while member states originally made use of the institutional framework of ASEAN to generate a common understanding of climate change mitigation and to apply it externally, the association is now becoming increasingly accepted for internal framing of adaptation processes.

In terms of institutional analysis, it became clear from the analysis that science should not only concern itself with how adaptation options can be facilitated, but should also include the question of how institutions themselves are able to adapt to changing circumstances. The underlying premise is that addressing those drivers that constrain sustainability will help societies to respond

87 Da Fonseca, Igor Ferraz, Marcel Bursztyn, and Benjamin S. Allen, "Trivializing Sustainability: Environmental Governance and Rhetorical Free-riders in the Brazilian Amazon," *Natural Resources Forum, Special Issue: Institutions for Sustainable Development* 36.1 (2012): 28–37.

88 Ibid., 90.

to climatic changes, but also to adverse trends more generally.[89] While this idea determines most of the academic literature on public institutions, it has not necessarily been reflected enough in the process of rule-making for sustainable adaptation so far. By combining the discussion on adaptation with the broader notion of sustainability, it becomes clear that impacts of economic growth are unsustainable, which puts into question the conventional formula of development in Southeast Asia. Hence there is a great point for arguing that the Association should play a stronger role in facilitating growth economies that are based on sustainability principles.

While the process of economic development has brought about 'winners' and 'losers' in terms of social and ecological stratification processes, climate change is expected to put another layer of risk on existing stresses, and to make development 'more difficult'.

The concept of institutions for sustainable development is much broader than that of institutions dedicated to sustainable development. Indeed, most of the challenges pertain to the mainstreaming of sustainable development thinking and principles into the daily routine of sectoral institutions.[90] There is good reason to support this argument from what we can track in the context of ASEAN's institutional framework.

However, we need to be cautious in interpreting the results, due to what Da Fonseca et al.[91] describe as 'gaps between rhetoric and reality'. While ASEAN has undoubtedly 'come to terms' with climate change, the notion of sustainability, which is a key concept for any adaptation strategy, remains on a broad rhetoric level only. In order to come up with more forward-looking decision-making, and to avoid the possibility that policies and development efforts run the risks of maladaptation (i.e. decision-making that leads to the long-term adverse consequences of climate change), more fundamental linkages between sustainability and the overarching topic of economic development should be sought.

Defining a 'well-developed' institution is problematic and subjective,[92] since the extent to which deliberate interventions need to be made by governors at national or supranational levels is still not clear.[93] But what also has become clear from the chapter is that 'adapting' means 'changing', in particular the

89 Riché et al., *Climate-related Vulnerability.*

90 LeBlanc, "Special Issue."

91 Da Fonseca et al., "Trivializing Sustainability."

92 Jones et al., *Responding to a Changing Climate.*

93 Berkhout, Huitema, and Termeer, "The Governance of Adaptation."

nature of decision-making in promoting social goals, self-organisation and, finally, of the capacity to adapt.[94]

Finally, it can be concluded that much will depend on whether ASEAN can play an active role in developing climate change-related actions and strategies that are stimulating adaptation policies based on sustainable development on the level of its member states. Therefore, it will be important to take a closer view on the level of national institutions in the future, and how they connect to related processes and dynamics on the regional level. Similarly, on the regional level, institutions that take into account both the aims of sustainable development and the prevailing political realities at the implementation level will be of paramount importance in this effort.

---

94 Adger, "Building Resilience," 2.

Bibliography

"5 Pemicu Kontroversi RUU Ormas [5 Initiators of the Controversy about the Mass Organisation Bill]." Published by Tempo.co, July 3, 2013. Accessed September 12, 2013. http://www.tempo.co/read/news/2013/07/03/078493040/5-Pemicu-Kontroversi-RUU-Ormas.

Acharya, Amitav. *Asia Rising: Who Is Leading?* Singapore: World Scientific, 2008.

Adger, W. Neil. "Institutional Adaptation to Environmental Risk under the Transition in Vietnam." *Annals of the Association of American Geographers* 90.4 (2000): 738–758.

———, "Building Resilience to Promote Sustainability." *IHDP Update* 2 (2003): 1–3.

Adger, W. Neil, Jouni Paavola, and Sameemul Huq. "Towards Justice in Adaptation to Climate Change." In *Fairness in Adaptation to Climate Change*, edited by W. Neil Adger, Jouni Paavola, Sameemul Huq, and M. J. Mace, 1–20. Cambridge, MA: MIT Press. 2006.

Adger, W. Neil, Katrina Brown, Jenny Fairbrass, Andrew Jordan, Jouni Paavola, Sergio Rosendo, and Gill Seyfang. "Governance for Sustainability: Towards a 'Thick' Analysis of Environmental Decision-making." *Environment and Planning A* 35 (2003): 1095–1110.

Adger, W. Neil, Nigel W. Arnell, and Emma L. Tompkins. "Successful Adaptation to Climate Change Across Scales." *Global Environmental Change* 15.2 (2005): 77–86.

Adger, W. Neil, Shardul Agrawala, and M. Monirul Qader Mirza. *Assessment of Adaptation Practices, Options, Constraints and Capacity: Climate Change 2007: Impacts, Adaptation and Vulnerability*. Contribution of Working Group II to the Fourth Assessment Report of the Intergovernmental Panel on Climate Change, chapter 17. Geneva: IPCC, 2007.

Adger, W. Neil, Suraje Dessai, Marisa Goulden, Mike Hulme, Irene Lorenzoni, Donald R. Nelsen, Lars Otto Naess, Johanna Wolf, and Anita Wrefold. "Are There Social Limits to Adaptation to Climate Change?" *Climate Change* 93 (2009): 335–354. doi: 10.1007/s10584-008-9520-z.

Agarwal, Anil. "An Indian Environmentalist's Credo." In *Social Ecology*, edited by Ramachandra Guha, 346–386. Berkeley: Oxford University Press, 1994.

Agarwal, Anil et al. *The Leap Frog Factor: Clearing the Air in Asian Cities*. New Delhi: Centre for Science and Environment, 2006.

Agassiz, Louis. *Etudes sur les glaciers*. Neuchatel: Jent et Gassmann, 1840.

Agrawal, Arun. "Local Institutions and Adaptation to Climate Change." In *Social Dimensions of Climate Change: Equity and Vulnerability in a Warming World*, edited by Robin Mearns and Andrew Norton, 173–197. Washington, D.C.: The World Bank, 2010.

Ahuja, Ravi, and Christiane Brosius, ed. *Mumbai, Delhi, Kolkata: Annäherungen an die Megastädte Indiens* [Mumbai, Delhi, Kolkata: Approaches to Megacities in India]. Heidelberg: Draupadi-Verlag, 2006.

Alagh, Yoginder K. "Energy and High Growth in India: Sustainability Issues." In *India's Economic Future: Education, Technology, Energy and Environment*, edited by Manmohan Agarwal, 219–246. New Delhi: Social Science Press, 2009.

Alley, Kelly D. "Idioms of Degeneracy: Assessing Gaṅgā's Purity." In *Purifying the Earthly Body of God: Religion and Ecology in Hindu India*, edited by Lance Nelson, 306–315. Albany: SUNY Press, 1998.

Amri, Ulil. "Divine Power: Faith-based Mass Groups Take on Climate Change." *Strategic Review* 1.1 (2011): 95–108.

Anuman Rajadhon, Phya. *Essays on Thai Folklore*. Bangkok: Thai Inter-Religious Commission for Development and Sathirakoses Nagapradipa Foundation, 1988.

Appadurai, Arjun. *Modernity at Large: Cultural Dimensions of Globalization*. Minneapolis: University of Minnesota Press, 1998.

———, "Spectral Housing and Urban Cleansing: Notes on Millennial Mumbai." *Public Culture* 12.3 (2000): 627–651.

Archibugi, Daniele, and Alberto Coco. "Measuring Technological Capabilities at the Country Level: A Survey and a Menu for Choice." *Research Policy* 34 (2005): 175–194.

Arnold, Edwin, trans., *The Gulistan: Being the Rose Garden of Shaikh Sa'di*. London, New York: Harper & Brothers, 1899.

Ashcroft, Bill. "The Multiplicity of Modernity: Post-colonial Studies and Globalization." Podcast published July 15, 2009 by Internationales Zentrum für Kultur- und Technikforschung, University of Stuttgart, lecture series "Kulturtheorien". Accessed August 13, 2013. http://www.uni-stuttgart.de/izkt2/2009-07-15-Bill_Ashcroft-The_Multiplicity_of_Modernity_Post-colonial_Studies_and_Globalization.mp3.

———, "Bollywood, Postcolonial Transformation, and Modernity." In *Travels of Bollywood Cinema: From Bombay to LA*, edited by Anjali Gera Roy and Chua Beng Huat. Oxford: Oxford University Press, 2012, 1–18.

Asian Development Bank. *The Economics of Climate Change in Southeast Asia: A Regional Review*. Manila: Asian Development Bank, 2009. Accessed January 28, 2012. http://www.adb.org/documents/books/economics-climate-change-sea/Economics-Climate-Change.pdf.

———, *Asia 2050: Realizing the Asian Century*. Manila: Asian Development Bank, 2011.

Assmann, Aleida. *Introduction to Cultural Studies: Topics, Concepts, Issues*. Berlin: Erich Schmidt Verlag, 2012.

Association of Academies of Sciences in Asia (AASA). *Towards a Sustainable Asia: Environment and Climate Change*. Berlin, Heidelberg: Science Press Beijing and Springer, 2011.

Association of Southeast Asian Nations (ASEAN). *ASEAN Declaration on the 13th Session of the Conference of the Parties to the UNFCCC and the 3rd Session of the CMP to the Kyoto Protocol*. Singapore: ASEAN, 2007.

———, *The ASEAN Charter*. Jakarta: ASEAN, 2008.

———, *Annual Report 2008–2009: Implementing the Roadmap for an ASEAN Community 2015*. Jakarta: ASEAN, 2009.

———, *ASEAN Joint Statement on Climate Change to the 15th Session of the Conference of the Parties to the United Nations Framework Convention on Climate Change and the 5th Session of the Conference of Parties Serving as the Meeting of the Parties to the Kyoto Protocol*. Cha-am Hua Hin: ASEAN, 2009.

———, *Fourth ASEAN State of the Environment Report*. Jakarta: ASEAN, 2009.

———, "ASEAN Vision 2020." Accessed May 31, 2012. http://www.aseansec.org/1814.htm.

Atteridge, Aaron. "Shifting Sands: India's New Approach to the Politics of Climate Change." Policy Brief. Published by Stockholm Environment Institute, 2010.

"Author Arundhati Roy: India's Economic Success a 'Lie'." Published by BBC News. Last modified June 2, 2011. Accessed May 23, 2011. http://www.bbc.co.uk/news/world-south-asia-13624077.

Azra, Azyumardi, Dina Afrianty, and Robert W. Hefner. "*Pesantren* and *madrasa*." In *Schooling Islam: The Culture and Politics of Modern Muslim Education*, edited by Robert Hefner and Muhammad Qasim Zaman, 172–198. Princeton: Princeton University Press, 2007.

Bäckstrand, Karin, and Eva Lövbrand. "Planting Trees to Mitigate Climate Change: Contested Discourses of Ecological Modernization, Green Governmentality and Civic Environmentalism." *Global Environmental Politics* 6.1 (2006): 50–75.

Bagla, Pallava. "Artificial Glaciers to Help Farmers." *Science* 282 (1998): 619.

———, "No Sign yet of Himalayan Meltdown, Indian Report Finds." *Science* 326 (2009): 924–925.

———, "Climate Science Leader Rajendra Pachauri Confronts the Critics." *Science* 327 (2010): 510–511.

Banjit Sairokham, and Thanat Baiya บานจิตร สายรอคำ และถนัด ใบยา. *Buat pa suep chata mae-nam kan jatkan pa lae nam choeng watthanatham* บวชป่า สืบชะตาแม่น้ำ การจัดการป่าและน้ำเชิงวัฒนธรรม [Ordaining Trees and River Life Prolonging (Ceremony): Cultural Management of Forest and Water]. Nan: Hak Mueang Nan Foundation, 2006.

Bano, Tarannum, Sachchidanand Singh, N. C. Gupta, Kirti Soni, R. S. Tanwar, S. Nath, B. C. Arya, and B. S. Gera. "Variation in Aerosol Black Carbon Concentration and its Emission Estimates at the Mega-City Delhi." *International Journal of Remote Sensing* 32.21 (2011): 6749–6764. doi: 10.1080/01431161.2010.512943.

Baso, Ahmad. *Pesantren Studies 2a. Buku II: Kosmopolitanisme Peradaban Kaum Santri di Masa Kolonial* [Pesantren Studies 2a. Book II: The Cosmopolitan Culture of Muslim Seminarians in the Colonial Era]. Jakarta: Pustaka Afid, 2012.

Batabyal, Somnath. *Making News in India: Star News and Star Ananda*. New Delhi: Routledge, 2012.

Bauman, Zygmunt. *Intimations of Postmodernity*. London, New York: Routledge, 1992.

Baviskar, Amita. "Between Violence and Desire: Space, Power and Identity in the Making of Metropolitan Delhi." *International Social Science Journal* 55.175 (2004): 89–98.

Bebbington, Anthony J., Samuel Hickey, and Diana C. Mitlin. "Introduction: Can NGOs Make a Difference? The Challenge of Development Alternatives." In *Can NGOs Make a Difference? The Challenge of Development Alternatives*, edited by Anthony J. Bebbington, Samuel Hickey and Diana C. Mitlin, 3–37. London, New York: Zed Books, 2008.

Beck, Ulrich. *World Risk Society*. Cambridge: Polity Press, 1998.

Bell, Martin. "Technology Transfer to Transition Countries: Are there Lessons from the Experience of the Post-War Industrialising Countries?" In *Technology in Transition*, edited by David Dyker, 63–94. Budapest: Central European University Press, 1997.

Bell, Martin, and Keith Pavitt. "Technological Accumulation and Industrial Growth: Contrasts between Developed and Developing Countries." *Industrial and Corporate Change* 2.2 (1993): 157–210.

Berkhout, Frans, Dave Huitema, and Katrien Termeer. "The Governance of Adaptation: An International Symposium." Paper presented in Amsterdam, March 22–23, 2012. Accessed May 31, 2012. http://www.ivm.vu.nl/en/Images/The%20Governance%20of%20Adaptation%20Call%20for%20Abstracts_tcm53-227922.pdf.

Betz, Joachim, and Melanie Hanif. *The Formation of Preferences in Two-level Games: An Analysis of India's Domestic and Foreign Energy Policy*. GIGA Working Papers No. 142. Hamburg: GIGA, 2010.

Bhabha, Homi. *The Location of Culture*. New York: Routledge, 1994.

Bhambri, Rakesh, and Tobias Bolch. "Glacier Mapping: A Review with Special Reference to the Indian Himalayas." *Progress in Physical Geography* 33.5 (2009): 672–704.

Bhan, Gautam. "'This Is No Longer the City I Once Knew': Evictions, the Urban Poor and the Right to the City in Millennial Delhi." *Environment and Urbanisation* 21.1 (2009): 127–142.

Bhattacharya, Arindam, Rahul Jain, and Amar Choudhary. *Green Manufacturing: Energy Products and Processes*. Boston: The Boston Consulting Group/Confederation of Indian Industry, 2011.

Bhattacharya, Subhes C., and Prasanta K. Dey. "Risk Management in a Large Rural Electrification Programme." *CEPMLP Internet Journal* 16.13 (2006): 1–24.

Bierbaum, Rosina, John Holdren, Michael MacCracken, Richard Moss, and Peter Raven. *Confronting Climate Change: Avoiding the Unmanageable and Managing the*

Unavoidable. Scientific Expert Group Report on Climate Change and Sustainable Development. New York: United Nations Foundation, 2007.

Billett, Simon. "Dividing Climate Change: Global Warming in the Indian Mass Media." *Climatic Change* 99 (2010): 1–16. doi: 10.1007/s10584-009-9605-3.

Bolch, Tobias, Anil V. Kulkarni, Andreas Kääb, Christian Huggel, Frank Paul, J. Graham Cogley, Holger Frey, Jeffrey S. Kargel, Koji Fujita, Marlene Scheel, Sagar Ratna Bajracharya, and Markus Stoffel. "The State and Fate of Himalayan Glaciers." *Science* 336 (2012): 310–314.

Boomgaard, Peter, ed. *Southeast Asia: An Environmental History*. Santa Barbara, Oxford: ABC-CLIO, 2007.

Boowa Ñāṇasampanno, Ācariya Mahā. *Venerable Ācariya Mun Bhūridatta Thera*. Translated by Bhikkhu Sīlaratano. Baan Taad: Wat Pa Baan Taad, 2003.

Bourdieu, Pierre. *Outline of a Theory of Practice*, translated by Richard Nice. Cambridge: Cambridge University Press, 1977.

———, *The Logic of Practice*, translated by Richard Nice. Stanford: Stanford University Press, 1990.

Bratton, Michael, and Nicholas van de Walle. *Democratic Experiments in Africa: Regime Transitions in Comparative Perspective*. Cambridge: Cambridge University Press, 1997.

Brosius, Christiane. "Die Vorstellung vom 'Neuen Indien': Anmerkungen zur unfassbaren Mittelklasse [The Image of 'New India': Annotations on the Incomprehensible Middle Class]." *Der Bürger im Staat* [The Citizen in the State] 59.3–4 (2009): 220–227. Accessed August 31, 2013. http://archiv.ub.uni-heidelberg.de/savifadok/volltexte/2010/1161.

———, *India's Middle Class: New Forms of Urban Leisure, Consumption and Prosperity*. New Delhi, London: Routledge, 2010.

Brown, Kevin. "Spectacle as Resistance: Performing Tree Ordination in Thailand." *The Journal of Religion and Theatre* 5.2 (2006): 91–103.

Bruinessen, Martin van. "*Pesantren* and *kitab kuning*: Maintenance and Continuation of a Tradition of Religious Learning." In *Texts from the Islands: Oral and Written Traditions of Indonesia and the Malay World* [*Ethnologica Bernica* 4], edited by Wolfgang Marschall, 121–146. Bern: University of Bern, 1994.

———, "*Ghazwul fikri* or Arabisation? Indonesian Muslim Responses to Globalization." In *Dynamics of Southeast Asian Muslims in the Era of Globalization*, edited by Ken Miichi and Omar Farouk, 47–70. Tokyo: Japan International Cooperation Agency Research Institute (JICA-RI), 2013.

Bruyninckx, Hans. "Sustainable Development: The Institutionalization of a Contested Policy Concept." In *Advances In International Environmental Politics*, edited by Michele Betsill, Kathryn Hochstetler and Dimitris Stevis, 265–298. Hampshire: Palgrave Macmillan, 2005.

Buckley, Brendan M., Kevin J. Anchukiatis, Daniel Penny, Roland Fletcher, Edward R. Cook, Masaki Sano, Wichienkeeo, Aruoonrut, Ton That Minh, and Truong Mai Hong. "Climate as a Contributing Factor in the Demise of Angkor, Cambodia." *Proceedings of the National Academy of Sciences* 107.15 (2010): 6748–6752.

Buddhadasa Bhikkhu พุทธทาสภิกขุ. *Phutthasatsanik kap kan-anurak thammachat* พุทธศาสนิกกับการอนุรักษ์ธรรมชาติ [Buddhists and the Conservation of Nature]. Bangkok: Komonkhimthong Foundation, 1990.

Burgess, Jacquelin. "The Production and Consumption of Environmental Meanings in the Mass Media: A Research Agenda for the 1990s." *Transactions of the Institute of British Geographers* 15.2 (1990): 139–161. Accessed December 22, 2012. http://www.jstor.org/stable/622861.

Burlingame, Eugene Watson. *Buddhist Legends.* Translated from the Original Pali Text of the Dhammapada Commentary. Part II, edited by Charles Rockwell Lanman, Delhi: Motilal Banarsidass Publishers, 2005.

Butcher, Melissa. *Transnational Television, Cultural Identity and Change: When Star Came to India*. New Delhi: Sage, 2003.

Cane, Mark A. "The Evolution of El Nino, Past and Future." *Earth and Planetary Science Letters* 230 (2005): 227–240.

Carey, Mark. "The History of Ice: How Glaciers Became an Endangered Species." *Environmental History* 12.3 (2007): 497–527.

Casimir, Michael J., ed. *Culture and the Changing Environment: Uncertainty, Cognition and Risk Management in Cross-cultural Perspective*. New York, Oxford: Berghahn Books, 2008.

Census of India. "Census 2011." Accessed May 18, 2011. http://censusindia.gov.in.

Centre for Science and Environment. *Comments on Interim Report of Expert Group on Low Carbon Strategies by the Planning Commission of India*. New Delhi: Centre for Science and Environment, 2011.

Chakravorty, Rangan, and Nandini Gooptu. "Imagi-Nation: The Media, Nation and Politics in Contemporary India." In *Cultural Encounters: Representing Otherness*, edited by Elisabeth Hallam and Brian V. Street, 89–118. London: Routledge, 2000.

"Chao ban tuen kho huai jau mae ชาวบ้านตื่นขอหวยเจ้าแม่ [Excited Folk Pray for Lotto Numbers from Mother Goddess]." Published by YouTube. Accessed September 25, 2012. http://www.youtube.com/watch?v=9vcMXrMYf3U.

Chapman, Graham, Keval Kumar, Caroline Fraser, and Ivor Gaber. *Environmentalism and the Mass Media: The North-South Divide.* London, New York: Routledge, 1997.

Charpentier, Johann von. *Essai sur les glaciers et sur le terrain erratique du bassin du Rhone*. Lausanne: Marc Ducloux, 1841.

Chatsumarn, Kabilsingh et al. (N.N.). *Buddhism and Nature Conservation*. Phra Nakorn: Thai Tibet Center, 2010.

Chaṭṭha Sangāyana. CD-ROM Version 3. Published by Vipassana Research Institute.

Chen, Dake, Mark A. Cane, Alexey Kaplan, Stephen E. Zebiak, and D.J. Huang. "Predictability of El Nino Over the Past 148 Years." *Nature* 428 (2004): 733–736.

Chikkatur, Ananth P., and Ambuj D. Sagar. *Positioning the Indian Coal-Power Sector for Carbon Mitigation: Key Policy Options*. White Paper Series. Arlington: PEW Center Global Climate Change, 2009.

"Chuan pai du mahatsajan kluai tani pli dang dokbua ชวนไปดู มหัศจรรย์ กล้วยตานี ปลีดั่งดอกบัว [Be Amazed by the Tani Banana Tree Whose Blossom is Like a Lotus Flower]." Oknation Blog posted by Khasemakhung คาเซมะคุง, 18 January 2011. Accessed December, 16, 2011. http://www.oknation.net/blog/kasemakung/2011/01/18/entry-1.

Climate Analysis Indicator Tool – CAIT. "Per Capita Emissions." Accessed May 19, 2012. http://cait.wri.org/.

Climate Change Agriculture and Food Security (CCAFS). *Open Call: A Discussion Paper Analyzing the Role of Rural Institutions in Enabling Adaptation to Climate Change in the Agricultural Sector*. Nairobi: CCAFS, 2011. Accessed May 31, 2012. http://ccafs.cgiar.org/news/careers-and-calls/call-proposal-discussion-paper-institutions-and-climate-change-adaptation.

Cogley, J. Graham. "Himalayan Glaciers in 2010 and 2035." In *Encyclopedia of Snow, Ice and Glaciers*, edited by Vijay P. Singh, Pratap Singh and Umesh K. Haritashya, 520–523. Dordrecht: Springer, 2011.

Cogley, J. Graham, Jeffrey S. Kargel, Georg Kaser, and Cornelis J. van der Veen. "Tracking the Source of Glacier Misinformation." *Science* 327 (2010): 522.

Cole, Luke W., and Sheila R. Foster. *From the Ground Up: Environmental Racism and the Rise of the Environmental Justice Movement*. New York, London: New York University Press, 2001.

Confederation of Indian Industry. *Renewable Energy: The Next Wave*. New Delhi: Confederation of Indian Industry, 2010.

Confederation of Indian Industry, and The Boston Consulting Group. *Green Manufacturing: Energy, Products and Processes*. New Delhi: Confederation of Indian Industry, The Boston Consulting Group, 2011.

Cook, Edward R., Kevin J. Anchukaitis, Brendan M. Buckley, Roseanne D. D'Arrigo, Gordon C. Jacoby, and William E. Wright. "Asian Monsoon Failure and Megadrought during the Last Millennium." *Science* 328 (2010): 486–489.

Cowen, Michael, and Robert W. Shenton. *Doctrines of Development*. London: Routledge, 1996.

Cribb, Robert. "Environmentalism in Indonesian Politics." In *Towards Integrated Environmental Law in Indonesia*?, edited by Adriaan Bedner and Nicole Niessen, 37–48. Leiden: CNWS, 2003.

Cruikshank, Julie. *Do Glaciers Listen?: Local Knowledge, Colonial Encounters, and Social Imagination*. Vancouver: UBC Press, The University of British Columbia, 2005.

Cruz, Rex Victor, Hideo Harasawa, Murari Lal, Shaohong Wu, Yurij Anokhin, Batima Punsalmaa, Yasushi Honda, Mostafa Jafari, Congxian Li, and Nguyen Huu Ninh. "Asia." In *Climate Change 2007: Impacts, Adaptation and Vulnerability: Working Group II Contribution to the Fourth Assessment Report of the Intergovernmental Panel on Climate Change*, edited by Martin Parry, Osvaldo Canziani, Jean Palutikof, Paul van der Linden and Clair Hanson, 469–506. Cambridge: Cambridge University Press, 2007.

Da Fonseca, Igor Ferraz, Marcel Bursztyn, and Benjamin S. Allen. "Trivializing Sustainability: Environmental Governance and Rhetorical Free-riders in the Brazilian Amazon." *Natural Resources Forum, Special Issue: Institutions for Sustainable Development* 36.1. (2012): 28–37.

Darlington, Susan M. "The Ordination of a Tree: The Buddhist Ecology Movement in Thailand." *Ethnology* 37.1 (1998): 1–15.

———, "Buddhism and Development: The Ecology Monks of Thailand." In *Action Dharma: New Studies in Engaged Buddhism*, edited by Christopher Queen, Charles Prebish and Damien Keown, 96–109. London: RoutledgeCurzon, 2003.

———, "The Spirit(s) of Conservation in Buddhist Thailand." In *Nature Across Cultures: Views of Nature and the Environment in Non-Western Cultures*, edited by Helaine Selin, 129–145. Dordrecht: Kluwer Academic Publishers, 2003.

———, "The Good Buddha and the Fierce Spirits: Protecting the Northern Thai Forest." *Contemporary Buddhism*, 8.2 (2007): 169–185.

———, "Translating Modernity: Buddhist Response to the Thai Environmental Crisis." In *TransBuddhism: Transmission, Translation, Transformation*, edited by Nalini Bhushan, Jay L. Garfield and Abraham Zablocki, 183–207. Amherst, MA: University of Massachusetts Press, 2009.

———, *The Ordination of a Tree: The Evolution of the Thai Buddhist Environmental Movement*. Albany, New York: State University of New York Press, 2012.

D'Arrigo, Roseanne, David Frank, Gordon Jacoby, and Neil Pederson. "Spatial Response to Major Volcanic Events in or about AD 536, 934 and 1258: Frost Rings and other Dendrochronological Evidence from Mongolia and Northern Siberia: Comment on R. B. Stothers, 'Volcanic Dry Fogs, Climate Cooling, and Plague Pandemics in Europe and the Middle East,' (*Climate Change* 42, 1999)." *Climate Change* 49 (2001), 239–246.

Day, Mary B., David A. Hodell, Mark Brenner, Hazel J. Chapman, Jason H. Curtis, William F. Kenney, Alan L. Kolata, and Larry C. Peterson. "Paleoenvironmental History of the West Baray, Angkor (Cambodia)." *Proceedings of the Natural Academy of Sciences* 109.4 (2012): 1046–1051.

"Death Toll from Months of Flooding in Thailand Rises Above 800." *Straits Times*, December 31, 2011. Accessed January 14, 2012. http://www.straitstimes.com/BreakingNews/SEAsia/Story/STIStory_750551.html.

DeCaroli, Robert. *Haunting the Buddha: Indian Popular Religions and the Formation of Buddhism*. Oxford, New York: Oxford University Press, 2004.

De Graaf, H. J., and Theodore G. Th. Pigeaud. *De eerste moslimse voorstendommen op Java: studien over de staatkundige geschiedenis van de 15de en 16de eeuw*. 'S Gravenhage: Nijhoff, 1974.

Delcore, Henry D. "Symbolic Politics or Generification? The Ambivalent Implications of Tree Ordination in the Thai Environmental Movement." *Journal of Political Ecology* 11 (2004): 1–30.

Delegation of the European Union to Indonesia, Brunei Darussalam and ASEAN. Accessed May 31, 2012. http://eeas.europa.eu/delegations/indonesia/index_en.htm.

Dembowski, Hans. "Environment and Politics in India." In *Taking the State to Court: Public Interest Litigation and the Public Sphere in Metropolitan India*, edited by Hans Dembowski, 63–82. Asia House's online version of the book originally published in 2001. Accessed November 28, 2012. http://www.asienhaus.de/public/archiv/taking_the_state_to_court.pdf.

Department of Non-Conventional Energy Sources (DNES). *Annual Report 1985–1986*. New Delhi: Ministry of Energy, 1986.

Depledge, Joanna. "The Opposite of Learning: Ossification in the Climate Change Regime." *Global Environmental Politics* 6.1 (2006): 1–22.

Deshpande, Satish. *Contemporary India: A Sociological View*. New Delhi: Penguin, 2003.

Dhavalikar, M. K. *Environment and Culture: A Historical Perspective*. Pune: Bhandarkar Oriental Research Institute, 2002.

Di Chiro, Giovanna. "Nature as Community: The Convergence of Environment and Social Justice." In *Uncommon Ground: Rethinking the Human Place in Nature*, edited by William Cronon, 298–531. New York: W.W. Norton & Co., 1996.

Diodato, Nazzareno, Gianni Bellocchi, and Gianni Tartari. "How do Himalayan Areas Respond to Global Warming?" *International Journal of Climatology* 32.7 (2012): 975–982.

Dobson, Andrew. *Justice and the Environment: Conceptions of Environmental Sustainability and Dimensions of Social Justice*. Oxford: Oxford University Press, 1999.

Donner, Wolf. *Thailand: Land zwischen Tradition und Moderne* [Thailand: Country between Tradition and Modernity]. München: Beck, 1996.

Dorsey, Michael K. "Climate Knowledge and Power: Tales of Skeptic Tanks, Weather Gods, and Sagas for Climate (in) Justice." *Capitalism Nature Socialism* 18.2 (2007): 7–21.

Dosi, Giovanni. "Technological Paradigms and Technological Trajectories: A Suggested Interpretation of the Determinants and Directions of Technical Change." *Research Policy* 11.3 (1982): 147–162.

Dosi, Giovanni, L. Marengo, and C. Pasquali. "How Much Should Society Fuel the Greed of Innovators? On the Relations between Appropriability, Opportunities and Rates of Innovation." *Research Policy* 11.3 (2006): 1110–1121.

Drennen, Thomas E., Jon D. Erickson, and Duane Chapman. "Solar Power and Climate Change Policies in Developing Countries." *Energy Policy* 24.1 (1996): 9–16.

Drew, Georgina. "A Retreating Goddess? Conflicting Perceptions of Ecological Change Near the Gangotri-Gaumukh Glacier." *Journal for the Study of Religion, Nature and Culture* 6.3 (2012): 344–362.

Dubash, Navroz K. *Toward a Progressive Indian and Global Climate Politics*. Working Paper No. 1. New Delhi: Centre for Policy Research, 2009.

Dubey, Salman. "The Middle Class." In *India Briefing*, edited by Leonard Gordon and Philip Oldenberg, 137–164. Boulder, London: Westview Press, 1992.

Dutt, Gautam. "Reaching a Climate Agreement: Beyond the Copenhagen Accord." *Economic and Political Weekly* 45.17 (2010): 32–37.

Earth-Mates Dialogue Center (EDMC) with the United Nations Development Program (UNDP) and the Alliance of Religions and Conservation (ARC). "The Muslim 7-Year Action Plan (M7YAP) To Deal with Global Climate Change." Last modified September 20, 2009. Accessed December 7, 2012. http://www.arcworld.org/downloads/Muslim-7YP.pdf.

Eberhard, Anton, Siv Tolke, Anna Vigh, Antonio del Monaco, Harald Winkler, and Stephen Danyo. *GEF Climate Change Program Study*. Washington: Global Environmental Facility Office of Monitoring & Evaluation, 2004.

Ebtekar, Massoumeh. "Leadership for Peace and Sustainable Development: The Spiritual and Feminine Factor." *Farzaneh* 15.5 (2008): n.p. http://en.farzanehjournal.com/index.php/articles/no-15/6-no-15-5-leadership-for-peace-and-sustainable-development-the-spiritual-and-feminine-factor.

Eckersley, Robyn. *The Green State: Rethinking Democracy and Sovereignty*. Boston: MIT Press, 2004.

Eckes, Thomas, and Six, Bernd. "Fakten und Fiktionen in der Einstellungs-Verhaltens-Forschung: Eine Meta-Analyse [Fact and Fiction in Attitude-Behavior Research: A Meta-Analysis]." *Zeitschrift für Sozialpsychologie* [Journal for Social Psychology] 15.4 (1994): 253–271.

El-Ashry, Mohamed. "Adaptation to Climate Change: Building Resilience and Reducing Vulnerability." In *Climate Change Policy: Recommendations to Reach Consensus*, edited by The Brookings Institution, 58–65. Washington: Brookings, 2009.

Ellis, Karen, Bryn Baker, and Alberto Lemma. *Policies for Low Carbon Growth*. ODI Research Report. London: Overseas Development Institute, 2009.

"El Niño-Southern Oscillation." Published by Wikipedia. Last modified July 23, 2012. Accessed May 3, 2012. http://en.wikipedia.org/wiki/El_Nino-Southern_Oscillation.

Emile-Geay, Julien, Mark A. Cane, Edward R. Cook, and Gerald H. Haug. "Volcanoes and ENSO over the Past Millennium." *Journal of Climate* 21 (2008): 3134–3148.

Emile-Geay, Julien, Richard Seager, Mark Cane, and Edward R. Cook. "The Volcanic Eruption of 1258 A.D. and the Subsequent ENSO Event." *Geophysical Research Letters* 321 (2006): 1–5.

Emirbayer, Mustafa, and Ann Mische. "What is Agency?." *American Journal of Sociology* 103.4 (January 1998): 962–1023.

Esteva, Gustavo. "Development." In *The Development Dictionary: Knowledge as Power,* edited by Wolfgang Sachs, 1–23. London, New York: Zed Books, 2010.

Eucker, Dennis, and Jonas Hein. *Klimawandel in Südostasien: Die ASEAN als Wegbereiter einer regionalen Klimapolitik? GIGA Focus Asien 4*. Hamburg: GIGA, 2010.

Europa Publications, ed. *South Asia 2013*. London: Routledge, 2012.

European Commission. *Stimulating Technologies for Sustainable Development: An Environmental Technologies Action Plan for the European Union*. Brussels, European Commission, 2004.

European Photovoltaic Industry Association (EPIA). *Global Market Outlook for Photovoltaics until 2015*. Brussels: EPIA, 2009.

Farrell, Diana, and Eric Beinhocker. "Next Big Spenders: India's Middle Class." *Newsletter International*, May 19, 2007, McKinsey Global Institute, accessed August 31, 2013.http://www.mckinsey.com/insights/mgi/in_the_news/next_big_spenders_indian_middle_class.

Fealy, Greg, and Virginia Hooker. *Voices of Islam in Southeast Asia: A Contemporary Sourcebook*. Singapore: Institute of Southeast Asian Studies, 2006.

Federation of Indian Chambers of Commerce and Industry (FICCI). *FICCI'S 100 Day Action Agenda for the New Government*. New Delhi: FICCI, 2009.

Fernandes, Leela. "The Politics of Forgetting: Class Politics, State Power and the Restructuring of Urban Space in India." *Urban Studies* 41.12 (2004): 2415–2430.

"Fifty-five Countries Pledge to Cut Greenhouse Emissions, UN Hails 'Important Step', but 137 Nations yet to Make Promise." *The Guardian*, February 2, 2010. Accessed May 31, 2012. http://www.guardian.co.uk/enviroronment/2010/feb/02/55-countries-greenhouse-emissions-pledge.

Finsterwalder, Richard. *Die geodätischen, gletscherkundlichen und geographischen Ergebnisse der Deutschen Himalaja-Expedition 1934 zum Nanga Parbat*. Berlin, Karl Siegismund Verlag, 1938.

Fisher, Julie. *The Road from Rio: Sustainable Development and the Nongovernmental Movement in the Third World*. Westport: Praeger, 1993.

Fletcher, Roland, Daniel Penny, Damien Evans, Christophe Pottier, Mike Barbetti, M. Kummu, Terry Lustig, and APSARA. "The Water Management Network of Angkor, Cambodia." *Antiquity* 82 (2008): 658–670.

"Flood Toll 813: Three Missing." *Bangkok Post*, January 10, 2012. Accessed January 14, 2012. http://www.bangkokpost.com/news/local/274506/flood-toll-813-three-missing.

Focus on the Global South. "Food, Livelihoods & Climate Change in the Mekong Region." Workshop Report. Bangkok, August 9–11, 2010. Accessed January 28, 2012. http://www.focusweb.org/sites/www.focusweb.org/files/Mekong%20Report.pdf.

Foltz, Richard C. "Islam." In *The Oxford Handbook of Religion and Ecology*, edited by Roger S. Gottlieb, 207–219. Oxford, New York: Oxford University Press, 2006.

Foucault, Michel. *The Order of Things: An Archaeology of the Human Sciences.* London: Tavistock, 1972.

Fransson, Niklas, and Tommy Garling. "Environmental Concern: Conceptual Definitions, Measurement Methods, and Research Findings." *Journal of Environmental Psychology* 19.4 (1999): 369–382.

Freeman, Christopher. "The Learning Economy and International Inequality." In *The Globalizing Learning Economy: Major Socio-Economic Trends and European Innovation Policy*, edited by Daniele Archibugi and Bengt-Åke Lundvall, 320–334. Oxford: Oxford University Press, 2001.

Freeman, Christopher, and Luc Soete. "Developing Science, Technology and Innovation Indicators: What We Can Learn from the Past." *UNU-MERIT Working Paper #2007-001.* Maastricht: UNU-MERIT, 2007.

Frickel, Scott, and Debra J. Davidson. "Understanding Environmental Governance: A Critical Review." *Organisation & Environment* 17.4 (2004): 471–492.

Frolking, Steve, Changsheng Li, Rob Braswell, and Jan Fuglestvedt. "Short- and Long-term Greenhouse Gas and Radiative Forcing Impacts of Changing Water Management in Asian Rice Paddies." *Global Change Biology* 10 (2004): 1180–1196.

Fu, Xiaolan, Carlo Pietrobelli, and Luc Soete. "The Role of Foreign Technology and Indigenous Innovation in the Emerging Economies: Technological Change and Catching-up." *World Development* 39 (2011): 1204–1212.

Fujiwara, Noriko. *The Political Economy of India's Climate Agenda.* CEPS Working Document 35. Brussels: Centre for European Policy Studies, 2010.

Fujiwara, Noriko, and Christian Egenhofer. "Understanding India's Climate Agenda." CEPS Policy Brief No. 206. Brussels: Centre for European Policy Studies, 2010.

Gabaude, Louis. "Note sur l''ordination' sans ordre des arbres et des forêts." *Aséanie* 25 (2010): 91–125.

Gade, Anna. "Tradition and Sentiment in Indonesian Environmental Islam." *Worldviews* 16 (2012): 263–285.

Gadgil, Madhav. *Ecological Journeys.* New Delhi: Orient Blackswan, 2005.

Gao, Chaochao, Alan Robock, Stephen Self, Jeffrey B. Witter, J. P. Steffenson, Henrik B. Clausen, Marie-Louise Siggard-Andersen, Sigfus Johnsen, Paul A. Mayewski, and Caspar Ammann. "The 1452 or 1453 A.D. Kuwae Eruption Signal Derived from

Multiple Ice Core Records: Greatest Volcanic Sulphate Event of the Past 700 Years." *Journal of Geophysical Research* 111 (2006): D12107.

Gautam, P. K. "Climate Change and India's Position." IDSA Strategic Comments. Published December 5, 2007, New Delhi.

Georg, C. "Climate Change and Social Vulnerability." Paper presented at the IHDP Open Meeting, World Conference Center, Bonn, April 26–30, 2009.

German Advisory Council on Global Change (WBGU). *Solving the Climate Dilemma: The Budget Approach.* Berlin: WBGU 2009.

———, *World in Transition: A Social Contract for Sustainability*. Berlin: WBGU, 2011.

Ghosh, Prodipto. "Climate Change: Is India a Solution to the Problem or a Problem to the Solution?" In *Climate Change Perspectives from India*, published by United Nations Development Programme, India, 17–36. New Delhi: UNDP India, 2009. http://www.in.undp.org/content/dam/india/docs/undp_climate_change.pdf.

Ginneken, Jac van. *Understanding Global News*. London, Thousand Oaks, New Delhi: Sage, 1998.

Gitlin, Todd. *The Whole World Is Watching: Mass Media in the Making & Unmaking of the New Left*. Berkeley et al.: University of California Press, 1980.

GlacierWorks. Accessed December 31, 2012. http://www.glacierworks.org/.

Global Environment Facility. *GEF Contribution to Agenda 21 – the First Decade*. Washington: Global Environment Facility, 2000.

———, *Report of the GEF to the Twelfth Session of the Conference of the Parties to the United Nations Framework Convention on Climate Change*. Washington: Global Environment Facility, 2006.

———, *GEF 5 Focal Area Strategies*. Washington: GEF, 2011.

Goh, Gillian. "The 'ASEAN Way': Non-Intervention and ASEAN's Role in Conflict Management." *Stanford Journal of East Asian Affairs* 3.1 (2003): 113–118.

Gombrich, Richard. *Theravāda Buddhism: A Social History from Ancient Benares to Modern Colombo*. Second Edition. London, New York: Routledge, 2006.

Gong, Gloria Jean. "What China Wants: China's Climate Change Priorities in a Post-Copenhagen World." *Global Change, Peace & Security* 23.2 (2011): 159–175.

Gopalakrishnan, Manasi. "Indian PM Criticized for Unsatisfactory Reshuffle." *Deutsche Welle Asia*, July 12, 2011.

Gosling, David. *Religion and Ecology in India and Southeast Asia*. London: Routledge, 2001.

Government of India. *Integrated Energy Policy: Report of the Expert Committee*. New Delhi: Government of India, 2006.

———, *National Environment Policy 2006*. New Delhi: Government of India, 2006.

———, *National Urban Transport Policy*. New Delhi: Government of India, 2006.

———, *National Action Plan on Climate Change*. New Delhi: Government of India, 2007.

———, *Jawaharlal Nehru National Solar Mission: Towards Building Solar India*. New Delhi: Government of India, 2009.

———, *Greenhouse Gas Emissions 2007*. New Delhi: Government of India, 2010.

———, *Low Carbon Strategies for Inclusive Growth: An Interim Report*. New Delhi: Government of India, 2011.

———, *Mid Term Appraisal for the Eleventh Five Year Plan 2007–2012*. New Delhi: Government of India, Planning Commission, 2011.

Government of India. (Ministry of External Affairs, Public Diplomacy Division). *The Road to Copenhagen: India's Position on Climate Change Issues*. New Delhi: Government of India, 2009.

———, (Ministry of Power). *National Mission for Enhanced Energy Efficiency. Draft Mission Document: Implementation Framework*. New Delhi: Government of India, 2008.

———, (Ministry of Power, Bureau of Energy Efficiency). *India: Addressing Energy Security and Climate Change*. New Delhi: Government of India, 2007.

———, (Planning Commission). *Eleventh Five Year Plan 2007–2012*. New Delhi: Government of India, Planning Commission, 2008.

Government of India, Ministry of Home Affairs, Office of the Registrar General & Census Commissioner, India. *Census 2001, Data on Language*. Accessed August 20, 2013. http://censusindia.gov.in/Census_Data_2001/Census_Data_Online/Language/data_on_language.html.

Grabowsky, Volker. *Kleine Geschichte Thailands* [A Concise History of Thailand]. München: Beck, 2010.

Greenpeace India. *Hiding behind the Poor*. Bangalore: Greenpeace, 2007.

Grodzins Gold, Ann. "Sin and Rain: Moral Ecology in Rural North India." In *Purifying the Earthly Body of God: Religion and Ecology in Hindu India*, edited by Lance Nelson, 165–195. Albany: SUNY Press, 1998.

Groslier, Bernard-Philippe. "La cité hydraulique angkorienne: exploitation ou surexploitation du sol." *Bulletin de l'école française d'extrême-orient* 66 (1979): 161–202.

Group Centennial. *India 2039: Unleashing an Energy Revolution*. Policy Paper 4. Manila: Asian Development Bank, 2009.

Grove, Jean M. *Little Ice Ages: Ancient and Modern*. 2nd Edition. London: Routledge, 2004.

Grove, R. H., Vinita Damodaran, and Stapal Sangwan, ed. *The Environmental History of South and Southeast Asia: Nature & the Orient*. Oxford: Oxford University Press, 1998.

Grove, Richard H., and J. Chappell. *El Niño, History and Crisis: Studies from the Asia-Pacific Region*. Cambridge: White Horse, 2000.

Guha, Ramachandra. *Social Ecology*. Berkeley: Oxford University Press, 1994.

Gurjar, B. R., A. Jain, A. Sharma, A. Agarwal, P. Gupta, A. S. Nagpure, and J. Lelieveld. "Human Health Risks in Megacities due to Air Pollution." *Atmospheric Environment* 44.36 (2010): 4606–4613. doi: 10.1016/j.atmosenv.2010.08.011.

Gurjar, B. R., J. A. van Aardeene, J. Lelieveld, and M. Mohan. "Emission Estimates and Trends (1990–2000) for Megacity Delhi and Implications." *Atmospheric Environment* 38.33 (2004): 5663–5681.

Guthrie, Elizabeth. "A Study of the History and Cult of the Buddhist Earth Deity in Mainland Southeast Asia." PhD diss., University of Canterbury, 2004.

Haberman, David L. *River of Love in an Age of Pollution: The Yamuna River of Northern India*. Berkeley: University of California Press, 2006.

Hacker, Paul. "Inklusivismus [Inclusivism]." In *Inklusivismus: Eine indische Denkform* [Inclusivism: An Indian Mode of Thought], edited by Gerhard Oberhammer, 11–28. Wien: Institut für Indologie der Universität Wien, Sammlung De Nobili, 1983.

"Hae kho huai mae tani rot tit lai kilomet แห่ขอหวยแม่ตานี รถติดหลายกม [People Pray for Lotto Numbers – Traffic Tailbacks of Many Kilometres]." Published by Thairat Newspaper Online. Originally published in ไทยรัฐฉบับพิมพ์ [Thairat Newspaper], January 15, 2011. Accessed May 20, 2011. http://www.thairath.co.th/today/view/141536.

"Hae kho lek det jau mae tani naen soi phetkasem แห่ขอเลขเด็ดเจ้าแม่ตานี แน่นซอยเพชรเกษม [People Flock to Mae Tani to Pray for Lucky Numbers, Phetkasem Soi Overcrowded]." Published by Webportal mthai.com. Accessed September 25, 2012. http://lotto.mthai.com/lotto-news/result-787.html.

Haleem, M. A. S. Abdel. *The Qur'an: A New Translation of M. A. S. Abdel*. Oxford: Oxford University Press, 2005.

Hall, Stuart. *Representation: Cultural Representations and Signifying Practices*. London: Sage, 1997.

Hannerz, Ulf. *Cultural Complexity: Studies in the Social Organization of Meaning*. New York: Columbia University Press, 1992.

Harding, Sandra. *The Feminist Standpoint Theory Reader: Intellectual and Political Controversies*. New York: Routledge, 2003.

Harris, Ian. "Getting to Grips with Buddhist Environmentalism: A Provisional Typology." *Journal of Buddhist Ethics* 2 (1995): 173–190.

Hart, Gillian. "Development Critiques in the 1990s: culs de sac and Promising Paths." *Progress in Human Geography* 25.4 (2001): 649–658.

Hartley, John. *Popular Reality: Journalism, Modernity, Popular Culture*. London: Arnold, 1996.

Haum, Rüdiger. "Energy Security, Rural Electrification, and Market-Based Climate Change Intervention by the Global Environment Facility in India: The Limits of Transformation." *International Energy Journal* 11.4 (2010): 115–125.

———, *Transfer of Low-Carbon Technology under the United Nations Framework Convention on Climate Change: The Case of the Global Environment Facility and Its Market Transformation Approach in India*. Brighton: University of Sussex, 2011.

———, "Project Based Market Transformation in Developing Countries and International Technology Transfer: The Case of the Global Environment Facility and Solar PV." In *Low-Carbon Technology Transfer to Developing Countries: From Rhetoric to Reality*, edited by David Ockwell and Alexandra Mallet, 185–210. London: Earthscan, 2012.

Haum, Rüdiger, Klaus Jacob, and Julia Hertin. "Integration von Umwelt-, Innovations- und Industriepolitiken: Eine Evaluation von europäischen Strategien und Maßnahmen." In *Nachhaltig regieren: Eine Bilanz zu Governance Prinzipien und Praktiken*, edited by Reinhard Steurer and Rita Trattnig, 97–122. München: Oekom, 2010.

Hayashi, Yukio. *Practical Buddhism Among the Thai-Lao: Religion in the Making of a Region*. Kyoto: Kyoto University Press, 2003.

Hein, Wolfgang. "Globalisierung und nachhaltige Entwicklung in den Ländern des Südens." In *Nachhaltigkeit*, edited by K. Rennings and O. Hohmeyer, 165–218. Baden-Baden: Nomos, 1997.

Heinrich Boell Foundation. *Towards Good Society? Civil Society Actors, the State, and the Business Class in Southeast Asia – Facilitators of or Impediments to a Strong, Democratic, and Fair Society?* Berlin: Heinrich-Böll-Stiftung, 2004.

Hewitt, Ken. "The Karakoram Anomaly? Glacier Expansion and the 'Elevation Effect', Karakoram Himalaya." *Mountain Research and Development* 25.4 (2005): 32–40.

———, "Glacier Change, Concentration and Elevation Effects in the Karakoram Himalaya, Upper Indus Basin." *Mountain Research and Development* 31.3 (2011): 1–13.

Hill, Christopher V. *South Asia: An Environmental History*. Santa Barbara, Oxford: ABC-CLIO, 2008.

Hobday, Mike. "East Asian Latecomer Firms: Learning the Technology of Electronics." *World Development* 23.7 (1995): 1171–1193.

Hobday, Mike, and Howard Rush. "Upgrading the Technological Capabilities of Foreign Transnational Subsidiaries in Developing Countries: The Case of Electronics in Thailand." *Research Policy* 36.9 (2007): 1335–1356.

Hornidge, Anna-Katharina, and Christoph Antweiler, ed. *Environmental Uncertainty and Local Knowledge: Southeast Asia as a Laboratory of Global Ecological Change*. Bielefeld: Transcript-Verlag, 2012.

Hörning, Karl H., and Julia Reuter, ed. *Doing Culture: Neue Positionen zum Verhältnis von Kultur und sozialer Praxis* [Doing Culture: New Viewpoints Regarding the Relation between Culture and Social Practice]. Bielefeld: Transcript-Verlag, 2004.

Howell, Julia Day. "Sufism and the Indonesian Islamic Revival." *Journal of Asian Studies* 60.3 (2001): 701–729.

Ibenholt, Karin. "Explaining Learning Curves for Wind Power." *Energy Policy* 30 (2002): 1181–1189.

Ignatow, Gabriel. "Globalizations and the Transformation of Environmental Activism: Turkey since the 1980s." *Globalizations* 5.3 (2008): 433–447.

"India, China Sign MoA on Climate Change." *The Hindu*, October 22, 2009.

Intergovernmental Panel on Climate Change (IPCC). *Climate Change 2007: Impacts, Adaptation and Vulnerability. Working Group II Contribution to the Fourth Assessment Report*. Cambridge, New York: Cambridge University Press, 2007.

———, *Climate Change 2007: The Physical Science Basis. Working Group I Contribution to the Fourth Assessment Report*. Cambridge, New York: Cambridge University Press, 2007.

International Energy Agency. *Analysis of the Scope of Energy Subsidies and Suggestions for the G-20 Initiative*. Paris: International Energy Agency, 2010.

———, *World Energy Outlook 2010*. Paris: International Energy Agency, 2010.

———, *World Energy Outlook 2011*. Paris: International Energy Agency, 2011.

International Finance Corporation. *India, Kenya, and Morocco: Photovoltaic Market Transformation Initiative (PVMTI)*. Project Document. Washington: International Finance Corporation, 1998.

International Monetary Fund (IMF). *India Sustainability Report 2011*. Washington: IMF, 2011.

Isager, Lotte. "History and People of North Thailand." In *Forest in Culture – Culture in Forest: Perspectives from Northern Thailand*, edited by Ebbe Poulson, Flemming Skov, Sureeratna Lakanavichian, Sornprach Thanisawanyangkura, Henrik Borgtoft, and Ole Høiri, 85–115. Tjele: Research Centre on Forest and People in Thailand, 2001.

Isager, Lotte, and Søren Ivarsson. "Contesting Landscapes in Thailand: Tree Ordination as Counter-Territorialization." *Critical Asian Studies* 34.3 (2002): 395–417.

Ishii, Yoneo. "Church and State in Thailand." *Asian Survey* 8.10 (1968): 864–871.

Ivarsson, Søren. "Man, Nature and Environmentalism in Thailand: The Role of Buddhism." In *Forest in Culture – Culture in Forest: Perspectives from Northern Thailand*, edited by Ebbe Poulsen, Flemming Skov, Sureeratna Lakanavichian, Sornprach Thanisawanyangkura, Henrik Borgtoft, and Ole Høiri, 33–53. Tjele: Research Centre on Forest and People in Thailand, 2001.

Ives, Jack, and Bruno Messerli. *The Himalayan Dilemma: Reconciling Conservation and Development*. New York: Routledge, 1989.

Jackson, Tim. "Prosperity without Growth? The Transition to a Sustainable Economy." Published by Sustainable Development Commission, 2009. Accessed May 31, 2012. http://www.sd-commission.org.uk/data/files/publications/prosperity_without_growth_report.pdf.

Jacob, Klaus, Marian Beise, Jürgen Blazejczak, Dietmar Edler, Rüdiger Haum, Martin Jänicke, Thomas Löw, Ulrich Petschow, and Klaus Rennings. *Lead Markets for Environmental Innovations*. Mannheim: Physica, 2005.

Jacoby, Hanan, Mariano Rabassa, and Emmanuel Skoufias. *Distributional Implications of Climate Change in India*. Policy Research Working Paper No. 5623. Washington, D.C.: World Bank, 2011.

Jaffe, Adam B., Richard G. Newell, and Robert N. Stavins. "A Tale of Two Market Failures: Technology and Environmental Policy." *Ecological Economics* 54.2–3 (2005): 164–174.

Jänicke, Martin. "Ecological Modernisation: New Perspectives." In *Environmental Governance in Global Perspective – New Approaches to Ecological Modernisation*, edited by Martin Jänicke and Klaus Jacob, 2–24. Berlin: FFU, 2006.

———, "Ecological Modernisation: New Perspectives." *Journal of Cleaner Production* 16 (2008): 557–565.

Jänicke, Martin, and Stefan Lindemann. "Governing Environmental Innovations." *Environmental Politics* 19.1 (2010): 127–141.

Jänicke, Martin, Philip Kunig, and Michael Stitzel. *Lern- und Arbeitsbuch Umweltpolitik: Politik, Recht und Management des Umweltschutzes in Staat und Unternehmen*. Bonn: Diez, 2002.

Jayaraman, T. "Will It Be a US Endgame at Copenhagen?" *Economic and Political Weekly* 44.50 (2009): 13–15.

Joas, Hans. *Die Kreativität des Handelns* [The Creativity of Action]. Frankfurt a. M.: Suhrkamp, 1996.

Jones, Lindsey. "Overcoming Social Barriers to Adaptation." Published by Overseas Development Institute, 2010. Accessed May 31, 2012. http://www.odi.org.uk/resources/download/4945.pdf.

Jones, Lindsey, Eva Ludi, and Simone Levine. *Towards a Characterisation of Adaptive Capacity: A Framework for Analysing Adaptive Capacity at the Local Level*. Background Note, December 2010. London: ODI, 2010.

Jones, Lindsey, Susanne Jaspars, Sara Pavanello, Eva Ludi, Rachel Slater, Alex Arnall, Natasha Grist, and Sobona Mtisi. *Responding to a Changing Climate: Exploring how Disaster Risk Reduction, Social Protection and Livelihoods Approaches Promote Features of Adaptive Capacity*. Working Paper 319. London: ODI, 2010.

Joyce, Patrick. *The Rule of Freedom: Liberalism and the Modern City*. London: Verso, 2003.

Kääb, Andreas, Etienne Berthier, Christopher Nuth, Julie Gardelle, and Yves Arnaud. "Contrasting Patterns of Early Twenty-first-century Glacier Mass Change in the Himalayas." *Nature* 488 (2012): 495–498.

Kalin, Ibrahim. *Knowledge in Later Islamic Philosophy: Mulla Sadra on Existence, Intellect and Intuition*. Oxford: Oxford University Press, 2010.

Kamala Tiyavanich. *Forest Recollections: Wandering Monks in Twentieth-Century Thailand.* Chiang Mai: Silkworm Books, 1997.

Karan, Pradyumna. "Environmental Movements in India." *Frontier Thinking in Sustainable Development and Human Well Being* 84 (1994): 32–41.

Kargel, Jeffrey S., J. Graham Cogley, Gregory J. Leonard, Umesh Haritashya, and Alton Byers. "Himalayan Glaciers: The Big Picture is a Montage." *Proceedings of the National Academy of Sciences of the United States of America (PNAS)* 108.36 (2011): 14709–14710.

Karlsson-Vinkhuyzen, Sylvia I. "From Rio to Rio via Johannesburg: Integrating Institutions across Governance Levels in Sustainable Development Deliberations." *Natural Resources Forum, Special Issue: Institutions for Sustainable Development* 36.1 (2012): 3–15.

Kasa, Sjur, Anne Gullberg, and Goril Heggelund. "The Group of 77 in the International Climate Negotiations: Recent Developments and Future Directions." *International Environmental Agreements* 8 (2008): 113–127.

Kates, Robert W., William C. Clark, Robert Corell, J. Michael Hall, Carlo C. Jaeger, Ian Lowe, James J. McCarthy, Hans Joachim Schellnhuber, Bert Bolin, Nancy M. Dickson, Sylvie Faucheux, Gilberto C. Gallopin, Arnulf Grübler, et al. "Sustainability Science." *Science* 292 (2001): 641–642.

Katz, Ruth Cecily. *Arjuna in the Mahabharata: Where Krishna is, there is Victory.* Columbia: University of South Carolina Press, 1989.

Kaufman, Steven, Richard Duke, Richard Hansen, John Rogers, Richard Schwartz, and Mark Trexler. *Rural Electrification with Solar Energy As a Climate Protection Strategy.* REPP Research Report No. 9. Washington: REPP, 2002.

Kawasaki, Jintana. "Thailand's Rice Farmers Adapt to Climate Change." Published by Our World 2.0, United Nations University. Last modified May 24, 2010. Accessed January 29, 2012. http://ourworld.unu.edu/en/climate-change-adaptation-for-thailands-rice-farmers/.

Kealhofer, Lisa. "Human-Environmental Relationships in Prehistory: An Introduction to Current Research in South and Southeast Asia." *Asian Perspectives* 35.2 (1996): 111–117.

Kemp, Rene. *Environmental Policy and Technical Change.* Cheltenham: Edward Elagr, 1997.

Keskitalo, E. Carina H., and Johanna Liljenfeldt. "Working with Sustainability: Experiences of Sustainability Processes in Swedish Municipalities." *Natural Resources Forum, Special Issue: Institutions for Sustainable Development*, 36.1 (2012): 16–27.

"Khlong Hok Wa Barrier Torn Down." *Bangkok Post*, November 17, 2011. Accessed January 16, 2012. http://www.bangkokpost.com/news/transport/266730/khlong-hok-wa-barrier-torn-down.

"Kiai NU se-Tuban Sepakat Lestarikan Hutan [Kiai of NU in Tuban Agree to Preserve the Forest]." Published by Shvoong. Last modified April 20, 2012. Accessed December 7, 2012. http://id.shvoong.com/society-and-news/opinion/2284646-kiai-nu-se-tuban-sepakat/.

Kick, Wilhelm. *Adolph Schlagintweits Karakorum-Forschungsreise 1856. Forschungsberichte, Band 6.* München: Deutscher Alpenverein, 1993.

———, "Forschung am Nanga Parbat: Geschichte und Ergebnisse." In *Forschung am Nanga Parbat: Geschichte und Ergebnisse, Band 8: Beiträge und Materialien zur Regionalen Geographie*, edited by Wilhelm Kick, 1–133. Berlin: Institut für Geographie der Technischen Universität Berlin, 1996.

King, David A. "Climate Change Science: Adapt, Mitigate, or Ignore?" *Science* 303 (2004): 176–177. Accessed December 8, 2012. http://www.sciencemag.org.

Kisner, Corinne. "Climate Change in Thailand: Impacts and Adaptation Strategies." Published by Climate Institute Washington, D.C. Last modified July, 2008. Accessed January 14, 2012. http://www.climate.org/topics/international-action/thailand.htm.

"Klai huai ok! Hae kho tani phetkasem naen ใกล้หวยออก! แห่ขอตานี เพชรเกษมแน่น [Shortly before the Lottery Draw! People Flock to Tani and Pray for Numbers – Phetkasem Overcrowded]." Published by Thairat Newspaper Online, January 14, 2011. Accessed September 25, 2012. http://www.thairath.co.th/content/region/141354.

Klostermaier, Klaus K. *A Survey of Hinduism*. Albany: SUNY Press, 2007.

Kohli, Vineeta. *The Indian Media Business*. New Delhi: Sage, 2006.

Kraft, Kenneth. "Nuclear Ecology and Engaged Buddhism." In *Buddhism and Ecology: The Interconnection of Dharma and Deeds*, edited by Mary Evelyn Tucker and Duncan Ryūken Williams, 269–290. Cambridge, MA: Harvard University Press, 1997.

Kula, E. "Islam and Environmental Conservation." *Environmental Conservation* 28 (2001): 1–9. Accessed October 10, 2012. doi 10.1017/S0376892901000017.

Laird, John. *Money Politics, Globalisation, and Crisis. The Case of Thailand: Exploring New Paths Towards Sustainable Development*. Singapore: Graham Brash Pte Ltd., 2000.

Lakha, Salim. "The State, Globalisation and Indian Middle-Class Identity." In *Culture and Privilege in Capitalist Asia*, edited by Michael Pinches, 251–274. London: Routledge, 1999.

Lal, Sumir. *Can Good Economics Ever Be Good Politics? Case Study of India's Power Sector*. World Bank Working Paper No. 83. Washington, D.C.: World Bank, 2006.

Lall, Sanjaya. "Promoting Technology Development: The Role of Technology Transfer and Indigenous Effort." *Third World Quarterly* 14.1 (1993): 95–108.

Lamoureux, Florence. *Indonesia: A Global Studies Handbook*. Santa Barbara, CA: ABC-Clio, 2003.

Landmann, Alexandra. "Endangered Guardians of the Sacred Forest." *The Newsletter* 65 (2013): 42. Accessed September 24, 2013. http://www.iias.nl/sites/default/files/IIAS_NL65_42.pdf.

"Langkah LPBI NU dalam penanggulangan bencana [Measures Taken by LPBI NU to Handle Disasters]." Published by NU Online. Last modified July 19, 2012. Accessed December 6, 2012. http://www.nu.or.id/a,public-m,dinamic-s,detail-ids,5-id,38862-lang,id-c,halaqoh-t,Langkah+LPBI+NU+dalam+Penanggulangan+Bencana-.phpx.

Lanjouw, J. O., and A. Mody. "Innovation and the International Diffusion of Environmentally Responsive Technology." *Research Policy* 25 (1996): 549–557.

Latour, Bruno. *We Have Never Been Modern*. Cambridge, MA: Harvard University Press, 1993.

Lebel, Louis, Tira Foran, Po Garden, and Jesse B. Manuta. "Adaptation to Climate Change and Social Justice: Challenges for Flood and Disaster Management in Thailand." In *Climate Change Adaptation in the Water Sector*, edited by Fulco Ludwig, Pavel Kabat, Henk van Schaik and Michael van der Valk, 125–142. London: Earthscan, 2009.

LeBlanc, David. "Special Issue on Institutions for Sustainable Development." *Natural Resources Forum, Special Issue: Institutions for Sustainable Development* 36.1 (2012): 1–2.

Lee, Seungjoo. *The Evolutionary Dynamics of Institutional Balancing in East Asia*. EAI Asia Security Initiative Working Paper. Seoul: EAI, 2012.

Legris, Pierre, François Blasco, and Tichit. *Notice de la Carte, Cambodge. Carte Internationale du Tapis Végétale Series.* Pondichéry: Institut français de Pondichéry, 1972.

Levine, Simone, Eva Ludi, and Lindsey Jones. *Rethinking Support for Adaptive Capacity to Climate Change: The Role of Development Interventions*. London: ODI, 2011.

Lian, Koh Kheng, and Lovleen Bhullar. "Adaptation to Climate Change in the ASEAN Region." Draft dated October 4, 2010. Accessed May 31, 2012. http://www.ucl.ac.uk/laws/environment/docs/hong-kong/Adaptation%20to%20CC%20ASEAN%20%28KL%20Koh%20AND%20Lovleen%20Bhullar%29.pdf.

Lim, Weida. "Climate Change Policies in Singapore: Whose 'Environments' Are We Talking About?" *Environmental Justice* 2.2 (2009): 79–83.

Löhr, Sabine. "Zwischen Ost und West: Ansätze einer Ökologischen Ethik im Modernen Buddhismus." PhD diss., Universität Hamburg, 2004.

"LPBI NU, garda depan penanganan bencana di NU [LPBI NU, the Avant Garde of Dealing with Disasters in NU]." Published by NU Online. Last modified June 28, 2010. Accessed December 4, 2012. http://www.nu.or.id/a,public-m,dinamic-s,detail-ids,5-id,23553-lang,id-c,halaqoh-t,LPBINU++Garda+Depan+Penanganan+Bencana+di+NU-.phpx.

Lupton, Deborah, ed. *Risk and Sociocultural Theory: New Directions and Perspectives*. Cambridge: Cambridge University Press, 1999.

Mahnke, Eva. "Muskelspiel der Schwellenländer." Published by Klimaretter.Info, December 7, 2011.

Majelis Lingkungan Hidup, Kementerian Lingkungan Hidup, ed. *Akhlaq Lingkungan: Panduan Berperilaku Ramah Lingkungan* [Environmental Ethics: A Guide to Environmentally Friendly Conduct]. Jakarta, Yogyakarta: Majelis Lingkungan Hidup PP Muhammadiyah, Kementerian Lingkungkan Hidup Republik Indonesia, 2011.

Mangunjaya, Fachruddin Majeri, and Jeanne Elizabeth McKay. "Reviving an Islamic Approach for Environmental Conservation in Indonesia." *Worldviews* 16 (2012): 286–305.

Mann, Michael E., Jose D. Fuentes, and Scott Rutherford. "Underestimation of Volcanic Cooling in Tree-ring-based Reconstructions of Hemispheric Temperatures." *Nature Geoscience*, Advance Online Publication (2012): 1–4. Accessed July 5, 2012. http://www.nature.com/ngeo/index.html.

Mann, Michael E., Mark A. Cane, Stephen E. Zebiak, and Amy Clement. "Volcanic and Solar Forcing of the Tropical Pacific over the Past 1000 Years." *Journal of Climate* 18 (2005): 447–456.

Mann, Michael E., Zhihua Zhang, Scott Rutherford, Raymond S. Bradley, Malcolm K. Hughes, Drew Shindell, Caspar Ammann, Greg Faluvegi, and Fenbao Ni. "Global Signatures and Dynamical Origins of the Little Ice Age and Medieval Climate Anomaly." *Science* 326 (2009): 1256–1260.

Marcinkowski, Christoph. "Aspects of Shi'ism in Contemporary Southeast Asia." *The Muslim World* 98 (2008): 36–71.

Marks, Danny. "Climate Change and Thailand: Impact and Response." *Contemporary Southeast Asia* 33.2 (2011): 229–258.

Markus, Hazel R., and Shinobu Kitayama. "Culture and Self: Implications for Cognition, Emotion, and Motivation." *Psychological Review* 98.2 (1991): 224–253.

Martinot, Eric. "Renewable Energy Markets and the Global Environment Facility." *Financial Times*, February 12, 2000.

Mass Communication Organization of Thailand (MCOT). "ASEAN Explores Tools to Address Climate Change." Accessed May 31, 2012. http://www.mcot.net/cfcustom/cache_page/159435.html.

Mawdsley, Emma. "India's Middle Classes and the Environment." *Development and Change* 35.1 (2004): 79–103.

McDaniel, Justin. *The Lovelorn Ghost and the Magical Monk: Practicing Buddhism in Modern Thailand.* New York: Columbia University Press, 2011.

McGurty, Eileen. *Transforming Environmentalism: Warren County, PCBs, and the Origins of Environmental Justice.* New Brunswick, London: Rutgers University Press, 2009.

Mearns, Robin, and Andrew Norton. "Equity and Vulnerability in a Warming World: Introduction and Overview." In *Social Dimensions of Climate Change: Equity and Vulnerability in a Warming World*, edited by Robin Mearns and Andrew Norton, 1–44. Washington, D.C.: The World Bank, 2010.

———, ed. *Social Dimensions of Climate Change: Equity and Vulnerability in a Warming World*. Washington, D.C.: The World Bank, 2010.

Meehl, Gerald A., Warren M. Washington, Julie M. Arblaster, and Aixue Hu. "Factors Affecting Climate Sensitivity in Global Coupled Models." *Journal of Climate* 17 (2006): 1584–1596.

"Meneguhkan Visi Kebangsaan Nahdlatul Ulama [Strengthening the Vision of Nationalism in Nahdlatul Ulama]," published by NU Online, last modified February 6, 2013. Accessed September 18, 2013. http://www.nu.or.id/a,public-m,dinamic-s,detail-ids,4-id,42323-lang,id-c,kolom-t,Meneguhkan+Visi+Kebangsaan+Nahdlatul+Ulama-.phpx.

Merli, Claudia. "Religious Interpretations of Tsunami in Satun Province, Southern Thailand: Reflections on Ethnographic and Visual Materials." *Svensk Religionshistorik Årsskrift* 14 (2005): 154–181.

Miller, Gifford H., Aslaug Geirsdottir, Y. Yufang Zhong, Darren J. Larsen, Bette L. Otto-Bliesner, Marika Holland, David A. Bailey, Kurt A. Refsnider, Scott J. Lehmann, John R. Southon, Chance Anderson, Helgi Bjornsson, and Thorvaldur Thordarson. "Abrupt Onset of the Little Ice Age Triggered by Volcanism and Sustained by Sea-Ice/Ocean Feedbacks." *Geophysical Research Let*ters 39 (2012): 1–5.

Miller, James D., Walter W. Immerzeel, and Gwyn Rees. "Climate Change Impacts on Glacier Hydrology and River Discharge in the Hindu Kush-Himalayas." *Mountain Research and Development* 32.4 (2012): 461–467.

Milne, R. S. "Technocrats and Politics in the ASEAN Countries." *Pacific Affairs: An International Review of Asia and the Pacific* 55.3 (1982): 403–429.

Ministry of Environment and Forests. "India: Climate-friendly Development." Published October 2002. Accessed November 28, 2011. http://www.envfor.nic.in/cc/cop8/moefbk/moefbk.pdf.

Ministry of External Affairs. "India and Global Challenges: Climate Change and Energy Security." Address by Shri Pranab Mukherjee at The Asia Society, New York, September 30, 2008. Accessed March 11, 2009. http://www.meaindia.nic.in/speech/2008/09/30ss03.htm.

———, "Prime Minister's Statement Prior to his Departure for Copenhagen." Published December 17, 2009. Accessed December 21, 2009. http://www.meaindia.nic.in/speech/2009/12/17ss01.htm.

———, "Remarks by PM at the Informal Plenary of HOS/Gs at the 15th COP at Copenhagen." Accessed December 21, 2009. http://www.meaindia.nic.in/speech/2009/12/18ss01.htm.

Ministry of Finance of Indonesia. *Ministry of Finance Green Paper: Economic and Fiscal Policy for Climate Change Mitigation in Indonesia*. Jakarta: Ministry of Finance and Australia Indonesia Partnership, 2009.

Ministry of Non-Conventional Energy Sources (MNES). *Annual Report 1993–1994*. New Delhi: MNES, 1994.

Miri, Seyyed Mohsen. "Prinsip-prinsip Islam dan filsafat Mulla Shadra sebagai basis etis dan kosmologis lingkungan hidup [Islamic principles and the philosophy of Mulla Shadra as ethical and cosmological basis for the environment]." In *Menanam sebelum kiamat: Islam, ekologi, dan gerakan lingkungan hidup* [Planting before Doomsday: Islam, Ecology, and the Environmental Movement], edited by Fachruddin M. Mangunjaya, Husain Heriyanto and Reza Gholami, 24–40. Jakarta: Conservation International, 2007.

Missbach, Andreas. *Das Klima zwischen Nord und Süd: Eine regulationstheoretische Untersuchung des Nord-Süd-Konflikts in der Klimapolitik der Vereinten Nationen*. Münster: Westfälisches Dampfboot, 1999.

Mitra, A. P., and C. Sharma, ed. *Global Environmental Changes in South Asia: A Regional Perspective*. Dortrecht: Springer Netherlands, 2010.

Mowery, David, and Nathan Rosenberg. "The Influence of Market Demand Upon Innovation: A Critical Review of Some Recent Empirical Studies." *Research Policy* 8.2 (1979): 102–153

Muchie, Mamo. "Old Wine in New Bottles: A Critical Exploration of the UN's Conceptions and Mechanisms for the Transfer of Environmentally Sound Technologies to Industry." *Technology in Society* 22 (2000): 201–220.

Mudur, Ganapati. "Atemlos in der Großstadt [Breathless in the City]." *MaxPlanckForschung* [MaxPlanckResearch] 4 (2010): 18–25.

Muhammad, Ahsin Sakho, Husein Mumammad, Roghib Mabrur et al., ed. *Fiqh lingkungan; fiqh al-bi'ah* [Jurisprudence of the Environment: Environmental Ethics]. 2nd edition. Jakarta: Conversation International, 2006.

Muhammad, Husein. "Manusia dan tugas kosmiknya menurut Islam [Humans and their Cosmic Duty According to Islam]." In *Menanam sebelum kiamat: Islam, ekologi, dan gerakan lingkungan hidup* [Planting before Doomsday: Islam, Ecology, and the Environmental Movement], edited by Fachruddin M. Mangunjaya, Husain Heriyanto and Reza Gholami, 3–8. Jakarta: Conservation International, 2007.

Mukherjee, Anit, and Xiaobo Zhang. "Rural Industrialization in China and India: Role of Policies and Institutions." *World Development* 35.10 (2007): 1621–1634.

Mukherjee, Pranab. "India and Global Challenges: Climate Change and Energy Security." Address at the Asia Society, New York, September 30, 2009). Accessed January 22, 2010. http://www.meaindia.nic.in/speech/2008/09/30ss03.htm.

Nair, K. Shadananan. "An Assessment of the Impact of Climate Change on the Megacities of India and of the Current Policies and Strategies to Meet Associated Challenges." Paper presented at Fifth Urban Research Symposium: Cities and Climate Change: Responding to an Urgent Agenda 2009, Marseille, June 28–30, 2009. Accessed

September 14, 2013. http://siteresources.worldbank.org/INTURBANDEVELOPMENT/Resources/336387-1256566800920/nair.pdf.

Najam, Adil, Saleemul Huq, and Youba Sokona. "Climate Negotiations Beyond Kyoto: Developing Countries Concerns and Interests." *Climate Policy* 3 (2003): 221–231.

Narain, Sunita. "Editorial." *Down to Earth*, March 1–15 (2010): 1.

Narayan, Vasudha. "'One Tree is Equal to Ten Sons': Hindu Responses to the Problems of Ecology, Population, and Consumption." *Journal of the American Academy of Religion* 65.2 (1997): 291–331.

Narlikar, Amrita. *New Powers: How to Become One and How to Manage Them*. London: Hurst & Company, 2010.

Nash, Roderick Frazier, ed. *American Environmentalism: Readings in Conservation History*. New York: McGraw-Hill, 1990.

Nasr, Seyyed Hossein. *An Introduction to Islamic Cosmological Doctrines. Conceptions of Nature and Methods Used for Its Study by the Ikhwan Al-Safa, Al-Biruni, and Ibn Sina.* Cambridge: Belknap Press, 1964.

———, *The Encounter of Man and Nature: The Spiritual Crisis of Modern Man*. London: Allen & Unwin, 1968.

———, "Masalah lingkungan di dunia Islam kontemporer [The Problem of the Environment in the Contemporary Muslim World]." In *Menanam sebelum kiamat: Islam, ekologi, dan gerakan lingkungan hidup* [Planting before Doomsday: Islam, Ecology, and the Environmental Movement], edited by Fachruddin M. Mangunjaya, Husain Heriyanto and Reza Gholami, 43–65. Jakarta: Conservation International, 2007.

Nelson, Richard. "Why Do Firms Differ?" *Strategic Management Journal* 12.2 (1991): 61–71.

Nidhi Eoseewong นิธิ เอียวศรีวงศ์. *Phutthasatsana nai khwamplianplaeng khong sangkhom* พุทธศาสนาในความเปลี่ยนแปลงของสังคม [Buddhism in the Context of Social Change]. Bangkok: Komonkhimthong Foundation, 2000.

Nilsson, Andreas, Chris von Borgstede, and Anders Biel. "Willingness to Accept Climate Change Strategies: The Effect of Values and Norms." *Journal of Environmental Psychology* 24.3 (2004): 267–277.

"No Fund Crunch for BRT Project: Delhi Govt." *The Indian Express*, October 16, 2008. Accessed December 5, 2012. http://www.indianexpress.com/news/no-fund-crunch-for-brt-project-delhi-govt/374191.

North, Douglass C. *Institutions, Institutional Change and Economic Performance*. Cambridge: Cambridge University Press, 1990.

Nugteren, Albertina. *Belief, Bounty, and Beauty: Rituals around Sacred Trees in India*. Leiden, Boston: Brill, 2005.

———, "From Cosmos to Commodity ... and Back: A Critique of Hindu Environmental Rhetoric in Educational Programs." In *Religion and Sustainable Development:*

Opportunities and Challenges for Higher Education, edited by Cathrien de Pater & Irene Dankelman, 159–168. Berlin: LIT Verlag, 2009.

Nüsser, Marcus. "Kilimanjaro and Mount Kenya: Colonized Mountains and their Rediscovery as Symbols of Global Climate Change." *Geographische Rundschau – International Edition* 5.4 (2009): 26–32 and supplement (map).

———, "Umwelt und Entwicklung im Himalaya: Forschungsgeschichte und aktuelle Themenfelder." *Geographische Rundschau* 64.4 (2012): 4–9 and supplement (map).

Nüsser, Marcus, Susanne Schmidt, and Juliane Dame. "Irrigation and Development in the Upper Indus Basin: Characteristics and Recent Changes of a Socio-hydrological System in Central Ladakh, India." *Mountain Research and Development* 32.1 (2012): 51–61.

Ockwell, David, Jim Watson, Gordon MacKerron, Prosanto Pal, Farhana Yamin, N. Vasudeva, and Parimita Mohanty. *UK–India Collaboration to Identify Barriers to the Transfer of Low Carbon Energy Technology*. Final Report. Falmer: SPRU – Science and Technology Policy, 2007.

Olson, Grant. "A Person-Centered Ethnography of Thai Buddhism: The Life of Phra Rajavaramuni (Prayudh Payutto)." PhD diss., Cornell University, 1989.

———, "Introduction." In *Buddhadhamma: Natural Laws and Values for Life*, by Phra Prayudh Payutto. Translated by Grant A. Olson, 1–33. Albany: SUNY Press 1995.

"On Commitments, Jairam Drops a Line, Eats Words." *Express India*, December 10, 2010.

Oppenheimer, Clive. "Ice Core and Palaeoclimatic Evidence for the Timing and Nature of the Great Mid-13th Century Volcanic Eruption." *International Journal of Climatology* 23 (2003): 417–426.

Oreskes, Naomi. "The Scientific Consensus on Climate Change." *Science* 306.3 (2004): 1686.

Organization for Economic Co-operation and Development (OECD). *India: OECD Economic Surveys.* Paris: OECD, 2007.

———, *OECD Science, Technology and Industry Scoreboard 2007*. Paris: OECD, 2007.

Orlove, Ben, Ellen Wiegandt, and Brian H. Luckman. "The Place of Glaciers in Natural and Cultural Landscapes." In *Darkening Peaks: Glacier Retreat, Science, and Society*, edited by Ben Orlove, Ellen Wiegandt, and Brian H. Luckman, 3–19. Berkeley: University of California Press, 2008.

Oxford Centre for Hindu Studies. "Bhumi Project: Hindu Groups Address Climate Change." Published January 6, 2011. Accessed November 28, 2011. http://ochs.org.uk/news/bhumi-project-hindu-groups-address-climate-change.

Pande, Varad. "India at Cancun: The Emergence of a Confident Dealmaker." *Economic and Political Weekly* XLVI.6 (2011): 14–15.

Parnwell, Michael. "Neolocalism and Renascent Social Capital in Northeast Thailand." *Environment and Planning D: Society and Space* 25 (2007): 990–1014.

Parnwell, Michael, and Martin Seeger. "The Relocalization of Buddhism in Thailand." *Journal of Buddhist Ethics* 29 (2008): 78–176.

Parry, M. L., O. F. Canziani, J. P. Palutikof, P. J. van der Linden, and C. E. Hanson, ed. *Climate Change 2007: Impacts, Adaptation and Vulnerability*. Oxford: Oxford University Press, 2007.

Pasuk Phongpaichit, and Chris Baker. "Thai Capital After the Asian Crisis." *Economic & Political Weekly* 42.50 (2007): 58–66.

Payutto, P.A. *Good, Evil and Beyond: Kamma in the Buddha's Teaching*. Third Impression. Translated by Bruce Evans. Bangkok: Buddhadhamma Foundation, 1996.

Payutto, Phra. "Ecology According to Buddhist Principles." Unpublished paper presented in 1993.

Payutto, Phra พระธรรมปิฎก (ป. อ. ปยุตฺโต). *Narok sawan samrap khon run mai* นรก – สวรรค์ สำหรับคนรุ่นใหม่ [Hells and Heavens for the People of the New Generation]. Bangkok: Buddhadhamma Foundation, 1995.

———, *Phutthatham* พุทธธรรม [Buddhadhamma]. Bangkok: Mahachulalongkorn University, 1995 CE/2538 BE.

———, *Kan phatthana thi yangyuen* การพัฒนาที่ยั่งยืน [Sustainable Development]. Second edition. Bangkok: Buddhadhamma Foundation, 1996.

———, *Phutthatham chabap doem* พุทธธรรม (ฉบับเดิม) [Buddhadhamma (Original Version)]. 23rd edition, n.p., 2003.

———, *Khon rak pa pa rak khon* คนรักษ์ป่า ป่ารักษ์คน [If Humans Protect the Forest, the Forest will Protect Humans]. Bangkok: Panyaprathip Foundation, 2011.

[Payutto, Phra] Phra Rajavaramuni (Prayudh Payutto). *Looking to America to Solve Thailand's Problems*. Translated by Grant A. Olson. Bangkok: Sathirakoses Nagapradipa Foundation, 1987.

Pedersen, Poul. "Nature, Religion, and Cultural Identity: The Religious Environmentalist Paradigm." In *Asian Perceptions of Nature: A Critical Approach*, edited by Ole Bruun, and Arne Kalland, 258–275. Richmond: Curzon Press, 1995.

Peleggi, Maurizio. *Thailand: The Worldly Kingdom*. London: Reaktion, 2007.

Perch, Leisa. "Inclusive and Sustainable Development: For Whom?" One Pager 126. Brasilia: UNDP International Policy Centre for Inclusive Growth (IPC-IG). Published 2011. Accessed May 31, 2012. http://www.ipc-undp.org/pub/IPCOnePager126.pdf.

"Pesantren and Community Involvement in Managing Disaster Risks in Coastal Areas through Mangrove Planting." Published by Mangroves for the Future. Accessed December 4, 2012. http://www.mangrovesforthefuture.org/assets/Repository/Documents/012-SG-ID-01-06-07.pdf.

Phaisan, Phra พระไพศาล วิสาโล. *Phra thammapidok kap anakhot khong phutthasatsana* พระธรรมปิฎกกับอนาคตของพุทธศาสนา [Phra Payutto and the Future of Buddhism]. Bangkok: Wutthitham, 1999.

Phaisan Visālo, Phra. "The Forest Monastery and its Relevance to Modern Thai Society." In *Radical Conservatism: Buddhism in the Contemporary World: Articles in Honour of Bhikkhu Buddhadasa's 84th Birthday Anniversary*, edited by Sulak Sivaraksa, Pracha Hutanuvatra, Nibhond Chaemduang, Santisukh Sobhanasiri and Nicholas P. Kholer, 288–300. Bangkok: Thai Inter-Religious Commission for Development/International Network of Engaged Buddhists, 1990.

———, *Phutthasatsana thai nai anakhot naenom lae thangok jak wikrit* พุทธศาสนาไทยในอนาคต แนวโน้มและทางออกจากวิกฤติ [Thai Buddhism in the Future: Trends and Solutions to the Crisis]. Bangkok: Sotsri-Saritwong Foundation, 2003.

Pigeaud, Theodore. G. Th. *Literature of Java: Synopsis of Javanese Literature 900–1900 A.D.*, vol. 1. The Hague: Nijhoff, 1967.

Piyaporn Wongruang. "Flood Mitigation Plan 'Unlikely to Be in Place in Three Years'." *Bangkok Post*, January 15, 2012. Accessed January 14, 2012. http://www.bangkokpost.com/news/local/275180/flood-mitigation-plan-unlikely-to-be-in-place-in-three-years.

Pollack, Henry. *A World Without Ice.* New York: Avery, 2009.

Prapod Assavavirulhakarn. *The Ascendancy of Theravāda Buddhism in Southeast Asia.* Chiang Mai: Silkworm Books, 2010.

Prasad, H. A. C., and J. S. Kochher. *Climate Change and India – Some Major Issues and Policy Implications.* New Delhi: Government of India, Ministry of Finance, 2009.

Prayun, Phra พระเมธีธรรมาภรณ์ (ประยูร ธมฺมจิตฺโต). *Thamma lae kan anurak singwaetlom* ธรรมะและการอนุรักษ์สิ่งแวดล้อม [The Dhamma and Environmental Conservation]. Bangkok: Buddhadhamma Foundation, 1995.

Prida Rueangwichathon ปรีดา เรืองวิชาธร. *Phlik fuen khuen chiwit chiwit lae ngan khong phra khru suphajarawat* พลิกฟื้นคืนชีวิต ชีวิตและงานของพระครูสุภาจารวัฒน์ [Revitalising Life: Life and Work of Phra Khru Suphajarawat]. Bangkok: Religious Council for Development, 1992.

Publications Division Ministry of Information and Broadcasting Government of India, ed. *India 2013: A Reference Annual.* New Delhi: Publications Division Ministry of Information and Broadcasting Government of India, 2013.

Puji, Siwi Tri. "MUI nyatakan dana lotere Greenpeace haram [MUI Declares Lottery Funds of Greenpeace Forbidden by Islam]." Published by *Republika.co.id*, August 4, 2011. Accessed August 12, 2013. http://www.republika.co.id/berita/nasional/umum/11/08/04/lpeec7-mui-nyatakan-dana-lotere-greenpeace-haram.

Putnam, Robert D. "Diplomacy and Domestic Politics: The Logic of Two-Level Games." *International Organizations* 42.3 (1988): 427–460.

Qomar, Mulamil. *Pesantren: Dari transformasi metodologi menuju demokrasi institusi* [Pesantren: From a Transformation of Methodology towards Institutional Democracy]. Jakarta: Erlangga, 2005.

Rai, Varun, and David G. Victor. "Climate Change and the Energy Challenge: A Pragmatic Approach for India." *Economic and Political Weekly* 44.31 (2009): 78–85.

Raina, Vijay Kumar. *Himalayan Glaciers: A State-of-Art Review of Glacial Studies, Glacial Retreat and Climate Change*. New Delhi: Ministry of Environment and Forests, Government of India, 2009.

Rajamani, Layanya. "India's Negotiating Position in Climate Change: Legitimate but Not Sagacious." Issue Brief No.2. New Delhi: Centre for Policy Research, 2007.

Ramana, M. V. *The Future of Nuclear Power*. Philadelphia: Center for the Advanced Study of India, India in Transition, 2010.

Rastogi, Namrata Pastodia. "Winds of Change: India's Emerging Climate Strategy." *The International Spectator* 46.2 (2011): 127–141.

REDD Indonesia. "Pidato Presiden Republik Indonesia, Susilo Bambang Yudhoyono di CIFOR tentang <kesetaraan pertumbuhan pembangunan berkelanjutan> [Speech by the President of the Republic of Indonesia, Susilo Bambang Yudhoyono in CIFOR about <Equal Growth for Sustainable Development>]." Last modified July 6, 2012. Accessed December 12, 2012. http://blog.cifor.org/10156/pidato-presiden-republik-indonesia-susilo-bambang-yudhoyono-di-cifor-tentang-kesetaraan-pertumbuhan-berkelanjutan/#.UMrz1Kx9QfY.

Rennings, Klaus. "Redefining Innovation – Eco-innovation Research and the Contribution from Ecological Economics." *Ecological Economics* 32 (2000): 319–332.

Reynolds, Frank E. and Mani B. Reynolds (trans.). *Three Worlds According to King Ruang: A Thai Buddhist Cosmology*. Berkeley: Asian Humanities Press, 1982.

Ṛgveda Saṁhitā. Edited by Theodor Aufrecht. Published as *Die Hymnen des Rigveda*. Two vols. Reprint Edition, Vienna: Otto Harrassowitz, 1988.

Riché, Beatrice, Excellent Hachileka, Cynthia B. Awuor, and Anne Hammill. *Climate-related Vulnerability and Adaptive Capacity in Ethiopia's Borana and Somali Communities*. Manitoba: International Institute for Sustainable Development, 2009.

Rigg, Jonathan, Anna Allott, Rachel Harrison, and Ulrich Kratz. "Understanding Languages of Modernization: A Southeast Asian View." *Modern Asian Studies* 33.3 (1999): 581–602.

Rist, Gilbert. "Development as a Buzzword." In *Deconstructing Development Discourse: Buzzwords and Fuzzwords*, edited by Andrea Cornwall and Deborah Eade, 19–28. Oxford: Practical Action Publishing, 2010.

Roberts, J. Timmons. "The International Dimension of Climate Justice and the Need for International Adaptation Funding." *Environmental Justice* 2.4 (2009): 185–190.

Roberts, J. Timmons, and Bradley C. Parks. *A Climate of Injustice: Global Inequality, North-South Politics, and Climate Policy*. Cambridge: The MIT Press, 2007.

Robock, Alan. "20 Reasons why Geoengineering May Be a Bad Idea." *Bulletin of Atomic Scientists*, May–June (2008): 14–18, and 59.

Rockström, Johan, Will Steffen, Kevin Noone, Åsa Persson, Stuart Chapin, Eric F. Lambin, Timothy M. Lenton et al. "A Safe Operating Space for Humanity." *Nature* 461 (2009): 472–475.

Roy, Arundhati. "Introduction." In *India: A Mosaic*, edited by R. Silvers and B. Epstein, xv–xxxiv. New York: New York Review of Books, 2000.

Sachs, Jeffrey D. *The End of Poverty: How We Can Make It Happen in Our Lifetime*. London: Penguin Books, 2005.

Sachs, Wolfgang. "Preface to the New Edition." In *The Development Dictionary: Knowledge as Power*, edited by Wolfgang Sachs, vi–xiv. London, New York: Zed Books, 2010.

"Sahasakha sueksa phuea phatthana chonnabot สหสาขาศึกษาเพื่อพัฒนาชนบท [Interdisciplinary Studies for the Development of Rural Areas]." Published by Chulalongkorn University Website. Accessed June 1, 2011. http://pioneer.chula.ac.th/~vsuntare/report/korn2.html.

Saniotis, Arthur. "Muslims and Ecology: Fostering Islamic Environmental Ethics." *Contemporary Islam* 6 (2012): 155–171.

Saranukromthai Ratchabanditayasathan สารานุกรมไทย ฉบับราชบัณฑิตยสถาน [The Thai Encyclopaedia of the Royal Institute], vol. 13. Bangkok: The Royal Institute, 1981.

Sastry, E. V. R. "The Photovoltaic Programme in India: An Overview." *Solar Energy Materials and Solar Cells* 47 (1997): 63–69.

———, "The Photovoltaic Programme in India." Paper presented at the IEA PVPS International Conference, Osaka, 2003.

Schedneck, Brooke. "Forest as Challenge, Forest as Healer: Reinterpretations and Hybridity within the Forest Tradition of Thailand." Unpublished Paper.

Schelling, Thomas. "Climate Change: The Uncertainties, the Certainties, and what they Imply About Action." Published by The Berkeley Electronic Press, 2007. Accessed May 31, 2012. http://are.berkeley.edu/courses/EEP131/fall2007/ClimateChange/Climate%20Change%20The%20Uncertainties...%20%28Schelling%29.pdf.

Schellnhuber, Hans Joachim. "Global Warming: Stop Worrying, Start Panicking?" *Proceedings of the National Academy of Sciences of the United States of America PNAS* 105 (2008): 38.

Scherler, Dirk, Bodo Bookhagen, and Manfred R. Strecker. "Spatially Variable Response of Himalayan Glaciers to Climate Change Affected by Debris Cover." *Nature Geoscience* 4 (2011): 156–159.

Schmidt, Guido. *Die Governance-Strukturen der ASEAN*. Norderstedt: Grin Verlag, 2009.

Schmidt, Susanne. "Der Himalaya bald ohne „Hima"? Was wir über die Gletscherentwicklung im Himalaya wissen." *Geographische Rundschau* 64.4 (2012): 10–16.

Schmidt, Susanne, and Marcus Nüsser. "Fluctuations of Raikot Glacier during the Last 70 Years: A Case Study from the Nanga Parbat Massif, Northern Pakistan." *Journal of Glaciology* 55.194 (2009): 949–959.

———, "Changes of High Altitude Glaciers from 1969 to 2010 in the Trans-Himalayan Kang Yatze Massif, Ladakh, Northwest India." *Arctic, Antarctic and Alpine Research* 44.1 (2012): 107–121.

Schmithausen, Lambert. "The Early Buddhist Tradition and Ecological Ethics." *Journal of Buddhist Ethics* 4 (1997): 1–74.

———, "Buddhism and the Ethics of Nature: Some Remarks." *The Eastern Buddhist* 32.2 (2000): 26–78.

Schneidewind, Uwe. "Technik allein bringt's nicht: Ohne soziale Innovationen wird der Klimawandel nicht zu beherrschen sein [Technology Alone Won't Do: Climate Change Cannot Be Managed without Social Innovations]." *Die Zeit* [The Times] 2, 2012, 25.

Schneidewind, Uwe, and Hanna Scheck. "Cities as 'Real World Laboratories' for System Innovations: Theories, Models and Empirical Designs." *Proceedings of the Third International Conference on Sustainability Transitions, August 29–31, 2012*, 361–379. Lyngby: Technical University of Denmark, 2012.

Schwencke, A. M. *Globalized Eco-Islam: A Survey of Global Islamic Environmentalism*. Leiden: Leiden Institute for Religious Studies, 2012.

Seeger, Martin. "Phra Dhammapiṭaka und die Pāli-Kanon-Debatte in Thailand: Ein Beitrag zur Untersuchung des modernen Buddhismus." PhD diss., Universität Hamburg, Staats- und Universitätsbibliothek Hamburg – Carl von Ossietzky, Hamburg, 2005. http://www.sub.uni-hamburg.de/opus/volltexte/2005/2342/.

———, "Against the Stream: The Thai Female Buddhist Saint Mae Chi Kaew Sianglam (1901–1991)." *South East Asia Research* 18.3 (2010): 555–595.

Sen, Amartya. "Climate Policy as Human Development, Special Contribution to the Human Development Report 2007/2008." In *Human Development Report 2007/2008*, edited by UNDP, 27–58. Hampshire, New York: Palgrave Macmillan, 2007.

Shanks, Edwin, Cecilia Luttrell, Tim Conway, Vu Manh Loi, and Judith Ladinsky. *Understanding Pro-poor Political Change: The Policy Process*. London: ODI, 2004.

Shiraishi, Takashi, and Patricio N. Abinales, ed. *After the Crisis: Hegemony, Technocracy and Governance in Southeast Asia*. Kyoto: Kyoto University Press, 2005.

Simamora, Adianto P. "Muslim Countries to Lead on Green Issues." *The Jakarta Post*, April 12, 2010. Accessed November 26, 2012. http://www.thejakartapost.com/news/2010/04/12/muslim-countries-lead-green-issues.html.

Sims, Holly. "Population, Urbanization, and the Environment." In *Understanding Contemporary India*, edited by Sumit Ganguly und Neil DeVotta, 157–189. New Delhi: Viva Books, 2003.

Singapore Government. *Singapore's National Climate Change Strategy*. Singapore: Singapore Government, 2008.

Sivam, Alpana. "Housing Supply in Delhi." *Cities* 20 (2003): 135–141.

Smit, I. Barry, and Olga Pilifosova. "Adaptation to Climate Change in the Context of Sustainable Development and Equity." In *Climate Change 2001: Impacts, Adaptation and Vulnerability*, edited by IPCC, 877–912. Cambridge: Cambridge University Press, 2001.

Soete, Luc. "International Diffusion of Technology, Industrial Development and Technological Leapfrogging." *World Development* 13.3 (1985): 409–422.

Somchai Jitsuchon. "Adaptations to Climate Change of Poor Thai Farmers." Paper presented at the conference "Environments of the Poor", Asian Development Bank, November 24, 2010. Accessed January 23, 2012. http://www.scribd.com/doc/43296343/Adaptations-to-Climate-Change-of-Poor-Thai-Farmers-paper.

Sonderforschungsbereich 700, ed. *Grundbegriffe der Governance-Forschung: Ein Beitrag aus dem Teilprojekt A1*. SFB Governance Working Paper Series 8. Published June 2009. Berlin: Freie Universität, 2009.

Sørensen, Henrik H. "Of Eco-Buddhas and Dharma-Roots: Views from the East Asian Buddhist Tradition." In *Nature, Environment and Culture in East Asia: The Challenge of Climate Change*, edited by Carmen Meinert. Leiden: Brill, 2013.

Sperling, Frank, ed. *Poverty and Climate Change – Reducing the Vulnerability of the Poor through Adaptation*. Washington, D.C.: AfDB, ADB, DFID, EC DG Development, BMZ, DGIS, OECD, UNDP, UNEP and the World Bank, 2003.

Sponsel, Les, and Poranee Natadecha-Sponsel. "Why a Tree is more than a Tree: Reflections on the Spiritual Ecology of Sacred Trees in Thailand." In *Santi Pracha Dhamma: Essays in Honour of the Late Puey Ungphakorn*, edited by Pipob Udomittipong, 364–373. Bangkok: Santi Pracha Dhamma Institute, 2001.

Sreberny, Annabelle. "Globalization and Me: Thinking at the Boundary." In *In Search of Boundaries: Communication, Nation States and Cultural Identities*, edited by Joseph Chan and Bryce T. Mcintyre, 293–309. Westport, London: Ablex, 2002.

Srivastava, Leena, and I.H. Rehman. "Energy for Sustainable Development in India: Linkages and Strategic Directions." *Energy Policy* 34 (2006): 643–654.

Stargardt, Janice. *Satingpra I: The Environmental and Economic Archaeology of South Thailand.* Oxford, Singapore: BAR International Series 158 in Association with ISEAS, 1983.

———, *The Ancient Pyu of Burma, Vol. I: Early Pyu Cities in a Man-Made Landscape*. Cambridge, Singapore: PACSEA in association with ISEAS, 1990.

———, *"Behind the Shadows: Archaeological Data on Two-Way Sea-Trade between Quanzhou and Satingpra, South Thailand, 10th–14th Century."* In *Emporium to the World: Quanzhou and Maritime Trade, 10th–14th Century*, edited by Angela Schottenhammer, 309–393. Leiden: Brill, 2001.

Stern, Nicholas. *The Economics of Climate Change: The Stern Review*. Cambridge: Cambridge University Press, 2006.

Stoker, Gerry. "Governance as a Theory: Five Propositions." *International Social Science Journal* 155 (1998): 17–28.

Stothers, Richard B. "Climatic and Demographic Consequences of the Massive Volcanic Eruption of 1258." *Climate Change* 45 (2000): 361–374.

Stott, Philip. "*Mu'ang* and *Pa*: Elite Views of Nature in a Changing Thailand." In *Thai Constructions of Knowledge*, edited by Manas Chitakasem and Andrew Turton, 142–154, London: SOAS, 1991.

Subramanian, Arvind, Nancy Birdsall, and Aaditya Mattoo. "India and Climate Change: Some International Dimensions." *Economic and Political Weekly* 64 (2009): 31; 43–50.

Sumedho, Bhikkhu. "Forest Tradition as a Challenge to the Modern World." In *Socially Engaged Buddhism for the New Millennium: Essays in Honor of the Ven. Phra Dhammapitaka*, edited by Pipob Udomittipong and Chris Walker, 470–482. Bangkok: Sathirakoses Nagapradipa Foundation and Foundation for Children, 1999.

Suphajarawat, Phra Khru. พระครูสุภาจารวัฒน์. *Botbat phra song nai chonnabot thamklang phawa wikrit* บทบาทพระสงฆ์ในชนบทท่ามกลาง ภาวะวิกฤติ [The Role of the *Saṅgha* in Rural Areas in the Middle of the Crisis]. Bangkok: Komonkhimthong Foundation, 1999 CE/2542 BE.

Susanto, Slamet. "Ministry Launches Eco-'pesantren' Program." *The Jakarta Post*, November 6, 2009. Accessed November 16, 2012. http://www.thejakartapost.com/news/2009/11/06/ministry-launches-ecopesantren039-program.html.

Sutton, Philip W. *The Environment: A Sociological Introduction.* Cambridge, UK: Polity Press, 2007.

Suwanna Kriengkraipetch. "Thai Folk Beliefs about Animals and Plants and Attitudes toward Nature." In *Culture and Environment in Thailand: A Symposium of the Siam Society*, edited by Siam Society, 195–211. Bangkok: Siam Society, 1989.

Suwanna Satha-anand. "Karma as Moral Justice in Thai Buddhism." *Manusya: Journal of Humanities*, Special Issue No. 4 (2002): 72–83.

Swearer, Donald. "The Hermeneutics of Buddhist Ecology in Contemporary Thailand: Buddhadāsa and Dhammapiṭaka." In *Buddhism and Ecology: The Interconnection of Dharma and Deeds*, edited by Mary Evelyn Tucker and Duncan Ryūken Williams, 21–44. Cambridge: Harvard University Press, 1997.

———, "An Assessment of Buddhist Eco-Philosophy." *Harvard Theological Review* 99.2 (2006): 123–137.

Swidler, Ann. "Culture in Action: Symbols and Strategies." *American Sociological Review* 51.2 (April 1986): 273–286.

"Taek tuen du tontan – muean luang pho to แตกตื่นดูต้นตาล-เหมือนรูปหลวงพ่อโต [People Get Excited About A Palm Tree that Looks like Luang Pho To]." Published by Webportal mthai.com. Accessed December 23, 2011. http://news.mthai.com/hot-news/14484.html.

"Taek tuen 'tontan' ok yot phayanak แตกตื่น "ต้นตาล" ออกยอดพญานาค [Excitement About the Sugar Palm that Sprouts into a Naga]." Khaosod Online Newspaper, December 19, 2009. Accessed December 23, 2011. http://www.khaosod.co.th/view_news.php?newsid=TUROd2NtOHhOREU1TVRJMU1nPT0.

Taele, B. M., K. K. Gopinathan, and L. Mokhuts'oane. "The Potential of Renewable Energy Technologies for Rural Development in Lesotho." *Renewable Energy* 35 (2007): 609–622.

Taittirīya Saṁhitā. 8 vols. Ānandāśrama Sanskrit Series, no. 42. Poona: Ānandāśrama, 1978.

Tambiah, S. J. *Buddhism and the Spirit Cults in North-East Thailand*. Cambridge: Cambridge University Press, 1970.

Tannenbaum, Nicola. "Protest, Tree Ordination, and the Changing Context of Political Ritual." *Ethnology* 39.2 (2000): 109–127.

Tarlo, Emma. *Clothing Matters: Dress and Identity in India*. London: C. Hurst & Co Publishers Ltd., 1996.

Taylor, J. L. *Forest Monks and the Nation-State: An Anthropological and Historical Study in Northeastern Thailand*. Singapore: Institute of Southeast Asian Studies, 1993.

Taylor, Jim. "Social Activism and Resistance on the Thai Frontier: The Case of Phra Prajak Khuttajitto." *Bulletin of Concerned Asian Scholars* 25 (1993): 3–16.

Taylor, Philip. "Poor Policies, Wealthy Peasants: Alternative Trajectories of Rural Development in Vietnam." *Journal of Vietnamese Studies* 2.2 (2007): 3–56.

"Technocrats – Minds like Machines." *The Economist*, November 19, 2011. Accessed May 31, 2012. http://www.economist.com/node/21538698.

Terwiel, Baas. "The Physical Transformation of the Central Thai Region in Premodern Times." In *A History of Natural Resources in Asia: The Wealth of Nature*, edited by Greg Bankoff and Peter Boomgaard, 41–60. Basingstoke : Palgrave Macmillan, 2007.

Terwiel, Barend Jan. "Die Naturkatastrophe aus der Sicht des thailändischen Buddhismus." In *Jahrbuch der Akademie der Wissenschaften zu Göttingen*, 59–68. Göttingen: Vandenhoeck & Ruprecht, 2005.

———, *Thailand's Political History: From Fall of Ayutthaya to Recent Times*. Bangkok: River Books, 2005.

———, *"Siam": Ten Ways to Look at Thailand's Past*. Großheirath: Ostasien Verlag, 2012.

Thanwa Jaithiang ธันวา ใจเที่ยง. "Phra ajan man phurithatto phu nam haeng kan-sang thammikasangkhom chao na lum nakhong พระอาจารย์มั่น ภูริทัตโต ผู้นำแห่งการสร้างธรรมิกสังคมชาวนาลุ่มน้ำโขง [Venerable Ajarn Man Bhuridatto, the Leader of a Dhammic Peasant Society in the Lowlands of the Mae Khong River]." Published 2006. Accessed December 28, 2011. http://61.47.2.69/~midnight/midnight2545/document95155.html.

The Energy and Resources Institute (TERI). *Report on Solar Photovoltaic Sector Review and Market Potential in India*. Delhi: TERI, 2006.

"The Islamic Perspective on the Environmental Crisis: Seyyed Hossein Nasr in Conversation with Muzaffar Iqbal." Transcript of a conversation between Professor Seyyed Hossein Nasr and Muzaffar Iqbal, recorded in Sherwood Park, Canada, on February 19, 2007. Published by TheFreeLibrary.com. Last modified June 22, 2007. Accessed November 20, 2012. http://www.thefreelibrary.com/The Islamic+perspective+on+the+environmental+crisis%3A+Seyyed+Hossein...-a0164596587.

"The Muslim Seven Year Action Plan on Climate Change 2010–2017 – Summary." Last modified September 20, 2009. Accessed December 7, 2012. http://www.bcca.org/ief/fl/WindsorARCMuslim_summary091020.pdf.

Thussu, Daya. *News as Entertainment: The Rise of Global Infotainment*. London: Sage, 2007.

Timmreck, Claudia, Stephan J. Lorenz, Thomas J. Crowley, Stefan Kinne, Thomas J. Raddatz, Mark Thomas, and Johann H. Jungclaus. "Limited Temperature Response to the Very Large AD 1258 Volcanic Eruption." *Geophysical Research Letters* 36 (2009): 1–5.

Tomalin, Emma. *Bio-divinity and Biodiversity: The Limits of Religious Environmentalism*. Aldershot: Ashgate, 2009.

Tompkins, Emma L., and W. Neil Adger. "Defining Response to Enhance Climate Change Policy." *Environmental Science and Policy* 8.6 (2005): 562–571.

Tompkins, Emma L., W. Neil Adger, and Katrina Brown. "Institutional Networks for Inclusive Coastal Zone Management in Trinidad and Tobago." *Environment and Planning A* 34.6 (2002): 1095–1111.

"Transition Research." Published by Wuppertal Institute for Climate, Environment and Energy. Accessed October 14, 2012. http://wupperinst.org/en/our-research/transition-research/.

Trenberth, Kevin, and Aiguo Dai. "Effects of Mt Pinatubo Volcanic Eruption on the Hydrological Cycle as An Analog of Geoengineering." *Geophysical Research Letters* 34 (2007): L15702.

Truman, Harry S. "Inaugural Address." Published by Gerhard Peters and John T. Woolley. *The American Presidency Project*, January 20, 1949. Accessed December 3, 2012. http://www.presidency.ucsb.edu/ws/?pid=13282.

Tsebelis, George. *Veto Players: How Political Institutions Work*. Princeton: Princeton University Press, 2002.

United Nations. "United Nations Framework Convention on Climate Change." Accessed December 3, 2012. http://unfccc.int/focus/adaptation/items/6999.php.

United Nations Development Program (UNDP). "Per Capita GDP at Current Prices – US Dollars." Accessed May 21, 2012. http://data.un.org/Data.aspx?d=SNAAMA&f=grID%3A101%3BcurrID%3AUSD%3BpcFlag%3A1.

United Nations Framework Convention on Climate Change. Berlin: UNFCCC Secretariat, 1992.

Uphoff, Norman T., and Louise Buck. *Strengthening Rural Local Institution Capabilities for Sustainable Livelihoods and Equitable Development*. Paper for Social Development Department. Washington, D.C.: The World Bank, 2007.

Vatsyayan, Kapila. "Ecology and Indian Myth." *India International Centre Quarterly* 19.1–2 (1992): 157–180.

Vedwan, Neeraj, and Robert E. Rhoades. "Climate Change in the Western Himalayas of India: A Study of Local Perception and Response." *Climate Research* 19.2 (2001): 109–117.

Veera Prateepchaikul. "Flood Victims Were Not Faceless People." *Bangkok Post*, January 10, 2012. Accessed January 16, 2012. http://www.bangkokpost.com/opinion/opinion/274536/flood-victims-should-not-die-in-vain.

Veron, Rene. "Remaking Urban Environment: The Political Ecology of Air Pollution in Delhi." *Environment and Planning A* 38.11 (2006): 2093–2109.

Vince, Gaia. "Glacier Man." *Science* 326 (2009): 659–661.

Visalo, Bhikkhu. "Buddhism for the Next Century: Toward Renewing a Moral Thai Society." In *Socially Engaged Buddhism for the New Millennium: Essays in Honor of the Ven. Phra Dhammapitaka (Bhikkhu P. A. Payutto) on His 60th Birthday Anniversary*, edited by Pipob Udomittipong and Chris Walker, 235–252. Bangkok: Sathirakoses Nagapradipa Foundation and Foundation for Children, 1999.

Viviroli, Daniel, and Rolf Weingartner. "'Water Towers' – A Global View of the Hydrological Importance of Mountains." In *Mountains: Sources of Water, Sources of Knowledge, Volume 31: Advances in Global Change Research*, edited by Ellen Wiegandt, 1520. Dordrecht: Springer, 2008.

Viviroli, Daniel, Hans H. Dürr, Bruno Messerli, Michel Meybeck, and Rolf Weingartner. "Mountains of the World, Water Towers for Humanity: Typology, Mapping, and Global Significance." *Water Resources Research* 43, W07447 (2007): 1–13. Accessed December 23, 2012. doi:10.1029/2006WR005653.

Wagner, Christian. "India: A Difficult Partner in International Climate Policy." In *International Climate Policy: Priorities of Key Negotiating Parties*, edited by Susanne Dröge, 17–36. Berlin: Stiftung Wissenschaft und Politik, 2010.

"Warta NU-Muhammadiyah Akan Deklarasikan Gerakan Moral Anti-Pornografi [Reports that NU and Muhammadiyah will Declare an Anti-pornography Morality Movement]." Published by NU Online. Last modified January 31, 2006. Accessed December 7, 2012. http://www.nu.or.id/a,public-m,dinamic-s,detail-ids,1-id,4077-lang,id-c,warta-t,NU+Muhammadiyah+Akan+Deklarasikan+Gerakan+Moral+Anti+Pornografi-.phpx.

Wat Nyanavesakavan วัดญาณเวศกวัน Official Website. Accessed December 27, 2012. http://www.watnyanaves.net/th/book_list.

Wat Nyanavesakavan วัดญาณเวศกวัน Official Website. Accessed December 27, 2012. http://www.watnyanaves.net/th/sound_album_list.

Watkins, Kevin. *Human Development Report 2007/2008: Fighting Climate Change: Human Solidarity in a Divided World*. New York: UNDP, 2007.

"West's Prescription No Panacea." *The Hindu*, December 6, 2011.

Wheeler, David and Saubrah Shome. *Less Smoke, More Mirrors: Where India Really Stands on Solar Power and Other Renewable*. Working Paper No. 204. Washington, D.C.: Center for Global Development, 2010.

White, Lynn jr. "The Historical Roots of Our Ecologic Crisis." *Science* 155 (1967): 1203–1207.

William, Marc. "The Third World and Global Environmental Negotiations: Interest, Institutions, and Ideas." *Global Environmental Politics* 5.3 (2005): 48–69.

Wissmann, Hermann von. *Die heutige Vergletscherung und Schneegrenze in Hochasien mit Hinweisen auf die Vergletscherung der letzten Eiszeit: Abhandlungen der Mathematisch-Naturwissenschaftlichen Klasse 14*. Mainz: Akademie der Wissenschaften und der Literatur, 1959.

Working Group II. "Impacts, Adaptation and Vulnerability." In *Fourth Assessment Report of the Intergovernmental Panel on Climate Change*, edited by Parry Martin L., Osvaldo F. Canziani, Jean P. Palutikof, P. J. van der Linden and Clair E. Hanson, 23–78. Cambridge: Cambridge University Press, 2007.

World Bank. *World Development Report 1999–2000: Entering the 21st Century*. Washington, D.C.: The World Bank, 2000.

———, *World Development Report 2010: Development and Climate Change*. Washington, D.C.: World Bank, 2010.

———, "Gross Development Product Data (Current US$)." Accessed January 16, 2012. http://data.worldbank.org/indicator/NY.GDP.MKTP.CD.

World Bank (South Asia Environment and Social Development Unit). *India: Strengthening Institutions for Sustainable Growth: Country Environmental Analysis*. Washington, D.C.: World Bank, 2008.

World Commission on Environment and Development (WCED). *Our Common Future, Report of the World Commission on Environment and Development*. Oxford: Oxford University Press, 1987.

World Wildlife Fund (WWF). "Mega-Stress for Mega-Cities: A Climate Vulnerability Ranking of Major Coastal Cities in Asia." Published 2010. Accessed September 14, 2013. http://assets.panda.org/downloads/mega_cities_report.pdf.

Wuppertal Institute for Climate, Environment and Energy. "CCS Global, Prospects of Carbon Capture and Storage Technologies (CCS) in Emerging Economies, Final Report to the German Federal Ministry for the Environment, Nature Conservation and Nuclear Safety (BMU), Part II: Country Study India". Published June 30, 2012. Accessed September 21, 2013. http://epub.wupperinst.org/frontdoor/index/index/docId/4582.

Young, Oran R., Leslie A. King, and Heike Schroeder, ed. *Institutions and Environmental Change: Principal Findings, Applications, and Research Frontiers*. Cambridge, London: The MIT Press, 2008.

Zuhaili, Wahbah. *Al-fiqh al-Islami wa adillathu* [Islamic Jurisprudence and Its Proofs]. Beirut: Dar al-Fikr, 1984.

Index

www.ingramcontent.com/pod-product-compliance
Lightning Source LLC
LaVergne TN
LVHW020633300926
849161LV00004B/684
9789004245884